DESIGN AND ANALYSIS OF MAGNETORESISTIVE RECORDING HEADS

DESIGN AND ANALYSIS OF MAGNETORESISTIVE RECORDING HEADS

EDGAR M. WILLIAMS
Read-Rite Corporation

A Wiley-Interscience Publication
JOHN WILEY & SONS, INC.
New York • Chichester • Weinheim • Brisbane • Singapore • Toronto

This book is printed on acid-free paper. ∞

Copyright 2001 by John Wiley & Sons, Inc. All rights reserved.

Published simultaneously in Canada.

No part of this publication may be reproduced, stored in a retrieval system or transmitted in any form or by any means, electronic, mechanical, photocopying, recording, scanning or otherwise, except as permitted under Sections 107 or 108 of the 1976 United States Copyright Act, witbout either the prior written permission of the Publisher, or authorization through payment of the appropriate per-copy fee to the Copyright Clearance Center, 222 Rosewood Drive, Danvers, MA 01923, (508) 750-8400, fax (508) 750-4744. Requests to the Publisher for permission should be addressed to the Permissions Department, John Wiley & Sons, Inc., 605 Third Avenue, New York, NY 10158-0012, (212) 850-6011, fax (212) 850-6008, E-Mail: PERMREQ@WILEY.COM.

For ordering and customer service, call 1-800-CALL-WILEY.

Library of Congress Cataloging-in-Publication Data:
Williams, Edgar M.
    Design and analysis of magneto-resistive recording heads/Edgar M. Williams.
        p. cm
    Includes index
    ISBN 0-471-36358-8 (cloth : alk. paper)
    1. Magnetic recorders and recording Heads. I. Title.

TK5984.W55 2001
621.389'32–dc21   00-035906

Printed in the United States of America.

10 9 8 7 6 5 4 3 2 1

*Put it before them briefly so they will read it,
clearly so they will appreciate it,
picturesquely so they will remember it
and, above all, accurately
so they will be guided by its light.*

Joseph Pulitzer

CONTENTS

FOREWORD xiii

PREFACE xv

1 Overview of Digital Recording Systems 1

Bit Error Rate of Digital Recording Systems, 2
Scaling Rules for Magnetic Recording Systems, 3
Signal Amplitude, 4
Areal Density and Magnetic Spacing, 8
The Written Transition, 10
Magnetic Field above a Written Transition, 12
Head and Preamplifier Noise, 13
Medium Noise (Jitter), 13
Total Noise and System SNR, 15
Nonlinear Interferences, 15
Top-down Design of a Recording System, 17
References, 18

2 Thin-Film Properties for MR Sensors 20

Ferromagnetic Thin Films, 21

Permanent Magnet Materials, 22

Induced Magnetic Anisotropy, 23

Stress Anisotropy and Magnetostriction, 24

Stoner–Wohlfarth Theory, 26

Anisotropic Magnetoresistance Effect, 28

Giant Magnetoresistance Effect, 32

Exchange Coupling With Antiferromagnetic Films, 36

    Dependence of Exchange Energy on Temperature, 37

The Basic Spin Valve, 39

Electrical, Insulating, and Thermal Properties of Thin Films, 42

    Fuchs–Sondheimer Theory of Film Resistivity, 42

    Thermal Properties of Films, 46

Thermal Analysis of MR Heads, 49

    Simple One-Dimensional Analysis of Temperature Rise, 50

    Two-Dimensional Analysis of Temperature Rise, 52

References, 58

3 Coupled Magnetic Films 62

Magnetization Distribution in Identical Films, 63

    Characteristic Length for Flux Propagation, 64

Magnetization Distribution in Nonidentical Films, 66

Influence of Exchange Energy in Coupled Films, 69

    Characteristic Length with Exchange Energy Included, 72

Saturation in Coupled Films, 74

Saturation Dependence on Stripe Height, 78

References, 80

4 Sense Element Biasing, Shielding, and Stabilization 82

Field of a Conducting Ribbon, 83

Field of a Current Sheet between Shields, 86
One Magnetic Film between Shields, 89
    Characteristic Length for a Shielded Film, 90
Two Magnetic Films between Shields, 92
Loss of Signal Flux to the Shields, 93
Shield Dimensions and Sensor Field with Uniform Excitation, 94
Sensor Stabilization with Exchange Pinning, 96
Sensor Stabilization with Permanent Magnet (Hard-Bias) Pinning, 99
    Field of a Uniformly Magnetized Slab, 100
    Magnetic Width of a Hard-Bias Stabilized Sensor, 105
References, 106

5 AMR Head Design and Analysis   108

Biasing the AMR Sensor for Signal Linearity, 109
    Self-Biasing (Shield Biasing), 109
    Shunt Biasing, 120
    Dual-Sensor MR (DSMR) Biasing, 123
    Soft Adjacent-Layer Biasing (Insulating Spacer), 132
SAL Biasing with a Conducting Spacer, 139
    Bias Fields, 140
    SAL and MR Layer Saturation, 141
    MR Layer Transfer Curves, 143
    Bias Curves: Signal and Asymmetry Versus Bias Current and MR/SAL Moment Ratio, 144
    Comparison between Experiments and Calculations, 146
    MR/SAL Moment Ratio and the Operating Current, 150
    Monte Carlo Analysis of Design and Process Control Requirements, 153
References, 158

6 GMR Head Design and Analysis   159

Basic Spin Valves: Top and Bottom Structures, 160

Shielded GMR Response (Transfer Curve), 164
  Sense Current Distribution and Bias Point Shift, 168
  Effective Anisotropy Field of the Free Layer, 171
  Flux Coupling between the Pinned and Free Layers, 172
  Sense Current Polarity for Top and Bottom Spin Valves, 176
  Interlayer Coupling Field, 177
  Output Signal and Asymmetry Versus Sense/Bias Current, 179
  Thermal Analysis of GMR Heads, 188
  AFM Layer and Distribution of Blocking Temperatures, 190
  Monte Carlo Analysis of Basic Spin Valves, 195
Critique of Basic Top and Bottom Spin Valve Designs, 205
Synthetic Spin Valves, 207
  Bias Point and Transfer Curves, 207
  Output Signal and Sense/Bias Curves, 211
  Monte Carlo Analysis of a SSV Design, 213
  Critique of SSV Designs, 214
Bias Compensation Layer Spin Valve, 215
Enhancement of GMR Ratio with Specular Reflection Layers, 216
Dual Synthetic Spin Valve, 217
References, 224

7 Microtrack Profiles 226

Ideal Track Profiles, 227
Nonideal Track Profiles, 228
Anisotropic Flux Propagation in a Sense Layer, 232
Read Width and Magnetic Center, 235
Microtrack Profiles of Narrow Read Tracks, 235
Experimental Microtrack Profiles with Spin Valves, 238
References, 245

8 Characterization of MR Device Function and Reliability   **247**

   Dynamic Functional Tests, 248

      Write Saturation Curves, 248

      Read Bias Curves, 251

      Output Stability Tests, 253

      Thermal Asperities, 260

   Quasi-Static Device and Wafer-Level Tests, 261

      QST Transfer Curves, 262

      Correlation between Quasi-Static and Dynamic Tests, 265

      Signal and Asymmetry Correlations, 266

   Electrostatic Discharge and Sensor Damage, 271

   Operating Temperature and Long-Term Reliability, 278

      Interdiffusion in Thin-Metal Films, 278

      Electromigration, 287

      Prediction of Sensor Lifetime, 291

   References, 297

Subject Index   **299**

FOREWORD

This book is being published at a time when our lives are changing rapidly by the growing use of the Internet. Its operation is critically dependent on the advancement of several technologies, including high-speed networks, client server technology, and the hard disk drive. The ubiquitous disk drive is now a key part of every personal computer sold, and it is the place where the Internet data servers store all of the information. Recently, the disk drive has made its first entry into the video consumer market as a new home video recorder that provides the customer with complete control of the timing and content of the video program material. This widespread adoption of the hard disk drive for on-line information storage is due to the remarkable reduction in the cost of storage on the disk drive in recent years. Ten years ago, a megabyte of data could be stored for about $6 on a hard disk drive. Today, due to the achievement of accelerating increases in the density of data storage on the drive, it costs about a penny to store a megabyte.

Magnetic recording density on disks depends on some critical dimensions in the recording components. In particular, the gap of the recording head and the spacing of the disk from the head must be reduced in order to increase the recording resolution. Also, the width of the recorded track must be decreased to provide more data tracks on the disk. These scaled dimensions, however, are insufficient to achieve higher data recording density unless a higher sensitivity reading head is also introduced. The magnetoresistive head provides the necessary gain in signal from the recorded disk to achieve substantial gains in storage density. Disk drive designers have increased the rate of storage density gains in recent years and have succeeded in widening the cost gap between disk and semiconductor storage, ensuring a continually increasing role for the inexpensive disk.

Early uses of magnetoresistance for magnetic field detection included the ill-fated magnetic bubble technology of the 1970s and the short-lived use as bar-code detectors. Throughout the 1970s, several companies worked on the more difficult development of a magnetoresistive reading head for magnetic recording. IBM succeeded in introducing the first head in a data tape drive in the mid-1980s and, by the early 1990s, also introduced the first such head in a disk drive. During the 1990s, the whole disk drive industry converted to the magnetoresistive head design.

This book presents a self-contained description of the design and analysis of magnetoresistive heads. Written by a leading physicist in the magnetic head business, it includes not only a thorough insight into the design evolution of the first generation of magnetoresistive heads but also the more recent giant magnetoresistance designs.

C. DENIS MEE

PREFACE

The computing, information processing, and networking industries have undergone rapid expansion in applications, all of which require improved data storage and retrieval systems. For many years hard disk drives have provided cost-effective on-line storage and retrieval of large quantities of data; hard drives are in the middle of the storage hierarchy because they are nonvolatile, cheaper than semiconductor memory, and faster than tape drive systems used as backup for other storage technologies. Demands for increased storage capacity, higher speed, increased reliability, and reduced cost have led to significant and impressive reductions in the size of the magnetic bit, or bit cell, in hard disk drives. The advent of the personal computer spurred reductions in the size of hard disk drives, which in turn led to smaller recording heads and disk diameters. Because data are retrieved in the form of electrical signals arising from magnetic transducers, recording head engineers and scientists have continuously improved these devices to provide useful signal levels in the presence of electrical noise as well as acceptable data integrity and long-term reliability.

The first hard disk drives were shipped to customers in 1957 at an areal density of 2000 bits/in.^{2}; the linear bit and track densities were 100 bpi and 20 tpi, respectively, so the size of the bit cell was about 250 µm long by 1250 µm wide. In the forty-plus years of hard disk drive history and evolution, volume production of inductive write and read heads has been pushed to areal densities approaching 5.0 Gbits/in.^{2}, and bit cells have concomitantly shrunk to about 0.10 µm long by 1.5 µm wide. Inductive technology has been demonstrated up to 5 Gbits/in.^{2}, but the signal-to-noise ratio and electrical bandwidth of inductive devices are limited, and technologies based on magnetoresistance effects [anisotropic magnetoresistance (AMR) and giant magnetoresistance (GMR)] have displaced the mature inductive technology in disk drive products operating at areal densities beyond 5 Gbits/in.^{2}. The rapid evolution in recording device physics and production processes has created significant emphasis on continued learning in the storage industry and academic institutions.

At present, there are a number of books covering the general subjects of magnetic recording, magnetic recording head technology, magnetic recording theory, and disk drive technology. In approximate chronological order they are Begun (1949), Hoagland (1963), Mee (1964), Camras (1985), White (1985), Mee and Daniel (1987), Mallinson (1987 and 1993), Cireanu and Gavrila (1990), Ruigrok (1990), Hoagland and Monson (1991), Bertram (1994), Jorgensen (1996), Mallinson (1996), Mee and Daniel (1996), Ashar (1997), Wang and Taratorin (1999), and Comstock (1999). However, the level and scope of these books are such that there exists a need for yet another book that addresses engineers and scientists working in the field of magnetic recording who want a deeper and practical understanding of specific technical challenges associated with designing or applying magnetoresistive transducers to hard disk drives operating at high areal densities. This book is written for this audience, and the level of presentation is appropriate for senior undergraduates, advanced-degree graduates, and data storage professionals who wish to maintain their skills in a rapidly changing field. I have assumed the reader has had courses in physics, electricity and magnetism, applied mathematics, and introductory magnetic recording subjects.

There is an old proverb roughly equivalent to the statement "Feed a hungry man and he will live for a day, but teach him to fish and he will feed himself for life." I have taken that attitude toward the readers of this book: magnetic recording concepts are introduced, the properties of materials used in recording heads are presented, the physics of magnetoresistive (MR) devices is developed in *closed-form mathematical language*, the mathematical tools are applied to a variety of MR recording head designs, and finally, back-of-the-envelope calculations are done with specific values of the design parameters to demonstrate use of the analytical tools. Experimental results are compared with the calculated curves to build confidence in the reader that the tools are indeed useful and sufficiently accurate for engineering work. The italics used above emphasizes the importance of closed-form solutions in revealing the relationships among many variables or parameters and in quantifying the relative importance of a given variable regarding its influence on device behavior. The physics of coupled magnetic films exhibiting AMR and GMR effects are developed using different methods, as there is much to learn by solving equivalent problems in different ways. Some design and analysis problems require micromagnetic finite-element numerical techniques, and these situations will be treated as necessary; however, the advantage of analytical techniques over finite-element treatments is that results can be quickly obtained and presented in forms that directly show the many variables involved and reveal how these design variables impact the function of MR heads in disk drives. There are, for example, over 30 individual parameters necessary to define an AMR or GMR head, so over a range of typical values, the electrical and thermal behavior can be quite complicated.

Chapter 1 gives a brief overview of digital recording systems, presents useful scaling rules for increasing the number of magnetic bits stored on disk in a typical hard disk drive, and builds the motivation for signal improvement gained with MR readback transducers. Chapter 2 discusses thin-film magnetic materials as well as the electrical and thermal properties required for use in MR devices. The physical and mathematical tools for analyzing the temperature rise of MR heads are developed in this chapter. The physics of coupled magnetic films and the magnetic fields associated with current flow in films are analyzed in Chapters 3 and 4. These tools are taken to the point where the material properties and geometries introduced in Chapters 1 and 2 may be substituted directly into the analytical equations such that the electrical and thermal behavior may be studied theoretically. In Chapters 5 and 6, these tools are applied to realistic design and analysis challenges found with AMR and GMR sensors, respectively. The side-reading behavior of MR devices is examined theoretically in Chapter 7, and experimental microtrack profiles are presented for consideration by the reader. Chapter 8 closes this book with a discussion of dynamic and quasi-static tests for assessing MR sensor function, stability, thermal damage, and long-term reliability. The relationships between bias current, operating temperature of a device, and slow changes in device structure arising from electromigration or interdiffusion of films are treated theoretically and experimentally. The book ends by showing how the lifetime of a sensor depends on the design parameters that influence head resistance and temperature. The reader will find the mathematical treatment of the physics and magnetics of MR heads, which requires approximately 390 equations, is supplemented with over 270 figures and graphs.

Units and dimensions in this book are centimeters-grams-seconds (cgs)–Gaussian for magnetic fields and properties and micrometers, nanometers, or angstrom units for linear dimensions of devices. Electrical signals are in volts or microvolts, currents are in amperes or milliamps, and resistivities are in microhm-centimeters. Thermal properties are in degrees kelvin per watt for thermal resistance, watts per meter-kelvin for thermal conductivity, parts per degree kelvin (or celsius) for temperature coefficient of resistance, and temperatures are given in degrees kelvin or celsius. These measures seem to be what many workers use and are accustomed to reading, notwithstanding the continued promulgation of International System (SI) of units in the world. Some of the data (kindly shared with me by co-workers) may be plotted with distance measured by equipment that reports results in microinches. Occasionally, a problem will be solved in SI units and then converted to the cgs–Gaussian system. This approach is sometimes helpful to readers familiar with only one system.

Many individuals helped me in the preparation of this book, and I have tried to make a complete list, but mistakes will happen and I may have overlooked someone who kindly provided me with information or other support; I offer my apologies to those individuals. This book owes its genesis to David Heim, who shared my belief that MR heads could be designed with closed-form analytic equations, guided me through the derivations of the equations for the physics of coupled magnetic films, and then gave me encouragement, inspiration, and generous support by reviewing Chapters 3–6. Special acknowledgment goes to C. Denis Mee, who exhorted me to write this book and provided me with the Foreword. The clarity of this book was improved by the critical reviews of various chapters and advice given by Subrata

Dey, Matthew Gibbons, David Hannon, Gordon Hughes, Marcos Lederman, Curtis Macchioni, John Mallinson, Denis Mee, Aric Menon, Stone Shi, and Joyce Thompson. Their interest, questions, critiques, and willingness to educate me have been most helpful. I received gracious support from many who cheerfully shared data and graphs and who took their time to teach me about the significance of their results. Thank you Geoff Anderson, Irmela Barlow, Jason Cain, Caleb Chang, Seila Chim, Vincent Doan, David Hannon, Yiming Huai, Chin-Ya (Stephanie) Hung, Chung Lam, Francis Liu, Ernest Louis, Curtis Macchioni, Aric Menon, Tue Ngo, Kroum Stoev, and Hua-Ching Tong. The executive staff of Read-Rite Corporation has been very supportive of this effort. In particular, Mark Re has closely followed the progress of each chapter and given me his full administrative support. In addition, Mark Re, Alan Lowe, and Cyril Yansouni have, in acknowledgment and support of my efforts at the time of my retirement, granted me the status of "Read-Rite Fellow Emeritus." Sherry McVicar took precious time from her busy schedule to help me prepare the index to this book. Special acknowledgement goes to my assistant, Joyce Travis, who prepared many of the figures in this book, handled the requests for permission to reproduce copyrighted material, and through her involvement made the task of writing a book more pleasant and less burdensome.

EDGAR M. WILLIAMS

Palo Alto, California
January 2000

REFERENCES

Ashar, K. G., *Magnetic Disk Drive Technology,* IEEE Press, New York, 1997.

Begun, S. J., *Magnetic Recording,* Murray Hill Books, New York, 1949.

Bertram, H. N., *Theory of Magnetic Recording,* Cambridge University Press, Cambridge, 1994.

Camras, M., *Magnetic Tape Recording,* Van Nostrand Reinhold, New York, 1985.

Cireanu, P. and Gavrila, H., *Magnetic Heads for Digital Recording,* Elsevier, Amsterdam, 1990.

Comstock, R. L., *Introduction to Magnetism and Magnetic Recording,* Wiley-Interscience, New York, 1999.

Hoagland, A. S., *Digital Magnetic Recording,* Wiley-Interscience, New York, 1963.

Hoagland, A. S. and Monson, J. E., *Digital Magnetic Recording,* 2nd ed., Wiley-Interscience, New York, 1991.

Jorgensen, F., *The Complete Handbook of Magnetic Recording,* 4th ed., TAB Books, an imprint of McGraw-Hill, New York, 1996.

Mallinson, J. C., *Foundations of Magnetic Recording,* 1st ed. (1987), 2nd ed., Academic, San Diego, 1993.

Mallinson, J. C., *Magneto-Resistive Heads, Fundamentals and Applications,* Academic, San Diego, 1996.

Mee, C. D., *Physics of Magnetic Recording,* North-Holland, Amsterdam, 1964.

Mee, C. D. and Daniel, E., *Magnetic Recording,* Vols. 1 and 2, McGraw-Hill, New York, 1987.

Mee, C. D. and Daniel, E., *Magnetic Storage Handbook,* 2nd ed., McGraw-Hill, New York, 1996.

Ruigrok, J. J. M., *Short-Wavelength Magnetic Recording,* Elsevier Advanced Technology, Oxford, 1990.

Wang, S. H. and Taratorin, A. M., *Magnetic Information Storage Technology,* Academic, San Diego, 1999.

White, R. M., *Introduction to Magnetic Recording,* IEEE Press, New York, 1985.

DESIGN AND ANALYSIS OF MAGNETORESISTIVE RECORDING HEADS

1

OVERVIEW OF DIGITAL RECORDING SYSTEMS

Data storage and retrieval systems are structured under engineering architectural rules. Hard disk drives place data in records that are easily addressed according to known locations at fixed radii and angular positions within a multitude of disk surfaces. Roughly one-third of the cost of a typical disk drive is found in the head and disk components, so an important motive is discovered for increasing the storage density (megabytes per disk) without increasing the number of components. In 1983, a 10-Mbyte 5.25-in. drive (four-disk) price was roughly $500, or $50 per megabyte. In 1987 storage capacity increased to about 400 Mbytes (eight disks) at a price of $2000, or $5 per megabyte. Today (late 1999) 12 Gbytes (12,000 Mbytes) on 3.5-in. drives (two disks) sell at less than $120, or roughly $0.01 per megabyte. Within this time frame, areal density, [bits per inch times tracks per inch (bpi · tpi)] has gone from a few megabits per square inch to about 5 Gbits/in^2, the bit cell has shrunk from roughly 3×70 μm^2 to about 0.09×1.4 μm^2, and data-sensing technology has switched from inductive to magnetoresistive (MR) heads. Over 140 million disk drives were produced in 1998 and virtually all of these drives exploited MR recording heads and longitudinal thin-film media. Manufacturers of recording heads have focused research-and-development (R&D) activities on MR technology, and factories have retooled to supply the industry with MR read heads combined with inductive write heads. Trend-FOCUS (1998) placed the total MR market at 441.5 million units in 1997, which was split 43.2% OEM (original equipment manufacturer) and 56.8% captive disk drive and head manufacturers and claimed that shipments of MR and giant magnetoresistance (GMR) heads totaled 636 and 50.7 million, respectively, in 1998. For 1999 Trend-FOCUS has followed the transition from anisotropic magnetoresistance (AMR) to GMR technology and projects markets of 82.1 and 752.9 million AMR and GMR heads, respectively.

2 OVERVIEW OF DIGITAL RECORDING SYSTEMS

This chapter examines magnetic storage system design requirements for acceptable data integrity at the output of a pulse detector, down through a preamplifier/write driver, and finally to the head and disk components. There are additional levels of error detection and correction hardware and firmware in disk drives that further improve data integrity, but these considerations go beyond the scope of this book. The discussion emphasizes signal, noise, linear and nonlinear effects, and the dependence of a raw (uncorrected) bit error rate (BER) on these factors. Because the writing process creates what is read by the MR sensor, we first briefly discuss the writing of transitions, magnetic properties of recording media, and several transition nonlinearities. As the bit cell shrinks, the requirement for an acceptable BER establishes strong motives for improving the signal amplitude levels and reducing sources of noise and interferences or distortions that degrade the BER to unacceptable levels.

BIT ERROR RATE OF DIGITAL RECORDING SYSTEMS

The BER in high-density disk recording is a subject of intense investigation, and identifying the various sources of error is a significant challenge. Yeh and Wachenschwanz (1997) have studied this problem in systems using sampling detectors known as PRML (partial-response maximum-likelihood) detection; they expanded the concept of signal-to-noise ratio (SNR) to include signal variations arising from linear and nonlinear intersymbol interference (ISI) in pseudorandom sequence (PRS) recording. With Gaussian noise added to random data sequences, the variance in readback amplitude becomes the sum of the noise variance (power) and the ISI variance; they call this the *mean-square error* (MSE) variance. That is,

$$\sigma_{MSE}^2 = \sigma_{Noise}^2 + \sigma_{ISI}^2, \qquad (1.1)$$

so with proper treatment, interference is "noise like" [see Williams, p. 253, in Arnoldussen and Nunnelley (1992)]. Yeh and Wachenschwanz (1997) discuss estimation of the BER using a complementary error function of the signal-to-MSE ratio, where the "signal" is normalized to unity. The relation given in their paper is

$$BER \sim 0.75 \; \text{erfc}\left(\frac{1}{2\sqrt{2}\sigma_{MSE}}\right), \qquad (1.2)$$

which is useful as an introduction to the analysis of root causes of bit errors. The analysis of amplitude variations in Viterbi detectors gives a more accurate estimate of BER; however, this subject goes beyond the scope of this book. Archival digital storage systems require extremely low error rates, so (1.2) is plotted on a loglinear basis in Fig. 1.1 to emphasize a raw BER in a design range of roughly 10^{-6}–10^{-10}. As a hypothetical example, assume a design center of $BER(z) = 10^{-8}$; then $z = 4.05$ approximately satisfies this requirement and a signal [zero-to-peak (0-p)] of

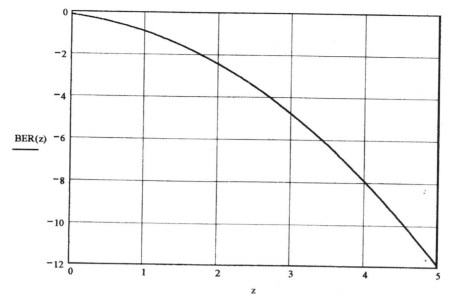

FIGURE 1.1 Log to base 10 of the BER.

$4.05 \times 2.828\sigma_{MSE}$ would be necessary. To take this idea one step further, assume the system has no ISI variance, or from (1.1), $\sigma_{MSE} = \sigma_{Noise}$. Achieving an error rate of 10^{-8} would be roughly equivalent to a signal-to-noise ratio [SNR(0-pk/RMS)] of 11.45, where RMS is the root-mean-square. Expressed in the conventional decibels, $SNR = 20 \log_{10}(11.45) = 21.2$ dB.

At very high bit densities, the nonlinear writing process creates magnetic transitions that are distorted in location and amplitude as a result of magnetic fields from adjacent magnetic transitions. These distortions are sometimes called *nonlinear transition shift* (NLTS) or *nonlinear distortion* (NLD) and *partial erasure* or *nonlinear amplitude loss*. In other words, ISI is composed of linear readback and nonlinear writing phenomena. The reader should consult the literature on magnetic recording signal processing for more on this subject. See, for instance, Yeh and Wachenschwanz (1997), Palmer et al. (1987), or Perkins and Keirn (1995). Extending the example above, if ISI is made equal to noise (perhaps by adding NLTS) the MSE increases on an RMS basis by the factor $2^{1/2}$, or 3 dB, and the BER degrades from 10^{-8} to an unacceptable level of $BER(4.05/1.414) = 10^{-4.3}$ unless SNR is increased to compensate for the additional NLTS.

SCALING RULES FOR MAGNETIC RECORDING SYSTEMS

Scaling rules of thumb are useful for estimating critical dimensions and relationships in recording systems. There is a correspondence between bit and track densities for given areal density (AD) points, and for many years disk drive producers have followed bpi/tpi ratios (the *bit cell aspect ratio*) R_{bc} in the approximate range of 15–25, and (AD) is simply given by the relation $AD = R_{bc} \cdot tpi^2 = bpi^2/R_{bc}$. As areal density increases beyond 5 Gbits/in^2, the bit cell aspect ratio appears to be moving to smaller values (R_{bc} values of 8–16 approximately). Justification for this trend arises from the observation that medium noise scales with the square root of track width whereas signal scales directly with width, such that high track densities will be preferred over high bit densities. The rule of thumb for bit cell aspect ratio facilitates exploration of product designs and assessment of consequences regarding track width tolerances for write and read heads and for device specifications. The disk drive industry has followed track width design rules based on the track-to-track distance, or track pitch:

$$\text{Write width} \simeq (0.70-0.9) \cdot \text{track pitch} \tag{1.3}$$

$$\text{Read width} \simeq 0.80 \cdot \text{write width} \simeq (0.60-0.80) \cdot \text{track pitch} \tag{1.4}$$

In recent technology demonstrations at 36 Gbits/in.^{2} the bit cell was 0.044 by 0.41 μm^2 ($R_{bc} = 580$ kbpi/62 ktpi = 9.3) and the GMR read width was approximately 0.30 μm, or 73% of the track pitch.

SIGNAL AMPLITUDE

Magnetoresistive heads exploit the AMR effect or the GMR effect. In either case, the change in resistance ΔR in a magnetic field divided by the resistance R in the zero-field state defines a figure of merit $\Delta R/R$ that is useful for comparing different technologies and devices. AMR devices commonly have a figure of merit of about 1.5–2%, while GMR/spin valve devices are roughly 4–15%, depending on design and complexity. These devices must be energized with a sense current to produce an electrical signal. That is,

$$\text{Signal} = I \, \Delta R(H) = I \frac{\Delta \rho(H)}{t} \frac{W}{h} \quad (V) \tag{1.5}$$

where I is the sense current, $\Delta \rho(H)$ is the resistivity change with magnetic field H, t is the film thickness, W is the sensor width, and h is its height. At high areal densities the track width W must reduce; in fact, the geometries and properties of the entire sensor must scale to maintain useful levels of signal. The task of this book is to delve into the details behind (1.5) and examine the design and analysis challenges of MR heads. AMR devices are inherently nonlinear, and one of the many design challenges is to bias the sensor for reasonably linear operation over a useful range of magnetic fields. GMR devices give more signal and have improved linearity but are more complicated and have a number of very thin critical layers in their construction. Because MR devices must be excited with a sense current, they heat up and thermal issues must be addressed. In addition, MR sense elements require magnetic shielding for operation at high bit densities.

The primary use of shields is to define a limited region of sensitivity to *external* flux for an MR sensor; however, the *internal* magnetic field environment for shielded conductors and magnetic sensors is altered as well. The unshielded sensor of Hunt (1971) produces a pulsewidth that is the geometrical mean of the element height h and its effective spacing $d + a$ from the surface of an arctangent magnetic transition $M_x(x) = (2M_r/\pi)\tan^{-1}(x/a)$,

$$\text{PW50} \simeq \sqrt{4(d+a)(d+a+h)}. \tag{1.6}$$

Mallinson (1996, p. 64) gives an equivalent expression $\text{PW50} = 2[d(d+h)]^{1/2}$ for a sharp transition. It is apparent an unshielded MR sensor would provide unacceptably wide pulses for useful heights $h \sim 1$ μm. Potter (1974) addressed this problem and analyzed a shielded MR sensor as two back-to-back Karlquist heads with the MR element of thickness t centered between shields ($G = g + t + g$). Bertram (1995) modified Potter's expression for the readback flux to include the medium thickness in the effective spacing $y = d + a + \delta/2$. Using a combination of Potter and Bertram, the flux entering a symmetrically shielded MR element for an arctangent transition is

$$\phi(\bar{x}, y) = \frac{2}{\pi} 4\pi M_r \, \delta W \cdot 10^{-8} \frac{y}{g} \left[f\left(\frac{\bar{x}+g+t/2}{y}\right) - f\left(\frac{\bar{x}+t/2}{y}\right) \right.$$
$$\left. + f\left(\frac{\bar{x}-g-t/2}{y}\right) - f\left(\frac{\bar{x}-t/2}{y}\right) \right], \tag{1.7}$$

where

$$f(z) = z\tan^{-1} z - 0.5 \log_e(1+z^2). \tag{1.8}$$

The flux is in webers for M_r in emu/cm^3 and all dimensions in centimeters (1.0 Maxwell = 10^{-8} Weber.) Normalized plots of (1.7) given by $F(x, y) = \Phi(x, y)/[4\pi M_r \, \delta W \cdot 10^{-8} y/g]$ and a simple Lorentzian function (1.9) are compared in Fig. 1.2. The Lorentzian is defined by the expression

$$L(x) = \frac{V_0}{1 + (2x/\text{PW50})^2}, \tag{1.9}$$

where V_0 is the zero-to-peak signal amplitude, x is the position along a track ($x = vt$, v is the disk surface velocity and t is time). The Lorentzian is physically incorrect as

OVERVIEW OF DIGITAL RECORDING SYSTEMS

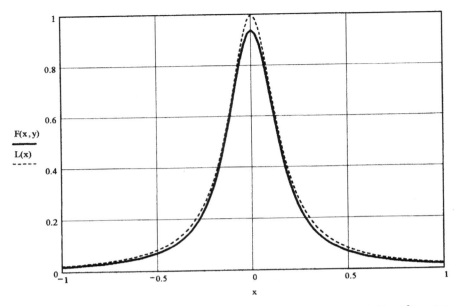

FIGURE 1.2 Comparison of pulses based on $F(x) = x \cdot a\tan(x) - 0.5 \ln(1 + x^2)$ and the Lorentzian $L(x) = 1/(1 + x^2)$.

a description of the readback process; however, it is simple and its shape approximates that of (1.7). The PW50 is approximated quite well by the relation

$$\text{PW50} \simeq \begin{cases} \left[\dfrac{G_{ss}^2 + t^2}{2} + 4(d+a)(d+a+\delta)\right]^{0.5} & \text{(single element),} \quad (1.10a) \\ \left[\dfrac{G_{ss}^2 + (g+2t)^2}{2} + 4(d+a)(d+a+\delta)\right]^{0.5} & \text{(dual element),} \quad (1.10b) \end{cases}$$

where G_{ss} equals $2g_1 + t$ or $2g_1 + g + 2t$ for single- and dual-element heads, respectively. Other approximations for PW50 are published; for example, see Smith (1991), Champion and Bertram (1995), or Bertram (1995). The parameters d, a, and δ refer to the magnetic spacing (bottom edge of the MR sensor to the magnetic surface of the disk), the length ($L = \pi a$) of the written magnetic transition, and the thickness δ of the magnetic layer on the disk surface, respectively. Each of these parameters will normally be scaled down to produce acceptably narrow pulses for high bit densities. Magnetic spacing d is fundamental to recording physics, and it emerges as a natural basis for scaling the other parameters; systems can be designed using the relations $G_{ss} \simeq 5d$, $a \simeq d$, and $\delta \simeq d/2$, for which PW50 $\simeq 5.7d$. That is, in a well-designed system, PW50 would scale with magnetic spacing. With $G_{ss} = 0.20$ μm, $d + a = 0.10$ μm, $\delta = 0.02$ μm, and stripe height $h = 1.0$ μm, estimates of shielded and unshielded pulsewidths for a single-element head are approximately 0.26 and 0.66 μm, respectively. That is, shielding can reduce PW50 by more than a factor of 2.

It is common to find that the literature on signal processing treats the readback signal with simple approximations for describing the amplitude and pulse shape. The Lorentzian pulse is especially useful because it agrees with experimental waveforms fairly well, and it has simple analytical properties. When data are stored, sequences of transitions are written, and the readback signal is a linear superposition of the responses to individual transitions. The total response for an infinite sequence of alternating polarity Lorentzians was published by Comstock and Williams (1973):

$$f(x, P, B) = \sum_{n=-\infty}^{\infty} (-1)^n V(x - nB) = V_0 \frac{\pi P}{2B} \frac{\sinh(\pi P/2B)\cos(\pi x/B)}{[\cosh^2(\pi P/2B) - \cos^2(\pi x/B)]} \quad (1.11)$$

where $P = $ PW50 and B is a constant bit cell length or distance between pulses (i.e., $1/B = $ bit density). This result is compact and useful in communicating the essential features of readback signals and amplitude dependence on recording density. Notice the argument scales with P/B, so pulsewidth P must be reduced along with reductions in bit cell length B to maintain useful levels of amplitude. Figure 1.3 shows a family of plots for $f(x, P, B)$ with fixed P and parameter B, while Fig. 1.4 shows the peak signal $f(x = 0, P, B)$ for variable B and parameter P.

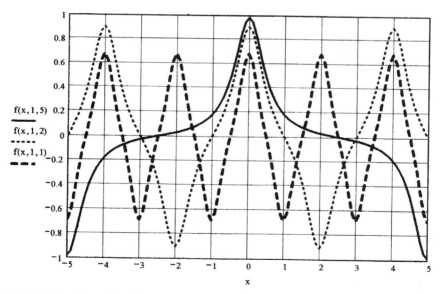

FIGURE 1.3 Normalized signal for infinite sum of alternating polarity Lorentzian pulses given by $f(x, P, B)$, where x is position along a track, P is the pulsewidth (constant at 1.0), and B is the distance between pulses ($B = 1, 2, 5$).

OVERVIEW OF DIGITAL RECORDING SYSTEMS

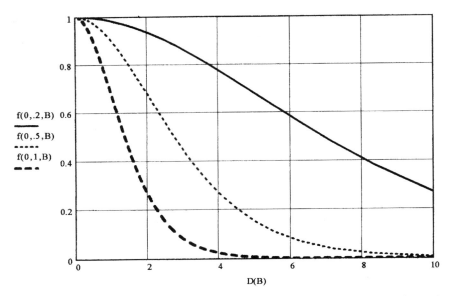

FIGURE 1.4 Normalized zero-to-peak signal for infinite sum of Lorentzians. Pulse peak at $x = 0$ for $P = 0.2, 0.5, 1.0$ units. Linear density $1/B$ is variable.

Signal processing engineers and the designers of detectors have defined the concept of *channel density* as a figure of merit for comparing channels. In the literature, this concept is identified with the relation

$$U = \text{channel density} = \frac{\text{PW50}}{B} = \text{PW50} \cdot \text{linear density}, \quad (1.12)$$

and today's PRML channels operate at useful error rates with U in the range of 2–3 or so. The relation given in (1.11) simplifies further when (1.12) is substituted for P and B:

$$\frac{f(0, P, B)}{V_0} = \frac{(\pi/2)U}{\sinh[(\pi/2)U]} \quad (1.13)$$

Detector channels operating at high user densities (say $U \geq 2.5$) place less burden on head and medium components because PW50 can be somewhat larger for a given linear density ($1/B$), and requirements can be relaxed for $G_{ss}, d, a,$ and δ.

AREAL DENSITY AND MAGNETIC SPACING

The history of rotating disk magnetic storage begins in 1957 with the IBM 350, which was the first production hard disk drive with movable recording heads. The IBM 350 stored data at 2000 bits/in.^{2}, and the magnetic head flew over the disk surface at a spacing of 800 µin., or 20 µm. Harker et al. (1981) review the first 25 years of disk drive technology and give some details regarding areal densities and critical geometrical parameters of IBM products. By 1981, the IBM 3380 drive was at about 12 Mbits/in.^{2} and heads were flying at 12 µin. (0.30 µm). One of the most critical parameters is the spacing between the recording head gap and the magnetic surface of the medium; this is called the *magnetic spacing*. If the magnetic disk and magnetic head have no protective coatings, then the magnetic spacing nearly equals the flying height of the head. When heads are mechanically lapped and polished, there exists a small recession (70 Å or less) of the head pole tips below the air-bearing surface of the head slider; this is called *pole tip recession* (PTR). Today, all recording heads and disks have a protective overcoat of diamondlike carbon (DLC) or other hard material applied to their surfaces, so the magnetic spacing and flying height are no longer identical.

The relationship of areal density to magnetic spacing is easily derived from PW50, channel density (U), bit cell ratio (R_{bc}), signal amplitude (at the highest density), and the scaling rules developed above. Reading from (1.9), (1.12), and (1.13), the areal density can be written

$$\text{Areal density} = \frac{\text{bpi}^2}{R_{bc}} = \frac{U^2}{R_{bc}\,\text{PW50}^2} = \frac{U^2}{R_{bc} K^2 d^2}, \qquad (1.14)$$

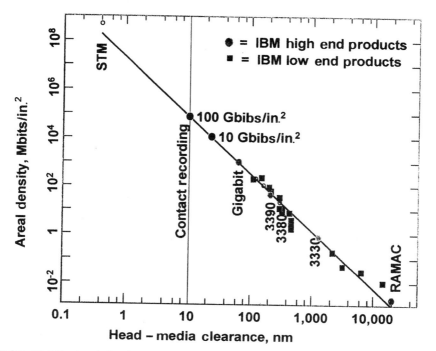

FIGURE 1.5 Areal density as a function of head–medium clearance [from Harker et al. (1981) and Ashar (1997)].

where the scaling rule PW50 = Kd is used. On the log–log plot of Fig. 1.5, (1.14) fits very well the published data of Harker et al. (1981) and Ashar (1997). With overcoats on head sliders and disk surfaces, the actual clearance between head and disk (the "flying height") is nearly $0.50d$. The hard coatings (e.g., DLC) are each normally designed to be roughly $0.25d$, and PTR is held to about $0.10d$. The 3σ tolerance on flying height variations will normally be about 15–20% of the nominal value.

THE WRITTEN TRANSITION

The highly nonlinear writing process can be treated with analytic approximations that give direct and immediate insight regarding the written transition, its magnitude, sharpness, and the dependence of these attributes on the recording head and properties of the magnetic medium. The Williams and Comstock (1971) model has become the archetype for this type of analysis, and their results are given here. They assume a magnetization transition of the form $M(x) = (2M_r/\pi)\tan^{-1}(x/a)$ in a medium whose magnetic properties are described by the remanence M_r, coercivity H_c, coercive squareness S^*, remanent coercivity $H_{cr} = H_c/r$, $r = (3 + S^*)/4$, and thickness δ. The transition is longitudinal and is written by the x-component of a Karlquist (1954) head,

$$H_x^h(x,y) = \frac{H_0}{\pi}\left[\tan^{-1}\frac{(x+g/2)}{y} - \tan^{-1}\frac{(x-g/2)}{y}\right] \quad (1.15)$$

at a position $H_x^h(x = x_0, y) = H_{cr}$, where the transition is centered; H_0 is the field at $x = y = 0$ and g is the gap length. The head field gradient is given by

$$\frac{dH_x^h}{dx} = -\frac{QH_c}{y} = \frac{-2x_0 H_0}{\pi g H_c}\sin^2\left(\frac{\pi H_c}{H_0}\right)\frac{H_c}{y} \quad (1.16)$$

Here, Q is the normalized write field gradient and is valid only for $H_0 \geq 2H_c$. This relation for Q from Bertram (1994, Chap. 8) is graphed in Fig. 1.6. Williams and Comstock (1971) calculate the intermediate transition (with parameter a_1) that exists at the trailing edge of the energized writing head, and then find the final transition (with parameter a) that exists after the medium has moved far away from the writing head (i.e., the field is turned off) and the transition has relaxed under the influence of its self-demagnetizing field:

$$\frac{a_1}{r} = \frac{y(1-S^*)}{\pi Q} + \left[\left(\frac{y(1-S^*)}{\pi Q}\right)^2 + \frac{2M_r\delta}{H_c}\frac{2y}{Qr}\right]^{1/2} \quad (1.17)$$

$$a = \frac{a_1}{2r} + \left[\left(\frac{a_1}{2r}\right)^2 + \frac{2\pi M_r\delta a_1}{(3+S^*)H_c}\right]^{1/2} \quad (1.18)$$

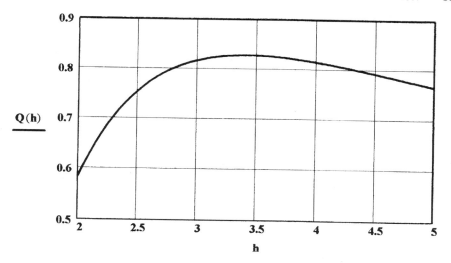

FIGURE 1.6 Normalized head field gradient [Eq. (1.16)] for Karlquist recording head field plotted as a function of normalized field $h = H_0/H_c$, where H_0 is the x-component of the field at $x = 0$, $y = 0$.

The final transition length ($L = \pi a$) is always greater than the intermediate one; that is, the written transition broadens (relaxes) as it moves away from the recording field. Figure 1.7 shows plots of intermediate and final transition parameters; the sharpest transition is obtained at the highest write field gradient. These calculations assume

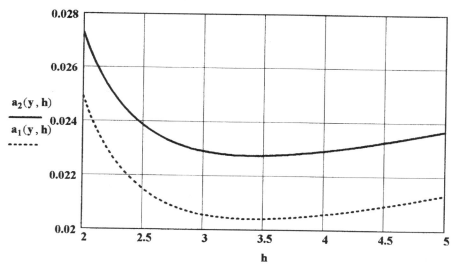

FIGURE 1.7 Williams–Comstock transition parameters for a transition in the presence of the recording field (a_1) and for the relaxed transition (a_2) in the absence of the recording field.

the write head never saturates magnetically, which is almost never the case in practice. When a head goes into saturation, the final transition parameter increases (perhaps by a factor of two- or three-fold in extreme cases) with increasing write current.

MAGNETIC FIELD ABOVE A WRITTEN TRANSITION

The sensing element of an MR head is excited by the magnetic field of written transitions. Potter (1970) analyzed the demagnetizing field for a single arctangent transition written longitudinally in a recording medium. The geometry of the head and medium system is shown in Fig. 1.8. The horizontal (H_x) and vertical (H_y) field components are given by the relations

$$H_x(x, y) = 4M_r \left\{ \tan^{-1}\left[\frac{x(\delta/2 + y)}{x^2 + a^2 + |\delta/2 + y|a}\right] \right.$$

$$\left. + \tan^{-1}\left[\frac{x(y - \delta/2)}{x^2 + a^2 + |\delta/2 - y|a}\right] \right\}, \quad (1.19)$$

$$H_y(x, y) = 2M_r \log_e \left[\frac{x^2 + (|\delta/2 + y| + a)^2}{x^2 + (|\delta/2 - y| + a)^2}\right]. \quad (1.20)$$

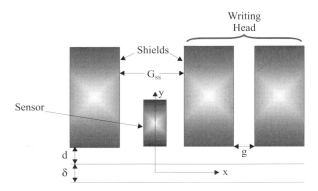

FIGURE 1.8 Geometry of a merged MR read, inductive write head system where the trailing write pole serves as one of the shield layers of the MR device.

The maximum field strength at a distance $y = d + \delta/2$ above the center of the medium is

$$H_y(x=0, y = d + \delta/2) = 4M_r \log_e\left[\frac{d+a+\delta}{d+a}\right] \simeq \frac{4M_r\,\delta}{d+a} \quad \text{Oe.} \quad (1.21)$$

This expression is a useful approximation for estimating the readback signal of an unshielded MR sensor and its dependence on medium properties and head–medium spacing; the injected sensor flux is approximately $\Phi_s \simeq 4\pi W t H_y(x, y)$. The analysis of Potter (1974), leading to the relation given in (1.7), must be used for signal calculations with shielded MR heads.

HEAD AND PREAMPLIFIER NOISE

The major contributions to noise in disk drives are from the magnetic head, the preamplifier, and spatial jitter of written transitions. Head and preamplifier noises arise from electronic thermal fluctuations and obey Gaussian statistics; their noise spectral densities (NSDs) are reasonably constant over the bandwidths of interest and the noise voltage is readily computed using Nyquist's theorem [see Williams, Chap. 7, in Arnoldussen and Nunnelley (1992)]:

$$e_n = \sqrt{4K_B T R(f)\,\Delta f} = \text{NSD}\sqrt{\Delta f} \quad \text{V, RMS,} \quad (1.22)$$

where $k_B = 1.3085 \times 10^{-23}$ J/K (Boltzmann's constant), T is the absolute temperature in kelvins, $R(f)$ is the noise-equivalent resistance (in ohms) which depends on frequency (f in hertz), and Δf is the noise bandwidth (in hertz) of the system. Magnetoresistive sensors are designed to meet customer requirements, and today the head designer is normally constrained to a maximum of about 60 Ω noise-equivalent resistance for the MR head (including the electrical connection network); at a temperature of 328 K, the head and connection NSD would be about 1.0 nV/$\sqrt{\text{Hz}}$. Preamplifier NSD varies somewhat, but it is normally in the range of 0.5–0.7 nV/$\sqrt{\text{Hz}}$. The combined electronic NSD from head and preamplifier would then be about 1.2 nV/$\sqrt{\text{Hz}}$, and with a system bandwidth of 100 MHz, the electronic noise voltage would be 12 μV (RMS).

MEDIUM NOISE (JITTER)

Thin-film disk transition jitter is a complicated subject; however, much insight has been developed since about 1983. A recording medium is composed of single-domain magnetic grains roughly 100 Å in diameter with nonmagnetic grain boundaries of 10 Å or less. As magnetic transitions are written along a track, statistical fluctuations in transition position and length arise from the switching properties of magnetic grains at a particular location; this is called *transition jitter*. Zhu in Chapter 6 of Arnoldussen and Nunnelley (1992) and Bertram (1994) are excellent references to the subject of noise in thin-film media, and many useful papers have appeared in recent years. Tarnopolsky and Pitts (1997) and Xing and Bertram (1997), for example, discuss medium-limited SNR; they show that medium noise power is composed of jitter in transition position and in width. Xing and Bertram show that the normalized noise power spectrum NP(k) for transition noise can be written as

$$\mathrm{NP}(k) = \frac{B}{|V_{\mathrm{iP}}(k)|^2} \mathrm{PSD}(k) \simeq \sigma_{\mathrm{jitter}}^2 k^2 + \sigma_{\mathrm{width}}^2 k^4, \quad (1.23)$$

where $V_{\mathrm{iP}}(k)$ is the Fourier transform of the isolated pulse and k is the wavenumber. The position and width variances depend on read track width W, transition parameter a and cross-track correlation length s:

$$\sigma_{\mathrm{jitter}}^2 = \frac{\pi^4 s a^2}{48 W}, \quad \sigma_{\mathrm{width}}^2 = \frac{\pi^4 a^2}{60} \sigma_{\mathrm{jitter}}^2. \quad (1.24)$$

Transition position and width jitter are related to the microstructure, grain size, intergranular coupling, and magnetic properties of the medium. Sato et al. (1996) and McKinlay et al. (1996), for example, discuss these subjects for specific magnetic alloys and relate them to medium noise. Small uniform grain diameters with low intergranular coupling (i.e., nonmagnetic grain boundaries and no exchange coupling between grains) give the best medium-limited SNR because these properties lead to reductions in switching variances at the level of individual grains in the transitions. Scaling rules start breaking down for grain diameters smaller than about 100 Å because thermal fluctuations in the disk magnetization lead to magnetic viscosity or relaxation effects (super-paramagnetism). This deep and broad subject goes beyond the scope of this book; however, a good entry point to a discussion is by Chikazumi and Charap (1978). Examining (1.23) and (1.24) shows if system SNR is limited by medium noise, improvements are bought only by reducing position and width jitter, and these must be scaled down with $B \times \mathrm{PW50}$ without creating reliability problems such as signal decay from thermal relaxation.

It is normal practice to measure medium noise on a spectrum analyzer by subtracting the head and preamplifier noises and integrating the remaining noise over the band of interest; this procedure gives the broad-band noise (voltage or power, depending on equipment setup). Since medium noise is in the transitions, the measured noise increases with linear density. The effective standard deviation of medium noise σ_n (composed of position jitter and "a" jitter) can then be extracted from the data. Typical numbers might be $V_n = \sqrt{P_n}$ (at $B = 100$ nm) $= 20$ μV (RMS), $V_0 = 300$ μV (0-p), and PW50 $= 250$ nm, for which $\sigma_n \simeq 6.9$ nm. Medium jitter is also measured with time-interval analyzers. In this case transitions are normally written at lower densities to facilitate the measurement; however, the jitter values are rather small (say 2–4 nm) compared to jitter at high densities (6–12 nm).

Medium noise is also studied with magnetic force microscopy (MFM), and noisy, high-density recordings show poorly defined transitions with significant percolation of switching clusters between transitions. These micromagnetic problems have been studied extensively by Zhu [see Chap. 6 of Arnoldussen and Nunnelley (1992)].

Xing and Bertram (1997) show that position jitter dominates at low densities and only at very high densities do width fluctuations become significant. In their experiments with thin-film media, the estimates of jitter for hyperbolic tangent and error function transition shapes give nearly the same values. For a disk with $M_r\delta = 0.8$ memu/cm^2 and $H_c = 2200$ Oe, they obtain values for position jitter of 3.7 nm, width jitter of 155 nm^2, cross-track correlation length $s = 18$ nm, and transition parameter $a = 32$ nm with an AMR head width $W = 2.7$ μm.

TOTAL NOISE AND SYSTEM SNR

Head, preamplifier and broad-band integrated medium noises are added to find the total RMS noise:

$$N_t = [N_{\text{head}}^2 + N_{\text{preamp}}^2 + N_{\text{medium}}^2]^{0.5}. \tag{1.25}$$

With our example of head, preamplifier, and medium noises of 10, 7, and 20 μV, respectively, the total RMS noise would be 23.4 μV, and the SNR (for low-density signal V_0 and high-density noise) would be 300/23.4, or 22.1 dB (0-p/RMS). The high-density signal (say, at $U = 2.3$) is estimated with (1.13) and $f(0, P = 2.3B, B) = 0.20 V_0$, or 60 μV (0-p).

Partial-response detectors (e.g., PR4 channels) use equalization to shape the incoming pulses in a manner that allows the Viterbi detector to eliminate most of the linear ISI. The signal and noise spectra of PRSs is substantially altered by the equalizer, such that the sampled data seen at the sequence detector differ from the RMS signal and noise. Yeh et al. (1998) show that the SNR of an N-bit PRS sequence is altered according to the relationship

$$\text{SNR}_{\text{sync}}^2 = \frac{2N}{N+1} \text{SNR}_{\text{RMS}}^2 \tag{1.26}$$

and a 31-bit PRS achieves an improvement of $\sqrt{62/32}$, or 2.87 dB, in SNR over the RMS value. Experimentally, they obtained about 25%, or about 2.5 dB, improvement in SNR.

NONLINEAR INTERFERENCES

Partial-response detectors shape the readback pulse to obtain equalized responses that reduce or eliminate much of the linear ISI at the sampling points. Nonlinear interferences or distortions associated with writing at high densities cannot be equalized out of the data patterns so they increase the MSE and the error rate suffers. These nonlinearities include NLTS, partial erasure, and overwrite. Bertram (1994) analyzes the physics of these issues, and he should be consulted for insight and details; for di-bit patterns NLTS (or displacement Δx) can be written in terms of its ratio to the bit cell length B:

$$\frac{\Delta x}{B} \simeq \frac{4a^2(d+\delta/2)^2}{B^4}. \quad (1.27)$$

Most of the literature on signal processing discusses NLTS or NLD in terms of percentage or decibels relative to the undistorted signal amplitude. Today it is common to find specifications for NLTS at -14 dB (i.e., 20%) or less for uncompensated data. Some compensation may be achieved during the writing process for data patterns having predictable shifting; this is called *write precompensation*, which removes perhaps half of the uncompensated shift, thus yielding about -20 dB (or 10%) NLTS.

Partial erasure arises when transitions are so closely spaced that small regions "percolate" over to adjacent transitions and reverse the magnetization with a concomitant loss of readback flux. The experiments of Lin et al. (1992) show that nonlinear amplitude losses become measurable at bit spacings $B \sim 4a$ and less. Zhu [Chap. 6 in Arnoldussen and Nunnelley (1992)] analyzes the physics of this micromagnetic problem.

In thin-film recording, "overwrite" is primarily the result of shifting "hard" transitions relative to "easy" transitions when writing new data over old data patterns. Hard transitions are those in which the magnetization is written in a direction opposite to the magnetization just entering the writing zone; easy transitions are those where the head field writes the magnetization in the same direction as the magnetization entering the write zone. Thus transition positions are modulated by the polarities of incoming old data, and this creates interesting phase modulation spectra. For special cases of new and old patterns (with imaging not included), Bertram (1994) shows that hard transition shifting Δ can be expressed as

$$\Delta \simeq \frac{M_r \, \delta(d+\delta/2)}{2x_0 \pi Q H_c}, \quad (1.28)$$

where $2x_0$ is the size of the writing zone and Q is defined in (1.16). Under most head and medium conditions, hard transition shifting will be less than a few percent of the high-density bit length B and overwrite will not substantially degrade the BER. Yeh and Wachenschwanz (1997) show -30 dB overwrite gives "less than a factor of 2 increase in BER, which is almost equal to the intrinsic uncertainty in BER measurement."

TOP-DOWN DESIGN OF A RECORDING SYSTEM

This introductory chapter ends with a high-level view of recording system design, starting with a BER requirement and spreading the noise and interference budgets over the system components. This "top-down" approach is discussed by Yeh et al. (1998), and their prescription includes noise and ISI reduction of 25% by sequence (Viterbi) detection. All noises and interferences are expressed as percentages of the sampled signal level (normalized to 1.0) in a partial-response channel, and (1.1) and (1.2) above are used to create the budget. Figure 1.9 is a flow diagram of the results, taking BER(4.32) = 10^{-9} and democratically splitting the noise and nonlinear ISI into equal parts of 5.8% each (after sequence detection.)

The noise budget is split equally between medium and system contributions (again, democracy in magnetic recording), or 5.5% each; that is, the medium SNR would be 25.2 dB (0-p/RMS), and the system noise is spread 4.5% for the head and 3.2% for the preamplifier electronics. In other words, the head SNR should be 26.9 dB (0-pk/RMS) after equalization in a PR4 channel. Figure 1.10 shows projections of the areal density achievable for a given readback sensitivity measured in millivolts [peak-peak (p-p)] per micrometer of sensor width. The curves are based on a low-density signal requirement of 1.00 mV before equalization; the solid curve allows the bit cell ratio (R_{bc}) to vary from 15:1 at low areal densities to 4.0:1 at high densities, whereas the dotted curve assumes $R_{bc} = 4.0$ at all densities. In Chapter 5, the reader will find analyses of soft adjacent-layer (SAL) biased AMR heads with sensitivites of 0.7–1.36 mV/μm that can support areal densities in the range of 3–6 Gbits/in.^{2} at bit cell ratios of approximately 15:1–18:1. Chapter 6 is devoted to the design and

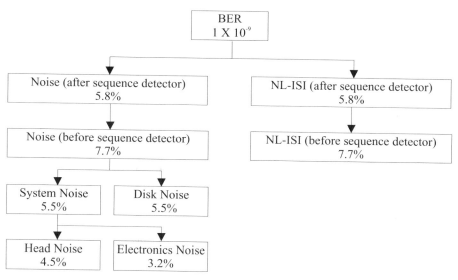

FIGURE 1.9 Top-down approach for a recording system designed to support a BER of 10^{-9} (from Yeh et al., 1998).

18 OVERVIEW OF DIGITAL RECORDING SYSTEMS

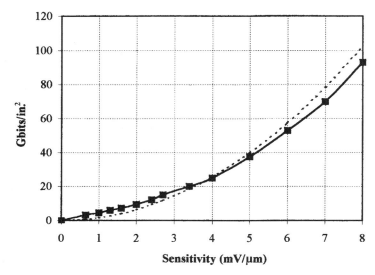

FIGURE 1.10 Areal density projections for read signal sensitivities in millivolts per micrometer of sensor width based on a low-density signal requirement of 1.0 mV (p-p) before equalization. Solid curve: the bit cell ratio R_{bc} varies from 15:1 at low areal density to 4:1 at high areal density. Dotted curve: $R_{bc} = 4.0$ at all areal densities.

analysis of GMR spin valve sensors with sensitivities at 0.8–5.4 mV/μm; the possibility of improving the sensitivity of dual synthetic spin valves (DSSVs) to about 8 mV/μm opens up the application to areal densities of 90 Gbits/in.^{2} or greater.

REFERENCES

Arnoldussen, T. and Nunnelley, L., Eds., *Noise in Digital Recording*, World Scientific, Singapore, 1992.

Ashar, K. G., *Magnetic Disk Drive Technology*, IEEE Press, New York, 1997.

Bertram, H. N., *Theory of Magnetic Recording*, Cambridge University Press, 1994.

Bertram, H. N., *IEEE Trans. Magn.*, **MAG-31**, 2573 (1995).

Champion, E. and Bertram, H. N., *IEEE Trans. Magn.*, **MAG-31**, 2461 (1995).

Chikazumi, S. and Charap, S., Chap. 15 in *Physics of Magnetism*, Krieger, Malabar, 1978.

Comstock, R. and Williams, M., *IEEE Trans. Magn.*, **MAG-9**, 342 (1973).

Harker, J. M., Brede, D. W., Pattison, R. E., Santana, G. R., and Taft, L. G., *IBM J. Res. Devel.*, **25**, 677 (1981).

Hunt, R. P., *IEEE Trans. Magn.*, **MAG-7**, 150 (1971).

Karlquist, O., *Trans. Roy. Inst. Tech., Stockholm*, **86**, 3 (1954).

Lin, H., Barndt, R., Bertram, N., and Wolf, J., *IEEE Trans. Magn.*, **MAG-28**, 3279 (1992).

Mallinson, J. C., *Magneto-Resistive Heads, Fundamentals and Applications*, Academic Press, San Diego, 1996.

McKinlay, S., Fussing, N., Sinclair, R., and Doerner, M., *IEEE Trans. Magn.*, **MAG-32**, 3587 (1996).

Palmer, D., Ziperovich, P., and Wood, R., *IEEE Trans. Magn.*, **MAG-23**, 2377 (1987).

Perkins, T. and Keirn, Z., *IEEE Trans. Magn.*, **MAG-31**, 1109 (1995).

Potter, R. I., *J. Appl. Phys.*, **41**, 1647 (1970).

Potter, R. I., *IEEE Trans. Magn.*, **MAG-10**, 502 (1974).

Sato, H., Nakai, J., Kikuchi, A., Mitsuya, H., Shimatsu, T., and Takahashi, M., *IEEE Trans. Magn.*, **MAG-32**, 3596 (1996).

Smith, R. L., *IEEE Trans. Magn.*, **MAG-27**, 4561 (1991).

Tarnopolsky, G. and Pitts, P., *J. Appl. Phys.*, **81**(8), 4837 (1997).

Williams, E., Chap. 7 in *Noise in Digital Recording*, Arnoldussen, T. C. and Nunnelley, L. L. (Eds), World Scientific, Singapore, 1992.

Williams, M. and Comstock, R., *Proc. 17th Ann. AIP Conf.*, **5**, 738 (1971).

Xing, X. and Bertram, N., *IEEE Trans. Magn.*, **MAG-33**, 2959 (1997).

Yeh, N. and Wachenschwanz, D., *IEEE Trans. Magn.*, **MAG-33**, 2962 (1997).

Yeh, N., Wachenschwanz, D., and Wei, L., *IEEE Trans. Magn.*, **35**, 776 (1999).

Zhu, J., Chap. 6 in *Noise in Digital Recording*, Arnoldussen, T. C. and Nunnelley, L. L. (Eds.), World Scientific, Singapore, 1992.

2

THIN FILM PROPERTIES FOR MR SENSORS

The anisotropic magnetoresistance (AMR) effect was discovered in ferromagnetic metals by Lord Kelvin (Thompson, 1857); 114 years later, Hunt (1971) published his proposal for and demonstration of NiFe thin films for MR sensors in digital recording systems. By 1974 the literature began reflecting increased interest in MR heads for storage applications, and McGuire and Potter (1975) published a 21-page review article on the AMR effect in ferromagnetic $3d$ alloys; they tabulated the properties of 30 binary and 7 ternary alloys. Magnetoresistive heads were developed for tape drives in the mid- to late 1970s (Shelledy and Brock, 1975), and commercially successful MR-based hard disk drive products began appearing in the early 1990s (Hannon, et al., 1994). For the sensing element, all of these products used Permalloy near the zero magnetostriction composition of 82% Ni and 18% Fe (see Bozorth, 1951, p. 649) and Hanazono et al. (1987)); at room temperature the MR figure of merit $\Delta\rho/\rho$ for thin films is about 1–3% depending on film thickness. Searches for other useful materials showing an improved AMR effect did not lead to any significant breakthroughs for AMR heads.

In 1988 the world of research in magnetism was alerted by the reports of Baibich et al. (1988), who showed large changes in $\Delta\rho/\rho$ (15–50%, depending on temperature) in very thin Fe/Cr multilayers. This effect was immediately called the "giant magnetoresistive," or GMR, effect, and Binasch et al. (1989) found the GMR effect in other ultrathin magnetic film systems. White (1992) discusses this early period of discovery and explains the physics of the GMR effect, which is quite different from that of the AMR effect. The magnetic fields required to achieve these changes were roughly 20 kOe; thus attention was focused on material systems that could lead to commercially useful GMR sensors. Grünberg (1990) filed a U.S. patent application in 1989 for a "magnetic field sensor with ferromagnetic thin layers having magnetically antiparallel polarized components," and the invention summary clearly describes three-layer GMR devices with $\Delta R/R \simeq 9\%$ that today are known as *spin valves*. In 1991, Dieny et al. (1991a) published on three-layer GMR systems exhibiting room temperature resistance changes between 4 and 9% at low fields (10–20 Oe). Neologist Virgil Speriosu, hearing John Slonczewski speak of "magnetic valves" in 1988, called these three-layer structures "spin valves" in 1989. This name quickly became synonymous with GMR heads.

This chapter introduces and discusses the properties of materials used in commercially successful AMR and GMR/spin valve devices. These include thin films of magnetically soft ferromagnetic alloys (Co, NiFe, NiCo, NiFeCo, NiFeZr, NiFeRh, CoZrMo, CoZrCr, CoNbTi), antiferromagnetic alloys (FeMn, FeMnRh, IrMn, NiMn, PtMn, RhMn, PdPtMn, CrPtMn), the antiferromagnetic oxide (NiO), nonmagnetic metals (Ta, Zr, Au, Cu), and electrically insulating oxides or nitrides. Materials like Zr and Ta promote favorable crystalline texture—the face-centered-cubic (fcc) [111] texture—in NiFe and Co layers, which lowers the coercivity and anisotropy fields. Magnetically hard materials (CoCrTa, CoCrPt) are used to stabilize MR sensors and maintain noise-free, single-domain behavior. The electrical, magnetic, thermal, and mechanical properties of thin films of these materials are briefly reviewed here, and a thermal analysis is developed to better understand the temperature and resistance behavior of MR heads. Discussion of diffusion and corrosion properties is deferred to Chapter 8, where reliability issues are treated.

FERROMAGNETIC THIN FILMS

Of the 37 AMR alloys tabulated by McGuire and Potter (1975), the alloys of NiCo and NiFe show useful levels of $\Delta \rho/\rho$ at room temperature. Of these two, only $Ni_{82}Fe_{18}$ has magnetic properties that are reasonably independent of mechanical stress and whose resistance to corrosion is acceptable. The alloy $Ni_{70}Co_{30}$ has $\Delta \rho/\rho$ of about 6% (thick films) at room temperature, but the magnetostriction coefficient $\lambda_s = -1.6 \times 10^{-5}$ (see Bozorth, 1951, p. 673), which is much too high for use in devices having residual mechanical stresses on the order of 10^8 dyn/cm^2 or more. Commercially successful AMR heads must be biased to achieve a reasonably linear operating region (a subject thoroughly covered in Chapter 5), and this is normally done by placing the AMR film in close proximity to another magnetic film called the soft adjacent layer (SAL). The magnetic and electrical properties of SAL films will generally be quite different from AMR layers. An ideal SAL film would have no AMR effect, high magnetization, high permeability, zero magnetostriction, and very high resistivity and would be chemically inert. Chen et al. (1991) reviewed the properties of NiFeX alloys (X = Al, Au, Nb, Pd, Pt, Si, and Zr) for SAL materials and showed NiFeNb and NiFeZr alloys are good candidates. Maruyama et al. (1988) and Yamada et al. (1988) reviewed the properties of CoZrMo alloys for SAL films. Table 2.1 lists relevant properties of various ferromagnetic materials used in thin films for AMR sensing layers, GMR free layers, and SAL biasing layers ($\Delta \rho$ is the AMR effect); sources of published data are indicated.

THIN FILM PROPERTIES FOR MR SENSORS

TABLE 2.1 Ferromagnetic Thin Films

Alloy, (at. %)/Use	$\Delta\rho$ ($\mu\Omega$–cm)	ρ ($\mu\Omega$–cm)	$4\pi M_s$ (G)	H_k (Oe)	λ_s	T_c (K)	Reference
$Ni_{82}Fe_{18}$ (100 Å)/AMR	0.64	30	10,000	3.5	-1×10^{-7}	805	Internal sources
$Ni_{66}Co_{34}$ (550 Å)/AMR	0.63	15	10,000	35	-5×10^{-6}	760	Simmons et al. (1989)
$Co_{90}Fe_{10}$ (20 Å)/GMR	—	19	19,400	32	2×10^{-6}	—	Internal sources
$Ni_{72}Fe_{18}Rh_{10}$ (100 Å)/SAL	0.17	66	9,100	3.5	7×10^{-7}	—	Internal sources
$Ni_{72}Fe_{20}Zr_5$ (300 Å)/SAL	0.2	100	7,000	4	4×10^{-6}	—	Chen et al. (1991)
CoNbTi (60 Å)/SAL	0.01	110	11,200	12	—	—	Internal sources
$Co_{82}Zr_6Mo_{12}$ (500 Å)/SAL	−0.028	140	6,000	4.4	1×10^{-7}	630	Yamada et al. (1988)
$Co_{90}Zr_9Cr_1$ (140 Å)/SAL	—	95	13,500	14	2.6×10^{-7}	—	Internal sources

The exchange constant A, (the "exchange stiffness" in ergs per centimeter) of a ferromagnetic material influences the magnetization distribution in thin films (see, e.g., Slonczewski (1966a,b)). Flux coupling between magnetic films is treated in more depth in Chapter 3, and that discussion requires estimates of the exchange constant. The values of A for various materials are difficult to pin down accurately; however, Tebble and Craik (1969) list values derived from electron spin resonance measurements that were summarized by Frait (1962). Apparently, A does not vary much above or below 10^{-6} erg/cm in ferromagnetic materials. The approximate values (all in units of 10^{-6} erg/cm) for some materials are for Fe, 1.5–2.0; for Co, 1.3–1.7; for Ni, 0.75–0.90; for NiFe (up to 60% Fe), 1.0; and for NiCu (up to 10% Cu), 0.5.

PERMANENT-MAGNET MATERIALS

Permanent-magnet (PM), or hard magnetic, materials are used in AMR and GMR heads for stabilizing sensor layers, maintaining single-domain structures, and enjoying signals free of Barkhausen jumps. The materials of choice are in the CoCr, CoNiCr, CoCrTa, CoPt, CoCrPt, and CoCrTaPt alloy families having a remanence M_r about 400–700 emu/cm^3, coercivity H_c about 800–2000 Oe, coercive squareness S^* about 0.80–0.95, and thickness δ in the range of 100–500 Å. According to Krounbi et al. (1991a,b), sensor stability is improved when PM flux ($M_r\delta$) matches the end-region flux ($M_s t$) of an MR layer, and a trade-off exists between sensitivity and stability; a ratio of "passive" to "active" flux between 1 and 2 defines a "suitable compromise." Xiao et al. (1998) discuss fabrication of PM junctions with MR sensors and studied the c-axis texture dependence on composition in the CoCrPt family. They matched the lattice of $Co_{81.8}Cr_{7.8}Pt_{10.4}$ films (100 to 500 Å) to a Cr underlayer (25 to 200 Å) to achieve large in-plane remanence and high coercivity; PM properties were $M_r = 703$ emu/cm^3, $H_c = 1100$ to 1950 Oe, and $S^* = 0.95$. Hard-bias (PM) stabilizing films are normally contiguous with the sensor layers and are part of the electrical contact between the connecting leads and the sensor element. Electrical resistance of the leads and junction are very sensitive to the design (see Xiao et al. 1998). The subject of hard-bias stabilization will be discussed with more depth in Chapter 4.

INDUCED MAGNETIC ANISOTROPY

Kelsall (1934) was the first investigator to show that the permeability of Ni-rich NiFe alloys could be significantly increased by annealing them in a magnetic field while cooling down from the Curie temperature. Bozorth and Dillinger (1935) studied this phenomenon extensively, and it is attributed to the directional diffusion ordering of Fe–Fe pairs in Ni, with the Fe pairs aligned along the field direction. The resultant material has induced uniaxial anisotropy, and it is now understood that pair-ordering

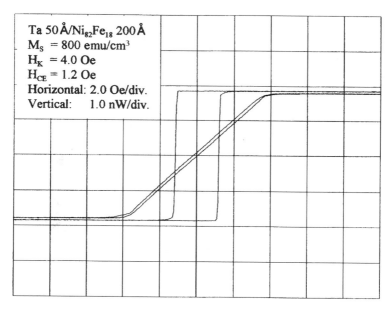

FIGURE 2.1 Hard- and easy-axis hysteresis loops for 200 Å NiFe on 50 Å Ta; $H_K = 4$ Oe, $H_{CE} = 1.2$ Oe.

anisotropy can be induced in thin films without high temperature if a magnetic field is applied during the deposition process. Figure 2.1 shows the magnetization (M) versus field (H) hysteresis behavior for a large sheet (3 in.) of Permalloy along the "easy" direction (parallel with the field direction during film deposition) and along the "hard" direction (perpendicular to the easy axis). The important magnetic properties are the intrinsic saturation induction $4\pi M_s$ (in gauss), easy-axis coercivity H_{CE} (in oersteds), hard-axis anisotropy field H_K (in oersteds) and the magnetostriction coefficient λ_S (which can be determined from changes in H_K with stress). The hard-axis permeability $\mu = 4\pi M_S/H_K$ is normally derived from the hysteresis loop. Along the easy direction, magnetization switches direction by domain wall processes (unless the sample is a single domain), and along the hard direction film magnetization rotates against the uniaxial anisotropy energy barrier (K_U in ergs/per cubic centimeter). The anisotropy field is a measure of the barrier height:

$$H_K = \frac{2K_U}{M_S} \quad \text{Oe} \quad (2.1)$$

For Permalloy with $H_K = 3.5$ Oe and $M_S = 800$ emu/cm^3, the barrier height K_U is 1.4×10^3 ergs/cm^3.

STRESS ANISOTROPY AND MAGNETOSTRICTION

Residual stress in thin films plays an important role in device behavior if the magnetostriction coefficient is not extremely small; this can be readily appreciated by a comparison of stress anisotropy and induced anisotropy energy densities. Cullity (1972) and Klokholm and Krongelb (1990) are useful references for this topic. Stress anisotropy is defined as

$$K_S = \tfrac{3}{2}\lambda_S(\sigma_y - \sigma_x) = \tfrac{3}{2}\lambda_S \Delta\sigma \quad \text{ergs/cm}^3 \quad (2.2)$$

and this energy density is positive if the product $\lambda_S \Delta\sigma$ is positive (positive $\Delta\sigma$ is an anisotropic tensile stress). Comparing (2.1) and (2.2), one can solve for the level of control on magnetostriction coefficient necessary to maintain useful levels of permeability (or H_K) at some residual stress. If the stress is $\Delta\sigma = 10^9$ dyn/cm^2, for example, and $H_K = 3.5$ Oe in a Permalloy film ($M_S = 800$), then

$$\lambda_S < \frac{M_S H_K}{3 \Delta\sigma} \approx 9 \times 10^{-7} \quad (2.3)$$

would be just sufficient to maintain control on the film anisotropy. Anisotropic stress creates uniaxial anisotropy, and it can overwhelm a competing anisotropy such as pair ordering. Cullity (1972, p. 245) discusses mixed anisotropies and shows that two anisotropies at right angles in a specimen give a resultant anisotropy energy E of

$$E = E_A + E_B = K_A \sin^2\theta + K_B \sin^2(\pi/2 - \theta) = K_B + (K_A - K_B)\sin^2\theta. \quad (2.4)$$

In other words, two equal anisotropies at right angles produce isotropic behavior ($E = K_B$), not biaxial anisotropy. If $K_A > K_B$, the energy minimum is along the A-axis, but if K_B is bigger, the easy direction shifts to the B-axis. Therefore, uncontrolled stresses produce interesting and complicated behavior in MR heads. Many thin films exhibit *isotropic* tensile or compressive stresses (depending on the fabrication process) for large-area film samples. The stress becomes *anisotropic*, however, after etching or removing portions of the film away to shape actual devices. This "patterning effect" is quite important in understanding the behavior of MR sensors, and Shiroishi et al. (1984) and Markham and Smith (1988) gave convincing evidence of the influence of pattern shape on the formation of magnetic domains and the resultant easy axis in thin films. Tatsumi et al. (1991) and Wang et al. (1992) discussed magnetostriction and its influence on the response of SAL-biased AMR sensors.

Shiroishi et al. (1984) determined the residual stress on NiFe films of various thicknesses with a length $L = 3a$ and width $W = a$. In all cases, the residual stress was tensile and directed along the length of the element. Sheet films were prepared by electron beam evaporation at 300°C (573 K) in a magnetic field, which induced pair-ordering uniaxial anisotropy; the glass substrates and Permalloy films had thermal expansion coefficients of 5.1×10^{-6}/K and 1.28×10^{-5}/K, respectively. In unpatterned sheet films, the residual *isotropic* tensile stress showed a maximum of about 1.75 GPa (1.75×10^{10} dyn/cm^2) at a thickness of 150 Å, with stress decreasing to a constant level of 0.35 GPa (3.5×10^9 dyn/cm^2) for thicknesses greater than 500 Å. They studied the dependence of the anisotropy field and shift of easy direction in patterned films with saturation magnetostriction constants λ_S between -2×10^{-6} and 6×10^{-6} and reached the following conclusions:

- Anisotropic stress is induced by patterning a thin film.
- The *long* direction of a pattern is the *axis* of stress anisotropy.
- If $\lambda_S \Delta\sigma$ is *positive*, the axis of stress is an *easy* direction.
- If $\lambda_S \Delta\sigma$ is *negative*, the axis of stress is a *hard* direction.
- The total anisotropy energy per cubic centimeter arising from pair-ordering, shape, and stress anisotropies is $K_T = K_u + (N_y - N_x)M_S^2/2 + 3\lambda_S \Delta\sigma/2$.
- The effective anisotropy field is $H_K = 2K_T/M_S$.

The shape demagnetizing factors N_x, N_y derived by Yuan (1992) are approximately $N_x = 8th/[W(W^2 + h^2)^{1/2}]$ and $N_y = 8tW/[h(W^2 + h^2)^{1/2}]$. For a Permalloy film with $W = 1.0$ μm, $h = 0.70$ μm, $M_S = 800$ emu/cm^3, $t = 0.01$ μm, $\lambda_S = -1.0 \times 10^{-6}$, $\Delta\sigma = 2.0 \times 10^9$ dyn/cm^2, and $K_u = 1.4 \times 10^3$ ergs/cm^3 (equivalent to $H_{Ku} = 3.5$ Oe), the total anisotropy energy density $K_T = 1.4 \times 10^3 + 1.5 \times 10^4 - 3.0 \times 10^3 = 1.36 \times 10^4$ ergs/cm^3 and the effective anisotropy field $H_K = 3.5 + 38 - 7.5 = 34$ Oe. Annealing the film to reduce the residual stress by a factor of 10 ($\Delta\sigma \approx 2.0 \times 10^8$ dyn/cm^2) would increase the anisotropy field to slightly over 40.8 Oe (an increase of about 20%). If the magnetostriction was initially $\lambda_S = +1.0 \times 10^{-6}$, annealing would shift H_K from 49 Oe down to nearly 42.3 Oe.

Data for magnetostriction of bulk NiFe alloys are published in Bozorth (1951, p. 649), and for sputtered and evaporated films, Klokholm and Aboaf (1981) may be consulted. The exact composition for zero magnetostriction is difficult to ascertain; however, Klokholm and Aboaf established the composition at about 81.5 wt. % Ni in Fe. The data are scattered; however the best linear fit for saturation magnetostriction as a function of Ni is approximately $\lambda_S = -1.4 \times 10^{-6}(\text{Ni-}81.5)$.

STONER–WOHLFARTH THEORY

Stoner and Wohlfarth (1948) created a theory that successfully explains many aspects of single-domain switching by magnetization rotation in uniaxial anisotropic systems. They treat shape, magnetocrystalline, and stress anisotropies. Their theory is now classical and is used for describing details of *M–H* curves at any angle between the applied field and easy axis and for explaining observations of AMR or GMR effects at various skew angles between current flow and the easy axis. Figure 2.2 defines the geometry of a uniaxial anisotropic thin-film system in which current flows along the easy axis, and an external field (at an angle α with respect to the hard axis) causes the magnetization to rotate toward the hard axis. This is the archetypical MR sensor with the easy axis aligned along the *x*-axis.

The energy of this system is composed of anisotropy energy E_a, magnetostatic potential energy E_p and demagnetizing energy E_d terms:

$$E = E_a + E_p + E_d = K_u \sin^2\theta - HM_S \cos(\alpha - \theta) + \tfrac{1}{2}(N_y - N_x)M_S^2 \sin^2\theta. \quad (2.5)$$

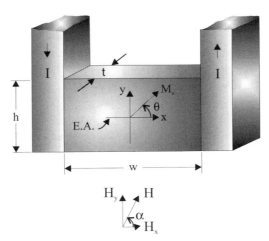

FIGURE 2.2 Geometry of a uniaxial anisotropic thin-film system with leads attached for current flow along the easy axis (EA). An external field *H* is directed at an angle α with respect to the EA.

The demagnetizing energy is of the same form as the anisotropy energy, so both terms in $\sin^2 \theta$ can be combined with a new coefficient $K_T = K_u + 0.5(N_y - N_x)M_S^2$ for the total effective anisotropy. The magnetization M_s establishes an equilibrium angle θ with respect to the easy axis, and this position is given by solutions to the equation

$$\frac{\partial E}{\partial \theta} = 2K_T \sin \theta \cos \theta - HM_S \sin(\alpha - \theta) = 0. \tag{2.6}$$

Figure 2.3 shows a family of plots for $m_y = M_y/M_S$ for selected values of $H_x = H \cos \alpha$ and $H_y = H \sin \alpha$; field components are normalized by the effective anisotropy field $H_K = 2K_T/M_S = H_{ku} + H_D$, where H_D is the demagnetizing field. These plots also show how well the solutions to (2.6) are approximated by the relation

$$m_y \cong \tanh\left(\frac{H_y}{H_{ku} + H_D + H_x}\right), \tag{2.7}$$

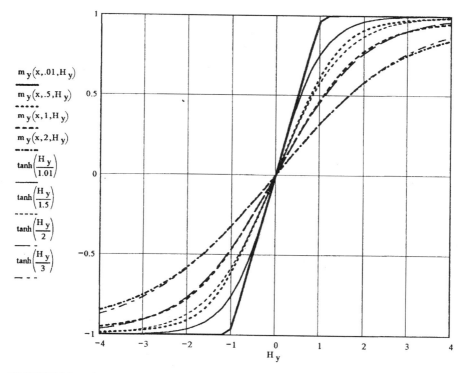

FIGURE 2.3 The y-component of magnetization versus H_y for a uniaxial anisotropic system with its EA along the x-direction for selected values of H_x. The numerical solutions are presented along with analytical solutions of the form $\tanh(H_y/H_x)$ for ease of comparison.

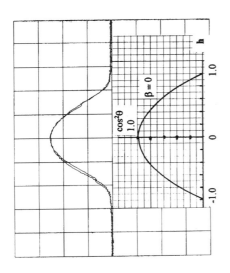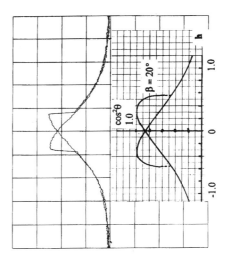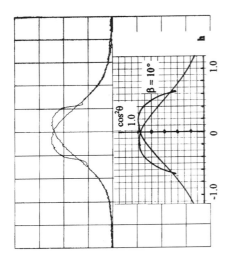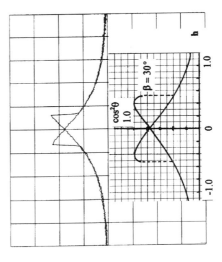

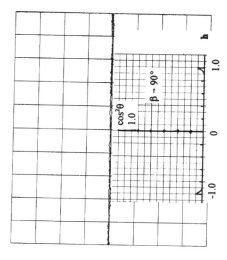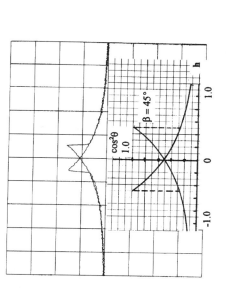

FIGURE 2.4 Experimental AMR transfer curves for NiFe 200 Å at various skew angles β between the easy axis and the current density direction compared with theoretical computations according to the Stoner–Wohlfarth model ($h = H/H_K$.) Vertical scale: $\Delta R = 1\%$/division; $\Delta R_{max} = 2.5\%$. Horizontal scale: 2.0 Oe/division; $H_K = 4.0$ Oe.

which is much easier for computational purposes. Biasing the sensor along the x-direction or increasing the demagnetizing field reduces the slope of the M_y-versus-H_y operating curve; that is, the sensitivity of the AMR head is reduced. In magnetic terms, it is the susceptibility $\chi = dM/dH$ that is reduced.

ANISOTROPIC MAGNETORESISTANCE EFFECT

Anisotropic magnetoresistivity is a measure of the difference in scattering probability of conduction electrons flowing parallel or perpendicular to the direction of magnetization M in a conductor. In the ferromagnetic materials of interest here, resistivity is always greatest for current flowing parallel to M and decreases a little when M is directed perpendicular to the current flow; this effect depends only indirectly on the magnetizing field H that aligns the magnetization. All nonmagnetic metals show an "ordinary galvanomagnetic effect" in which the electrical resistance increases quadratically along the direction of H; this is opposite to the AMR effect. Theoretical explanations of these effects go beyond the scope of this book, but they will be found in McGuire and Potter (1975), where their discussion involves spin–orbit coupling, d-band splitting, and scattering of conduction electrons from the Fermi surface of the relevant material. The approach here will be phenomenological, where the measured anisotropic magnetoresistivity is

$$\Delta\rho = \rho_\| - \rho_\perp. \tag{2.8}$$

In multidomain magnetic materials, where M is not necessarily aligned in a preferred direction, the average resistivity in the demagnetized state is given as

$$\rho_{av} = \tfrac{1}{3}\rho_\| + \tfrac{2}{3}\rho_\perp. \tag{2.9}$$

The anisotropic magnetoresistivity ratio is defined as $\Delta\rho/\rho_{av}$; in a single-domain sample, M lies along the easy axis, so the average resistivity would not change when a field is applied in the easy direction, but it does change by the amount $\Delta\rho$ when the field is along the hard direction. Experimentally, a four-point probe is placed on a film sample along the easy axis, the sample is magnetized along the hard axis, and only the *reduction* in resistivity associated with $\Delta\rho$ is plotted as a function of field. Figure 2.4 shows experimental results for NiFe (200-Å Permalloy) at selected skew angles between current direction and the easy axis. Each plot also has a small inset showing the Stoner–Wohlfarth theoretical curve (with $h = H/H_K$) for that skew angle; the qualitative agreement is striking. The curve for zero skew angle follows a simple mathematical relation,

$$\rho = \rho_\perp + \Delta\rho \, \cos^2 \theta \tag{2.10}$$

and for H along the hard axis ($\alpha = \pi/2$), the solution to (2.6) reduces to $\sin\theta = H/H_K$, where H_K is the anisotropy field $2K_T/M_S$ (in oersteds). Using the identity $\cos^2\theta = 1 - \sin^2\theta = 1 - (H/H_K)^2$, one immediately discovers the simple parabolic nature of the AMR effect shown in Fig. 2.5 for zero skew angle. The curve $\Delta\rho$ versus H is called the "transfer curve" for MR sensors in devices.

The AMR sensors are quite small, and for use at an areal density of 2 Gbits/in.^{2} (10,000 tpi and 200 kbpi), for example, the dimensions in Fig. 2.2 would be roughly $W = 2.0$ µm, $h = 1.3$ µm, and $t = 150$ Å. With these dimensions, the maximum change in resistance (reading directly from Table 2.1) would be $\Delta R_{\text{sensor}} = \Delta\rho\, W/th = 0.64\,\Omega$, and with a sense current $I_{\text{MR}} = 10$ mA flowing through the MR film, the maximum voltage change would be 6.4 mV. In practice, less than half of this voltage swing can be used because transfer curve nonlinearities cause problems in signal detection electronics. The transfer curve depends on a number of factors such as induced anisotropy, stress anisotropy, and shape and thickness of the MR element, all of which are subsumed in the effective total anisotropy field $H_K = 2K_T/M_S$. Figure 2.5 shows transfer curves for several values of effective anisotropy. Anderson et al. (1972) compared predicted curves with experiment and discussed how demagnetizing effects dominate devices at short stripe heights h (see Fig. 2.2). Hunt (1971), Smith (1987), and Bhattacharyya and Simmons (1994) have discussed signal output and its dependence on stripe height.

Device physicists and recording engineers have attacked the problem of transfer curve nonlinearity and sensitivity to an external field in various ways; this effort has led to successful techniques for biasing the sensing element along the H-axis to an operating point that produces useful error rates in a digital recording channel. These biasing schemes receive detailed discussions in Chapter 5.

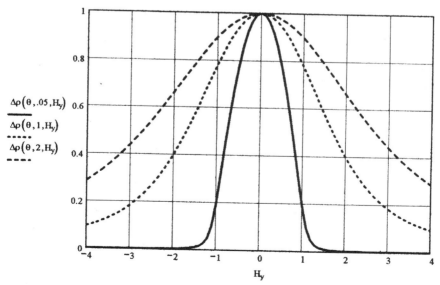

FIGURE 2.5 AMR transfer curves for various values of effective anisotropy field.

GIANT MAGNETORESISTANCE EFFECT

White (1992) has written a primer on the GMR effect, and it is difficult to improve on the clarity of his presentation, so his approach will be followed closely here. The GMR effect appears when there are at least two very thin layers of ferromagnetic material separated by a very thin nonmagnetic metal and when there is a way of changing the relative orientations of the magnetization in the adjacent magnetic layers. The thickness of each of these layers must be a fraction of the mean free path of an electron in the layered array. The large changes in resistance of the GMR effect arise from differences in conduction electron scattering when its spin orientation is parallel or antiparallel to the magnetic moment of a ferromagnetic conductor. The mean free path is apparently short for antiparallel spins and long for parallel spins, so spin-up and spin-down electrons are differentially scattered. A number of reviews on magnetoresistance, materials, and magnetic coupling are now available: Hathaway et al. (1994), Dieny (1994), Coehoorn et al. (1998), and Kools (1996) are good places to start. In the early work on Fe/Cr multilayers, the GMR effect was assigned to differential scattering at the material interfaces, whereas in their work on spin valve systems, Dieny et al. (1991b) interpreted spin-dependent scattering as a *bulk* phenomenon with a mean path of about 80 Å; they studied the spin valve effect in NiFe/X/NiFe and in Co/X/Co sandwiches, where X was a noble metal (Cu, Ag, or Au). The data shown in Figs. 2.6a–c are from a sample whose structure is two sandwiches of Ta 50 Å/NiFe 62 Å/Cu 22 Å/NiFe 40 Å/FeMn 70 Å deposited on a Si substrate and a final capping layer of Ta 50 Å. This structure is called a "top spin valve" because the antiferromagnetic (AFM) pinning layer is deposited on top. They found that, over a range of one to five multiple sandwiches, the relative GMR response $\Delta R/R$ was independent of the number of periods; the GMR room temperature responses were 4.1% for NiFe/Cu/NiFe at 10 Oe and 8.7% for Co/Cu/Co at 20 Oe excitation.

Figure 2.6a shows the combined hysteresis loops of two NiFe layers; at a field of −200 Oe both NiFe films are aligned along the negative field direction. At slightly positive fields, the pinned layer stays aligned in the negative direction and the free layer switches in the positive direction. At large positive fields (500–600 Oe) the pinned layer switches against the exchange anisotropy field of the FeMn layer and both NiFe films are now aligned in the positive direction. Figure 2.6b shows the magnetoresistance state for the parallel (low-resistance), antiparallel (high-resistance), and parallel (again low resistance) alignments of magnetization with the applied field. In Fig. 2.6c, a magnified plot of the GMR response in the −50- to +50-Oe range shows a shift of about 6 Oe of the NiFe free layer, which is evidence of ferromagnetic coupling between the NiFe layers through the Cu layer. If this coupling is not strong, the magnetization in the first magnetic layer is free to rotate in response to an external magnetic field, while the magnetization in the second magnetic layer is constrained, or "pinned," by exchange coupling to the antiferromagnetic FeMn layer on top of it. In their review of GMR materials, Kools et al. (1995) showed that the Cu layer serves to decouple the ferromagnetic layers, and if the Cu layer is less than about 17 Å thick, the ferromagnetic coupling is so strong,

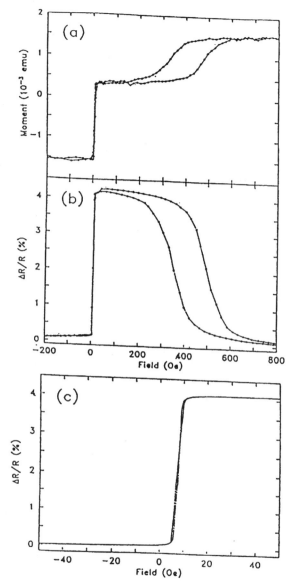

FIGURE 2.6 (*a*) Room temperature hysteresis loop, (*b*) magnetoresistance, and (*c*) GMR response for fields swept between ±50 Oe. Reprinted with permission from B. Dieny, V. S. Speriosu, S. Metin, S. S. P. Parkin, B. A. Gurney, P. Baumgart, and D. R. Wilhoit, *J. Appl. Phys.*, **69**, 4774. Copyright 1991 by the American Institute of Physics.

presumably from bridging sites (pin holes) between the ferromagnetic layers, that the two layers cannot rotate independently. Dieny et al. (1991b) showed the GMR effect attenuates exponentially with thickness of the nonmagnetic layer, so an optimal Cu thickness is found at 20–25 Å where the GMR effect maximizes at about 4–5% for NiFe films. The coupling behavior described by Kools et al. (1995) is plotted in Fig. 2.7. The solid line in this figure is computed from the Néel (1962) model of correlated waviness between two magnetic layers; this is called "orange-peel" or "topological" coupling. The coupling energy per unit area [in centimeter-gram-second (cgs) units] is

$$E_{\text{topology}} = \frac{\pi^2}{\sqrt{2}} \frac{h^2}{\lambda} M_1 M_2 \exp\left(\frac{-2\pi\sqrt{2}t_{\text{Cu}}}{\lambda}\right) \quad \text{ergs/cm}^2, \quad (2.11)$$

and is about 0.005 erg/cm^2 for 22 Å of Cu spacing between Permalloy films with $h \sim 8$ Å and $\lambda \sim 150$ Å. The interaction field, which depends on the ferromagnetic (FM) layer thickness and magnetization, is derived from the surface energy,

$$H_{\text{coupling}} = \frac{E_{\text{topology}}}{M_S t_{FM}} \quad \text{Oe} \quad (2.12)$$

and is about 10 Oe for a NiFe layer 60 Å thick. Leal and Kryder (1996) also discussed interlayer coupling for NiFe/Cu/NiFe/FeMn sandwiches and showed an

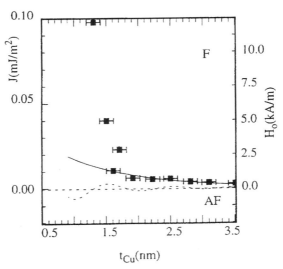

FIGURE 2.7 Measured interlayer coupling energy as a function of copper thickness (squares). Solid line = fit with the Néel model. Dashed line: oscillating exchange coupling for fit parameters as given in the text. Reprinted with permission from J. C. S. Kools, *IEEE Trans. Magn.*, **MAG-32**, 3165. Copyright 1996 by IEEE.

oscillatory interaction combined with topological coupling; in very smooth films the oscillatory behavior dominated, and with rougher films, the topological influence was stronger. The oscillatory (RKKY-like) interaction (Ruderman–Kittel–Kasuya–Yosida theory) can be expressed as

$$E = \frac{E_0}{(k_0 t_{Cu})^2} \sin\left(\frac{2\pi t_{Cu}}{\Lambda} + \phi\right). \tag{2.13}$$

The observed wavelength for NiFe was $\Lambda = 13.5$ Å; the surface energy E_0, wavenumber k_0, and phase angle ϕ were not specified. Positive values of E yield FM coupling and negative values yield AFM coupling; Parkin et al. (1990) discussed enhancement of the GMR effect with superlattice structures of Co/Ru, Fe/Cr, and Co/Cr systems. These synthetic spin valve designs rely on a good understanding of oscillatory interactions.

The GMR effect and its dependence on free-layer thickness and materials was studied by Dieny et al. (1991b) and later by Rijks (1996) for Co, $Ni_{80}Fe_{20}$, and $Ni_{66}Fe_{16}Co_{18}$ alloys; these references each show a broad maximum in the GMR effect at thicknesses in the range of 60–110 Å and both show that Co gives the largest room temperature response. Cobalt and Fe are nearly immiscible with Cu, so spin valve systems employing sandwiches with CoFe/Cu/CoFe show improved thermal stability [e.g., see the system $Ti/Ir_{20}Mn_{80}/Co_{90}Fe_{10}/Cu/Co_{90}Fe_{10}/Ni_{80}Fe_{20}$ described by Yoda et al. (1996)]. The review by Coehoorn et al. (1998) brings out important information about magnetic properties and processing details in spin valve systems. The microstructure of an AFM layer and the adjacent pinned FM layer can depend on the order in which the layers are deposited; for this reason they distinguish between a "top" spin valve and a "bottom" spin valve. They also note that a few monolayers of a second FM metal (e.g., Co) at the interface between the free and the nonmagnetic layer can enhance the GMR ratio by *improving the transmission* of spin-up electrons through the interface. Buffer layers can modify the crystal orientation of deposited films, and they recommend Ta as a material that promotes a strong [111] crystalline columnar texture to $Ni_{80}Fe_{20}/Cu/Ni_{80}Fe_{20}/Fe_{50}Mn_{50}$ and similar spin valve systems. The [111] texture improves the free layer by reducing its coercivity, increases exchange biasing with the FeMn AFM layer, and improves the GMR ratio by lowering the sheet resistance of the system.

In a definitive series of interface "dusting" experiments, Parkin (1993) inserted layers of Co at the interfaces of the Cu layer in sandwiches of a top spin valve system of the form $Si/NiFe(53$ Å$-t_i)/Co(t_i)/Cu(32$ Å$)/Co(t_i)/NiFe(22$ Å$-t_i)/FeMn(90$ Å$)/Cu(10$ Å$)$. The GMR response $\Delta R/R\,(\%) = 2.9 + 3.5[1 - \exp(-t_{Co}/2.3$ Å$)]$ doubled with about 8–10 Å of Co; a plot is given in Fig. 2.8a. He then took 5-Å Co layers, initially positioned at the Cu interfaces, and gradually moved these Co layers a distance d_{NiFe} to the interior of the NiFe layers and measured the GMR response. The response $\Delta R/R\,(\%) = 3.1 + 2.5\exp(-d_{NiFe}/2.3$ Å$)$ dropped nearly a factor of 2 after moving the Co layers about

THIN FILM PROPERTIES FOR MR SENSORS

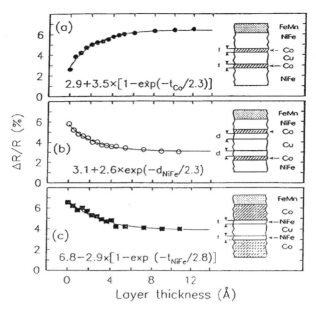

FIGURE 2.8 Dependence of room temperature saturation magnetoresistance on (*a*) Co interface layer thickness t_{Co}, in sandwiches of the form Si/Ni$_{81}$Fe$_{19}$(53 − t_i)/Cu(32)/Co(t_i)/Ni$_{81}$Fe$_{19}$(22 − t_i)/FeMn(90)/Cu(10); (*b*) distance of a 5-Å-thick Co layer from the Ni$_{81}$Fe$_{19}$/Cu interfaces in sandwiches of the form Si/Ni$_{81}$Fe$_{19}$(49 − *d*)/Co(5)/Ni$_{81}$Fe$_{19}$(*d*)/Cu(30)/Ni$_{81}$Fe$_{19}$(*d*)/Co(5)/Ni$_{81}$Fe$_{19}$(18 − *d*)/FeMn(90)/Cu(10); and (*c*) Ni$_{81}$Fe$_{19}$ interface layer thickness, t_i, in sandwiches of the form Si/Co(57 − t_i)/Ni$_{81}$Fe$_{19}$(tNi$_{81}$Fe$_{19}$)/Cu(24)/Ni$_{81}$Fe$_{19}$(t_i)/Co(29 − t_i)/FeMn(100)/Cu(10). Note layer thicknesses are in angstroms. Reprinted with permission from S. S. P. Parkin, *Phys. Rev. Lett.*, **71**, 1641. Copyright 1993 by the American Physical Society.

8–10 Å from the Cu interface; these results are plotted in Fig. 2.8*b*. If any doubts remained about the *interfacial* nature of the GMR effect, Parkin's final series cleared them up: he prepared Co/Cu/Co sandwiches and introduced NiFe layers of a thickness t_{NiFe} at the Co/Cu interfaces. The GMR $\Delta R/R$ (%) = 6.8 − 2.9[1 − exp(−t_{NiFe}/2.8 Å)] reduced nearly a factor of 2 for NiFe layers about 8–10 Å thick; these final results are plotted in Fig. 2.8*c*. In each of these experiments, the characteristic length (2.3–2.8 Å) necessary to establish the magnitude of the GMR response is very short, and for this reason Parkin claims that the GMR effect originates at the nonmagnetic metal interface.

EXCHANGE COUPLING WITH ANTIFERROMAGNETIC FILMS

The magnetization direction of a an FM can be pinned by exchange biasing the FM layer to an AFM film. This will normally be done by field annealing the FM–AFM system in a modestly high magnetic field (2 kOe) at a temperature above the "blocking temperature" T_B of the AFM layer. The blocking temperature is defined as the point where the exchange field has reduced from its room temperature value to zero; this is normally below the Néel temperature of the AFM film. Annealing will be accomplished in a vacuum oven for several hours, and the maximum temperature, which depends on T_B, could be 250–275°C. Nogués and Schuller (1999) review the exchange bias phenomenon and compile the surface energy, blocking temperature, and Néel point for numerous materials. Lederman (1999) reviews Mn-based metallic AFM materials for use in spin valves, and Lin et al. (1995) discuss their studies with FeMn, NiMn, and NiO AFM layers. Table 2.2 summarizes many of the relevant technical properties of AFM films as published in Lederman (1999), Lin et al. (1995), and Nogués and Schuller (1999). Exchange coupling is extremely sensitive to any conditions existing at the FM–AFM interface; thus top and bottom spin valve systems can have different properties depending on the exchange field, film thicknesses, topography, annealing temperatures, times, and other details with which device processing scientists routinely cope. The literature must be consulted, as these processing considerations go beyond the scope of this book. The main criteria for selecting an AFM layer are high exchange field with thin FM layer, thin AFM layer, high blocking temperature, high resistivity for the AFM material, little or no annealing requirement, low diffusion between FM and AFM layers, and good corrosion resistance of the AFM layer. For example, PtMn layers have demonstrated better corrosion resistance than FeMn or IrMn materials in recording devices.

Dependence of Exchange Energy on Temperature

With a given AFM/FM structure, the exchange energy E_{ex} and field H_{ex} are related by the expression $E_{ex} = H_{ex}M_S t_{FM}$, so it is common practice (see Nogués and Schuller, 1999) to assume that E_{ex} depends primarily on the saturation magnetization

TABLE 2.2 Antiferromagnetic Thin Films: Room Temperature Data

Alloy (at. %)	t_{AFM} (Å)	T_B (K)	E_{ex} (ergs/cm^2) with NiFe	H_{CE} (Oe) with NiFe	E_{ex} (ergs/cm^2) with CoFe	H_{CE} (Oe) with CoFe	ρ (μΩ-cm)
FeMn$_{50}$Rh	100	420	0.12	40	0.084	35	210
Ir$_{20}$Mn$_{80}$	70	470	0.11	40	0.13	35	325
Rh$_{20}$Mn$_{80}$	140	300	0.12	109	0.15	77	200
PdPtMn$_{50}$	250	620	0.09	110	—	—	185
Ni$_{50}$Mn$_{50}$	350	690	0.16	180	—	—	210
CrPtMn$_{50}$	350	570	0.06	32	—	—	360
PtMn	250	620	0.02–0.32	—	0.24	—	150
Fe$_{50}$Mn$_{50}$	80	420	0.05–0.47	—	—	—	130
Co$_x$Ni$_{1-x}$	—	430	0.04–0.06	—	—	—	—
NiO	350	470	0.05–0.29	—	—	—	10^8

but not the type of FM material. For this reason, the exchange field is then estimated from the energy using the relation

$$H_{ex} = \frac{E_{ex}}{M_S t_{FM}} \quad \text{Oe.} \qquad (2.14)$$

At room temperature E_{ex} is roughly 0.02–0.47 erg/cm^2, depending on choice of materials. The exchange energy depends on temperature (T), and the blocking temperature T_B describes how exchange biasing drops with increasing T as magnetic order is lost in the AFM layer. Exchange loss is an important consideration in analyzing GMR heads because FM layers will become unpinned with heating. Unpinning from transients [surge currents from electrostatic discharge (ESD), plugging a head into an energized sense/bias circuit, etc.] and repinning the FM layer will be a relevant concern. This important subject is discussed in Chapters 6 and 8. Kools (1996) shows that for some disordered AFM materials (e.g., FeMn, FeMnRh) the loss in exchange is approximately linear in T, while the ordered materials PdPtMn, IrMn, and NiMn discussed in Lederman (1999) and Fuke et al. (1997) are not linear in T. The exchange energy of these AFM systems empirically fit a distribution function normally identified with systems obeying Fermi statistics. That is,

$$E_{ex}(T) = \frac{E_{ex0}}{\exp\left[(T-T_0)/\Delta T\right]+1} \quad \text{ergs/cm}^2, \qquad (2.15)$$

where the parameters T_0 and ΔT play the role of defining the temperature distribution of exchange energy such that T_0 locates the mean and ΔT quantifies the distribution width; (2.15) has an inflection point at $T = T_0$ and the blocking temperature T_B is approximately equal to $T_0 + 2\,\Delta T$, which is found by linear extrapolation through the inflection point to the zero-energy intercept. The ordered and disordered AFM materials correlate with narrow and wide distribution widths, respectively ($\Delta T \sim 20$ K for ordered, $\Delta T \sim 100$ K for disordered). Plots of two hypothetical AFM systems calculated with (2.15) are shown in Fig. 2.9; the parameters are $E_{ex0} = 0.25$ erg/cm^2, $T_0 = 150°C$, and $\Delta T = 75°C$ for a "disordered" archetype and $E_{exo} = 0.10$ erg/cm^2, $T_0 = 350°C$, and $\Delta T = 20°C$ for an "ordered" archetype. The disordered curve is quasi-linear over most of the displayed temperature range, whereas the ordered curve is fairly constant up to about 300°C with a sharp reduction in exchange surface energy at higher temperatures. Experimental results for the exchange field $H_{ex}(T)$ of four different AFM systems and empirical curve fits are given in Chapter 6, Table 6.2 and Fig. 6.19.

An interpretation of the apparent agreement of experiment with Fermi–Dirac statistics is as follows: in analogy with the derivation of the Fermi–Dirac distribution by Kittel (1966, p. 620), FM atoms interact with AFM sites having two energy states (0 or Δ) and the probability of an FM atom occupying either state is $p(0)$ or $p(\Delta)$ for which the FM atom has an energy ε or $\varepsilon + \Delta$, respectively. The probability that an AFM(0) site is occupied is $f(\varepsilon)$, and the probability that an AFM(Δ) site is vacant is

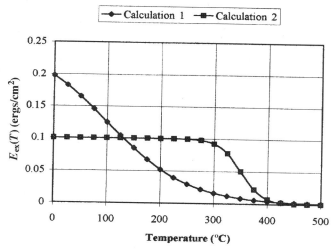

FIGURE 2.9 Hypothetical curves of exchange surface energy versus temperature for AFM systems obeying Fermi–Dirac statistics. Calculation 1 and calculation 2 show what are conventionally referred to as disordered and ordered AFM systems.

$1 - f(\varepsilon + \Delta)$. At thermal equilibrium the population factors are equal, so $f(\varepsilon)p(\Delta)[1 - f(\varepsilon + \Delta)] = f(\varepsilon + \Delta)p(0)[1 - f(\varepsilon)]$. The probabilities $p(0)$ and $p(\Delta)$ follow a Boltzmann distribution, that is, $p(\Delta)/p(0) = \exp(-\Delta/k_B T)$, so $f(\varepsilon + \Delta)/[1 - f(\varepsilon + \Delta)] = [p(\Delta)/p(0)]f(\varepsilon)/[1 - f(\varepsilon)]$. Kittel shows a solution exists for all T if the relation $[1 - f(\varepsilon)]/f(\varepsilon) = \exp[(\varepsilon - \mu)/k_B T]$, where μ is a constant energy independent of ε, thus the Fermi–Dirac distribution law becomes $f(\varepsilon) = 1/\{\exp[(\varepsilon - \mu)/k_B T] + 1\}$. Feller (1957, p. 39) is even more succinct in his statement of the basis for Fermi–Dirac statistics. He says, "Fermi–Dirac statistics is based on these hypotheses: (1) it is impossible for two or more particles to be in the same cell, and (2) all distinguishable arrangements satisfying the first condition have equal probabilities." Feller continues to observe that for r particles and n cells (where in this present chapter $r =$ FM atoms and $n =$ AFM sites) with $r \leq n$, there are $\binom{n}{r} = n!/[r!(n-r)!]$ ways of choosing the n cells for arranging the r particles, and the probability of each arrangement is $\binom{n}{r}^{-1}$. According to Feller, a good example of Fermi–Dirac statistics is the distribution of misprints in a book where r misprints are distributed among n letters.

THE BASIC SPIN VALVE

The basic spin valve sensor is sketched in Fig. 2.10, and the view is exploded to show the relative orientation between the magnetization vectors in the free and pinned layers. This "top" spin valve design, discussed by Heim et al. (1994), has a

40 THIN FILM PROPERTIES FOR MR SENSORS

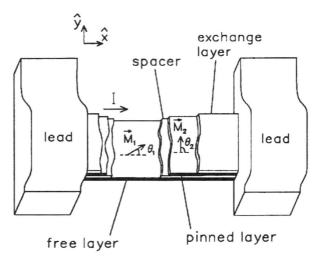

FIGURE 2.10 Unshielded spin valve sensor. Reprinted with permission from D. E. Heim, R. E. Fontana, Jr., C. Tsang, V. S. Speriosu, B. A. Gurney, and M. L. Williams, *IEEE Trans. Magn.*, **MAG-30**, 316. Copyright 1994 by IEEE.

free layer deposited with an induced anisotropy easy axis aligned along the x-axis; the intent is to have the magnetization M_1 align with the easy axis when no field is present and to rotate up or down in response to a signal field. A nonmagnetic spacer is deposited, followed by another ferromagnetic layer that is pinned in the transverse (y) direction by an AFM layer. The change in resistance from the GMR effect follows the relation

$$\Delta R = \left(\frac{\Delta \rho_{\text{GMR}} W}{t_{\text{FM}} h}\right) \frac{1 - \cos(\theta_1 - \theta_2)}{2} \qquad (2.16)$$

which was described by Dieny et al. (1991a), and this leads directly to the "ideal" GMR transfer curve discussed by Heim et al. (1994); the factor of 2 normalizes this relation over the total range from -1 to $+1$. When the pinned FM layer is as shown above ($\theta_2 = \pi/2$), (2.16) reduces to a resistance change proportional to $\sin \theta_1$, and the Stoner–Wohlfarth model applies to the idealized uniform rotation of M_1 in a uniform field along the hard axis of the free FM layer. The ideal transfer curve is linear in the applied field, and this is shown in Fig. 2.11 along with the simple hyperbolic tangent function used to approximate the Stoner–Wohlfarth solutions when a field component H_x exists [see (2.7) above and Fig. 2.3].

The free FM layer could have an AMR effect that cannot be ignored, so the total change in resistance of a spin valve device would be the sum of the GMR and AMR effects. In this more general case, (2.16) is altered to show the GMR effect (with

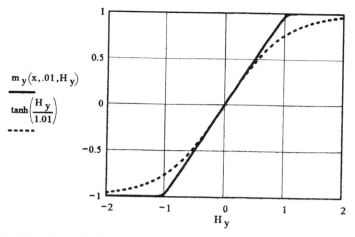

FIGURE 2.11 Ideal GMR transfer curve shown with a hyperbolic tangent function to approximate Stoner–Wohlfarth solutions when a field H_x exists.

difference in orientations $\Delta\theta = \theta_1 - \theta_2$) combined with an AMR effect that depends only on θ_1:

$$\Delta R = \frac{W}{t_{FM}h}\left(\Delta\rho_{GMR}\frac{1-\cos\Delta\theta}{2} + \Delta\rho_{AMR}\cos^2\theta_1\right) \quad (2.17)$$

Coehoorn et al. (1998) and Tsang et al. (1998) discuss the annoying presence of the AMR effect in spin valves, and Gill (1998) describes a technique wherein the sense current flows at 45° with respect to the free-layer magnetization to enhance the combined MR effect by about 25%. The normalized transfer curve for a total resistance change of a free layer having AMR effect equal to 25% of the GMR effect is shown in Fig. 2.12.

In actual spin valve devices it is not possible to achieve an ideal bias state where M_1 is uniform and lies along the easy axis. Heim et al. (1994) explain that the free layer is influenced by an FM coupling field [see (2.12)], a field arising from the sense current flowing through all layers of the sensor, and the demagnetizing field from the pinned FM layer. The challenge of designing spin valve sensors to achieve useful linearity and dynamic operating range is covered in Chapter 6; top and bottom spin valve designs will be analyzed, spin valves with two pinned layers (the "dual" spin valve) will be discussed, questions regarding current splitting among very thin conducting layers (where the thicknesses are less than the mean free path for conduction) are addressed, and thermal considerations such as unpinning and repinning the pinned layer will be examined.

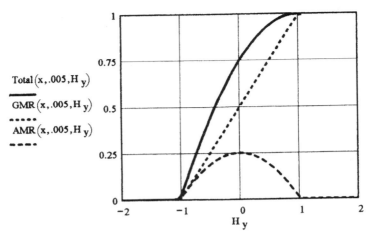

FIGURE 2.12 A GMR transfer curve containing 25% relative AMR effect. The dotted lines show the GMR and AMR effects separately.

ELECTRICAL, INSULATING, AND THERMAL PROPERTIES OF THIN FILMS

Fuchs–Sondheimer Theory of Film Resistivity

The resistivity of thin films depends on film thickness; as films become thin relative to the mean path length for bulk scattering, conduction electrons scatter with increasing probability from film surfaces; thus resistivity increases. This behavior is known as the size effect, and it was studied by Fuchs (1938) and Sondheimer (1952). Mayadas et al. (1974) showed that the thickness dependence of resistivity for annealed polycrystalline evaporated Permalloy films was consistent with the Fuchs–Sondheimer surface scattering model combined with additional scattering at grain boundaries; this is shown in Fig. 2.13a for evaporated Permalloy films with thicknesses between 100 Å and 2500 Å. The AMR effect is a bulk phenomenon (it is independent of surface scattering), so $\Delta\rho$ is constant with film thickness; this important point is illustrated in Fig. 2.13b because the electrical signal from MR recording heads depends on $\Delta\rho$ and is independent of ρ. The behavior of ρ with thickness has important consequences for current sharing in multilayered films and for I^2R heating in very thin films.

The Fuchs–Sondheimer theory uses the Boltzmann equation to derive the flow of electrons in metallic films under the assumption that electron scattering is entirely diffuse at each boundary. Electrons drift away from film surfaces toward the film interior with diffusion velocities $v_z = v\cos\theta$ that are either $v_z \geq 0$ or $v_z \leq 0$ depending on the surface $z = 0$ or $z = t$. The current density is written

$$J(z) = \frac{4\pi\varepsilon^2 m^2 \tau \bar{v}^3}{h^3} E \int_0^{\pi/2} \sin^3\theta \left[1 - \exp\left(-\frac{t}{2\lambda \cos\theta}\right) \cosh\left(\frac{t-2z}{2\lambda \cos\theta}\right)\right] d\theta,$$

(2.18)

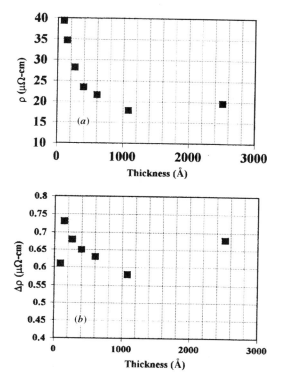

FIGURE 2.13 (a) Resistivity versus film thickness for Permalloy films; (b) the AMR effect of Permalloy versus film thickness. Data from McGuire and Potter (1975).

where ε, m, and $\lambda = \tau\bar{v}$ are the electron charge, effective mass, and mean free path, respectively, h is Planck's constant, E is the electric field, and t is the film thickness. With E along the x-direction, $J_x = \sigma E_x$ is averaged over the film thickness ($z = 0$ to $z = t$) to obtain the effective conductivity σ. That is,

$$\sigma = \frac{1}{Et}\int_0^t J_x(z)\,dz = \sigma_0\left[1 - \frac{3\lambda}{2t}\int_0^{\pi/2} \sin^3\theta\,\cos\theta\left[1 - \exp\left(-\frac{t}{\lambda\cos\theta}\right)\right]d\theta\right], \quad (2.19)$$

where σ_0 is the bulk conductivity. The integral in (2.19) is replaced by a convenient form for numerical tabulation,

$$\frac{\sigma_0}{\sigma} = \frac{\rho}{\rho_0} = \frac{\Phi(k)}{k}, \quad (2.20)$$

where the resistivity $\rho = 1/\sigma$, $k = t/\lambda$, and after repeated integration by parts, the function $\Phi(k)$ is expressed as

$$\frac{1}{\Phi(k)} = \frac{1}{k} - \frac{3}{4}\left(1 - \frac{k^2}{12}\right)Ei(-k) - \frac{3}{8k^2}(1 - e^{-k}) - \left(\frac{5}{8k} + \frac{1-k}{16}\right)e^{-k}. \quad (2.21)$$

The factor $Ei(-k)$ is a tabulated integral

$$-Ei(-u) = \int_u^\infty \frac{e^{-x}}{x}\, dx. \quad (2.22)$$

Sondheimer gives very useful limiting forms of $\rho/\rho_0 = \sigma_0/\sigma$ for large k (thick film) and small $k = t/\lambda$ (thin-film) regimes:

$$\frac{\rho}{\rho_0} = \begin{cases} 1 + \dfrac{3}{8k} & \text{(large } k\text{)}, \quad (2.23a) \\ \dfrac{4}{3k\,\log(1/k)} & \text{(small } k\text{)}. \quad (2.23b) \end{cases}$$

The complete solution to (2.19) and the limiting forms (2.23a,b) are plotted in Fig. 2.14.

The large-k approximation is about 10% greater than the complete solution for $k = 0.05$, is about 8% lower at $k = 0.50$, and is within 0.5% for $k > 5$. The small-k

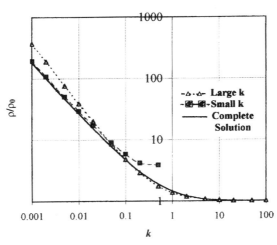

FIGURE 2.14 Normalized resistivity (solid line) versus normalized film thickness ($k = t/\lambda$) for complete Fuchs–Sondheimer theory, along with limiting forms for thick-film (large-k, triangles) and thin-film (small-k, squares) regimes.

TABLE 2.3 Sheet Resistance of Thin Films Versus Thickness

	Resistance (Ω/sq)				
Thickness (Å)	Cu	CoFe	NiFe	Ru	PtMn
35	537.7	439.7	246.3	142.7	460.3
45	113.9	179.6	128.9	—	—
50	57.4	127	106.8	78.7	355.6
65	20.5	58.5	65.3	—	—
100	8.6	28.3	33.5	30.5	175.7
150	4.65	14.9	19.5	18.4	113.6
200	2.7	10.0	13.4	12.6	85.8

approximation is always greater than the complete solution: at $k = 0.05$ the small k approximation is 15.5% high with improving agreement as k approaches 0.001 (6% high). Sondheimer argues that, lacking a completely rigorous theory of conductivity, the mean free path appears as a parameter that can be experimentally determined. The unpublished data of Huai and Macchioni (1999) for the resistivity of sputtered thin films of Cu, CoFe, NiFe, Ru, and PtMn on oxide-coated Si are shown in Table 2.3.

Macchioni fitted these data with the Fuchs–Sondheimer large-k relation (2.23a) using ρ_0 and λ as adjustable parameters; his results are found in Table 2.4. Note that the mean free path for PtMn is only 0.3 Å with this fit, and it is not clear what physical meaning (if any) can be attached to such a small distance. Also note the differences in resistivity of the sputtered NiFe films when compared with the evaporated film data in Fig. 2.13a. Dieny et al. (1991b) measured the sheet conductance of Si/Co 70 Å/t_{cu}/NiFe 47 Å/FeMn 78 Å/Cu 15 Å structures over a range of Cu thicknesses from about 12 Å up to 150 Å; their data (Fig. 7 of their publication) fit the Fuchs–Sondheimer large-k relation with the values $\rho_0 = 4.1$ μΩ-cm and $\lambda = 101$ Å. That is, using the Dieny et al. terminology, their sheet conductivity follows the relation $G_T(t_{Cu}) = G_{rest} + t_{Cu}/\rho_{Cu}(t_{Cu})$ where $G_{rest} = 0.045\ \Omega^{-1}$ is the sheet conductance of the rest of the structure and $\rho_{Cu}(t_{Cu}) = \rho_0(1 + 3\lambda/8t_{Cu})$. The thin-film Cu results of Dieny et al. (1991b) and Huai and Macchioni (1999) agree within about 10%. In spin valve structures, the Cu spacer is normally about 25 Å thick, so, depending on the process conditions, the resistivity of this layer could be $\rho \simeq (4.5\ \mu\Omega\text{-cm})(1 + 0.375 \times$

TABLE 2.4 Resistivity and Mean Free Path Parameters for Thin Films

Parameter	Cu	CoFe	NiFe	Ru	PtMn
λ, Å	111	95	74	58	0.3
ρ_0, μΩ-cm	4.5	16.7	23.3	23.3	167

TABLE 2.5 Electrical, Insulating and Thermal Properties of Films

Material	Temperature Coefficient (parts/K)			Thermal Conductivity (W/m-K)	Resistivity (μΩ-cm)	Notes
	Of ρ	Of $\Delta\rho$ AMR	Of $\Delta\rho$ GMR			
NiFe	4×10^{-3}	-3×10^{-3}	—	~20–50	~20–50	Film
NiFe/Cu/NiFe/ IrMn	5×10^{-4}	—	-2.3×10^{-3}	~30		Film
CoCrPt	~3×10^{-3}	—	—	~15	~75	Film
Cr	~2×10^{-3}	—	—	~20	~50	Film
Cu	3.2×10^{-3}	—	—	~60–180	~4–12	Film
Ta	3.8×10^{-3}	—	—	~5	~200	Film
Al_2O_3	—	—	—	~1.0	~10^{20}	Film
AlN	—	—	—	~3.0	~10^{15}	Film

111 Å/25 Å) $\simeq$ 12.0 μΩ-cm, which is nearly a factor of 3 greater than the resistivity of much thicker films. The resistivities for films of Cu, NiFe, and CoFe are estimated from the data in Table 2.5 and are plotted as a function of film thickness in Fig. 2.15.

These results will be used in Chapter 6 for estimating the current density in the various thin layers of spin valve structures.

Thermal Properties of Films

A tabulation of relevant electrical, insulating, and thermal properties of the various layers found in AMR and GMR recording heads is given in Table 2.5. In SAL-biased AMR heads, one finds the common practice of depositing a thin film of Ta between the SAL and MR layers, both of which are metals and possibly good electrical

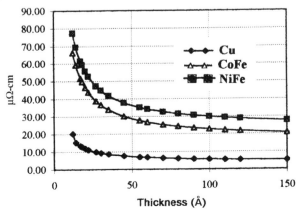

FIGURE 2.15 Resistivity of thin films of Cu, NiFe, and CoFe versus film thickness.

conductors. The Ta layer is a metal that promotes good [111] texture to the MR layer and has a resistivity high enough to provide controlled shunting of current between SAL and MR layers. Commercially successful MR heads are shielded from external magnetic fields by very thick magnetic shields (about 2–3 µm of Permalloy or Sendust), and the thin sensor layers require good electrical insulation from the thick shields. It is common to find sputtered alumina Al_2O_3 or aluminum nitride AlN in the range of 500–900 Å thick serving in this role. As areal density increases, one can trust that the thickness of insulating films will reduce to follow the scaling rules discussed in Chapter 1 and that materials engineers and scientists will continue to fight the reliability issues that arise from electrical current flowing in thin metal films and the dissipation of the heat thus generated. The data are collected from a number of sources (see Guo and Ju, 1997; Ishiwata et al., 1996; Lide, 1998; Glocker and Shah, 1995; and unpublished internal measurements) or estimated where necessary.

The Lorentz number $L = \kappa/(\sigma T) = (\pi^2/3)(k_B/e)^2 = 2.45 \times 10^{-8}$ $(V/K)^2$, which comes from the Wiedemann–Franz law (see Mott and Jones, 1958, p. 307; k_B is Boltzmann's constant and e is the electronic charge), is used to estimate the thermal conductivity (κ) from the resistivity of a film ($\rho = 1/\sigma$). This relation is apparently good for most metals as long as the temperature and the resistance are not too high. As an example, a NiFe film 3 µm thick would have $\rho \simeq 20$ µΩ-cm [$\sigma \simeq 5.0 \times 10^6$ $(\Omega$-m$)^{-1}$], and at $T = 300$ K, the thermal conductivity would be approximately $\kappa = L\sigma T \simeq 37$ W/(m-K); in their thermal analysis, Guo and Ju (1997) use $\kappa = 35$ for thick NiFe shields. The estimates of κ for Ta and CoCrPt may not be as reliable as for the lower resistivity metals. The temperature coefficient of κ is about -5×10^{-4} for Cu and Fe over the temperature range of 300–900 K (see, e.g., *the Handbook of Chemistry and Physics*, CRC Press, Boca Raton, FL).

Hasegawa (1996) discusses a theory of temperature dependence of the GMR effect and shows the GMR ratio $\Delta R/R$ versus T/T_C, where T_C is the Curie temperature of the FM layers. It is difficult to extract the $\Delta R(T)$ behavior (which produces the electrical signal) from the normalized $\Delta R/R(T/T_C)$ because $R(T/T_C)$ increases with temperature. Dieny et al. (1992) discussed the temperature dependence of $\Delta R/R$ for spin valves of various free-layer thicknesses. Their data, shown in Figs. 2.16a,b, were fit with second-order polynomials and extrapolated to zero $\Delta R/R$ at a temperature $T_{0SV} \simeq 515$ K for spin valve systems of NiFe/Cu/NiFe and Co/Cu/NiFe. Dieny et al. (1998) also published results for the temperature behavior of Co/Cu/Co spin valve systems for various free-layer thicknesses in which they obtained the extrapolated result $T_{0SV} \simeq 640$ K. Data on the AMR effect are extracted from Bozorth (1951, pp. 758–761) for Ni alloys, and the fitted curves shown in Fig.2.16c are calculated with (2.24b), where the coefficient $\alpha_{AMR} \simeq -0.0055$ for the bulk samples of NiCo, NiFe, and NiCu alloys from about -200 up to $400°C$. In the range from room temperature up to about $150°C$, the resistivity of thin-film samples of NiFe increases almost linearly with temperature. In this book, the resistivity and magnetoresistivity for AMR devices will be described by the relations

$$\rho(T) = \rho_0(1 + \alpha_\rho \, \Delta T) \quad (2.24a)$$

$$\Delta\rho_{AMR}(T) \simeq \Delta\rho_{0AMR} \exp(\alpha_{\Delta\rho} T). \quad (2.24b)$$

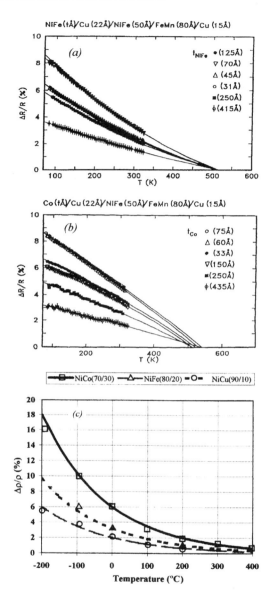

FIGURE 2.16 (*a*) Thermal variation of the magnetoresistance of samples with the structure glass/NiFe (t Å)/Cu (22 Å)/NiFe (50 Å)/FeMn (80 Å)/Cu (15 Å). The lines are second-order polynomial fits. (*b*) Thermal variation of the magnetoresistance of samples with the structure glass/Co (t Å)/Cu (22 Å)/NiFe (50 Å)/FeMn (80 Å)/Cu 15 Å). The lines are second-order polynomial fits. (*a*, *b*) Reprinted with permission from B. Dieny, P. Humbert, V. S. Speriosu, S. Metin, B. A. Gurney, P. Baumgart, and H. Lefakis, *Phys. Rev. B*, **45**, 806. Copyright 1998 by the American Physical Society. (*c*) AMR effect of NiFe (80/20), NiCo (70/30), and NiCu (90/10) alloys versus temperature. Data extracted from Bozorth (1951, pp. 758–761).

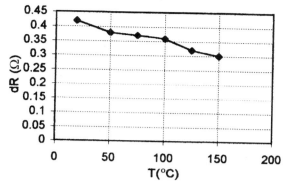

FIGURE 2.17 GMR effect versus temperature for a synthetic spin valve structure.

The temperature coefficients of resistivity and magnetoresistivity are given in columns 2 and 3 of Table 2.5. The AMR effect is also present in the free layer of a spin valve, and it is difficult to separate it from the GMR effect. In GMR devices, elevated temperatures above about 200°C may produce irreversible changes in the transport properties (see McMichael et al., 1997).

Unpublished data for a synthetic spin valve GMR $\Delta\rho(T)$ are plotted in Fig. 2.17; for the limited temperature range, the trend is approximately linear with a negative coefficient, but more complete temperature studies are indicated.

THERMAL ANALYSIS OF MR HEADS

Magnetoresistive heads are activated with a sense/bias current, and the power generated ($P = I^2R$) heats the device. This heat is dissipated through insulating layers to thick magnetic shields on each side of the sense element and through the shields to the surrounding environment. Compared to the size of a sense element, the entire structure of a recording head is huge and amounts to an infinite heat sink connected to the environment by conductive and convective pathways that maintain the assembly at a temperature close to the ambient inside the disk drive enclosure (15–25°C above room temperature). The temperature rise of a sense element above the drive ambient is an important consideration regarding signal amplitude and long-term reliability issues related to diffusion, electromigration, or other unwanted effects. The element resistivity has a positive temperature coefficient, and the AMR and GMR effects each have a negative temperature coefficient; thus the resistance increases and the signal reduces with device heating. The thermal models discussed here will give insight regarding the dependence of temperature rise and its relation to device geometry and materials properties.

Anisotropic and giant MR elements are composed of a number of metallic layers in a sandwich that is electrically insulated between thick magnetic shields. Each layer in the sensor is a resistor whose resistance is given by $R = \rho W/th$ (see Fig.

2.2), and when many layers are built one upon the other, the resistance is that of a parallel network. To keep this analysis tractable, the validity of Ohm's law is assumed, however, size effects must be included in extremely thin layers, and one would normally resort to experimentally determined resistivities for sheets of multilayered thin films. For two, three, or four resistances in parallel, the total resistance is

$$R_t = \frac{R_1 R_2}{R_1 + R_2}, \quad (2.25a)$$

$$R_t = \frac{R_1 R_2 R_3}{R_1 R_2 + R_1 R_3 + R_2 R_3}, \quad (2.25b)$$

$$R_t = \frac{R_1 R_2 R_3 R_4}{R_1 R_2 R_3 + R_1 R_2 R_4 + R_1 R_3 R_4 + R_2 R_3 R_4}, \quad (2.25c)$$

but it is easier to work with the conductance $G_t = \sum 1/R_i$ for many parallel resistors.

Simple One-Dimensional Analysis of Temperature Rise

In a simplified model, power flows from hot to cold bodies following the rule

$$Q = \frac{P}{A} = -\kappa \, \nabla T \quad \text{W/cm}^2, \quad (2.26)$$

where P is power in watts, A is the area of the boundary for heat flow in square centimeters, κ is the thermal conductivity of the surrounding material in watts per centimeter-kelvin, and ∇T is the temperature gradient between hot and cold bodies in kelvins per centimeter. Assuming a constant gradient across the hot–cold interface of thickness g in centimeters, $\nabla T = (T_h - T_c)/g$ and (2.26) is rewritten and solved for the temperature change in terms of power and device geometry:

$$\Delta T = T_h - T_c = \frac{g}{\kappa A} P \quad \text{K}. \quad (2.27)$$

The factor $g/\kappa A$ is the "thermal resistance" in kelvins per watt. In MR heads, heat flows from the element through two insulating gaps (g_1 and g_2) to shields 1 and 2, so the thermal resistance is that of two parallel thermal resistors,

$$R_\tau = \frac{g_1 g_2}{\kappa A(g_1 + g_2)} \quad \text{K/W}, \quad (2.28)$$

and the paths are assumed to have identical thermal conductivities and areas. With equal gaps, (2.28) reduces to the simpler relation $R_\tau = g/(2\kappa A)$ where $A \simeq Wh$. More complex models would include heat flow through the connection leads and three-dimensional aspects of the design. Figure 2.18 gives top and end views of the

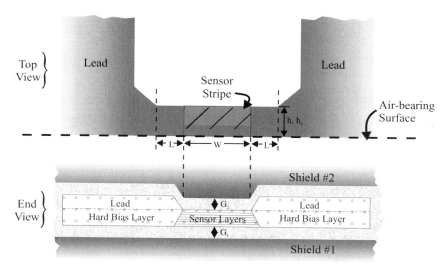

FIGURE 2.18 Geometry for a two-dimensional thermal analysis of AMR and GMR sensors.

geometry involved in this analysis; the effective area A is greater than Wh because heat flows through the sensor leads and the film sandwich is etched at an angle (the "junction angle" varies from perhaps 30° up to 80°, depending on the process used). The resistance depends on temperature in a simple manner,

$$R(T) = R_0(1 + \alpha \, \Delta T), \qquad (2.29)$$

where α is the experimental temperature coefficient of resistance (see Table 2.5). With a little algebra, the temperature rise can be expressed in terms of the resistance R_0 at ambient temperature, α, R_τ and the total current I flowing through all layers,

$$\Delta T \simeq \frac{R_\tau I^2 R_0}{1 - \alpha R_\tau I^2 R_0}. \qquad (2.30)$$

This approximation is useful up to temperatures where heat damage (including melting) occurs.

Some ball-park numbers will be helpful at this point in the analysis: assume a three-layer sensor with MR, Ta, and SAL thicknesses of 130, 55, and 85 Å and resistivities (in micro-ohm-centimeters) of 32, 160, and 60, respectively. Let the stripe length (track width) $W = 1.5$ μm, height $h = 1.0$ μm, and element-to-shield gaps at $g_1 = g_2 = 0.075$ μm with equal thermal conductivities $\kappa = 1.0$ W/m-K. (This geometry is appropriate for application at 2 Gbits/in.² areal density.) The temperature coefficient of resistance of the trilayer sandwich is set at $\alpha = 3 \times 10^{-3}$ for this example. Using the tools developed so far, the resistances are 37, 436, and

106 Ω for the MR, Ta, and SAL layers, respectively, and the total parallel resistance (2.25c) is 25.8 Ω at 300 K. From (2.28) the thermal resistance $R_\tau = 2.5 \times 10^4$ K/W, so 10 mA of sense/bias current generates a power sufficient to raise the temperature by 80 K, which increases the resistance to 32 Ω; thus the power dissipation at T = ambient + 80 K is 3.2 mW. Additional resistance arising from sensor leads is ignored here; depending on design rules, the leads could add another 5–15 Ω to the device. Figure 2.19 shows temperature rise and resistance as a function of bias current for this simple model and hypothetical MR head.

The AMR effect $\Delta\rho = \Delta\rho_{0AMR} \exp(\alpha_{AMR} \Delta T)$, where from Table 2.5, $\alpha_{AMR} = -3 \times 10^{-3}$, so at 10 mA and a temperature rise of 82 K, the output signal would drop 22%. The main conclusions are that temperature rise (2.30) goes somewhat greater than I^2 because the resistance increases with temperature, temperature rise is inversely proportional to the area A of dissipation, temperature rise is inversely proportional to the thermal conductivity κ of the gap material, and signal drops rapidly with heating. Jander et al. (1996) describe an approximate analytical model and show trends for head geometry and material properties; one key area for improvement is the thermal conductivity of the insulating gap material. As areal density increases, A scales with Wh, so it behooves the design engineer to find new materials with increased κ for controlling device operating temperatures and to create biasing schemes that reduce the required sense/bias current. As will be seen in Chapters 5 and 6, the history of MR recording heads follows this evolutionary path.

Two-Dimensional Analysis of Temperature Rise

In their work on electromigration in thin films, d'Heurle and Ho (1978) discuss the temperature distribution along thin-film conductors in good thermal contact with a

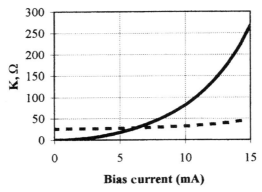

FIGURE 2.19 Temperature rise and resistance versus bias current for a simple hypothetical MR head.

substrate and cite an analysis of heat flow by Chaung and Huang (1976). The steady-state heat flow equation for a uniform conductor of thickness t_f is given as

$$\kappa_f \frac{d^2 T}{dx^2} + J^2 \rho - \frac{H}{t_f}(T - T_{\text{amb}}) = 0, \tag{2.31}$$

where κ_f is the thermal conductivity of the film, $J = I/ht_f$ is the current density (in amperes per square centimeter), ρ is the film resistivity, and $(H/t)(T - T_{\text{amb}})$ is the heat loss to the substrate. Using a transmission line model, Guo and Ju (1997) derive an analytical formula for MR heads of the same form as (2.31), and they include heat flow through leads, gaps with thermal conductivity κ_0 and thickness g between the film and two shields. Heat generation in the leads is excluded from their analysis. Gibbons (1999) extends the Guo and Ju (1997) analysis by including heat generation in the connecting leads, and his analysis is followed here. At equilibrium, the power density $Q = P/A$ generated in the sense element and leads equals the dissipation to the shields. That is,

$$Q(x + dx)th - Q(x)th + Q_{\text{out}} h \, dx = \rho(T)J^2 th \, dx \tag{2.32}$$

where

$$\rho(T) = \rho_0[1 + \alpha T(x)],$$

$$Q(x + dx) = -\kappa \frac{dT(x + dx)}{dx},$$

$$Q(x) = -\kappa \frac{dT(x)}{dx},$$

$$Q_{\text{out}} = 2\kappa_0 \frac{T(x)}{g}.$$

Upon substitution and collection of terms, the equation for temperature rise of the MR film becomes (with $J_f = I/(t_f h)$)

$$-\frac{d^2 T(x)}{dx^2} + \frac{1}{\lambda_f^2} T(x) = \frac{\rho_f J_f^2}{\kappa_f}, \quad \text{where } \lambda_f = \sqrt{\frac{\kappa_f g t_f}{2\kappa_0 - \alpha_f \rho_f J^2 g t_f}}, \tag{2.33}$$

and for the conducting leads (with $J_L = I/t_L h_L$) the equation is

$$-\frac{d^2 T}{dx^2} + \frac{1}{\lambda_L^2} T = \frac{\rho_L J_L^2}{\kappa_L}, \quad \text{where } \lambda_L = \sqrt{\frac{\kappa_L g t_L}{2\kappa_0 - \alpha_L \rho_L J_L^2 g t_L}}. \tag{2.34}$$

The solution to (2.33) is

$$T_f(x) = Ae^{x/\lambda_f} + Be^{-x/\lambda_f} + C, \quad \text{where } C = \frac{\lambda_f^2 \rho_f J^2}{\kappa_f}, \quad (2.35)$$

and for the left- and right-lead regions, the solutions to (2.34) are

$$T_L(x) = De^{(x \pm W/2)/\lambda_L} + Ee^{-(x \pm W/2)/\lambda_L} + F, \quad \text{where } F = \frac{\lambda_L^2 \rho_L J_L^2}{\kappa_L}. \quad (2.36)$$

The parameters λ_f and λ_L each have the dimension of (centimeters) and are known as the characteristic decay lengths for the system. The MR element extends along x from $-W/2 \leq x \leq W/2$, and the leads have a length L extending beyond the MR film. The parameter $T(\pm W/2)$ is the same for leads and MR element and $T(\pm L) = 0$ (i.e., the lead ends are at ambient). With these boundary conditions, the solution for temperature rise of the MR film is symmetric about $x = 0$, so $A = B$ and (2.35) becomes

$$T_f(x) = 2A \cosh\left(\frac{x}{\lambda_f}\right) + C. \quad (2.37)$$

At the end of the left lead $T_L(-L) = 0$, and substituting these conditions into (2.36) gives

$$D = -Ee^{-2(L-W/2)/\lambda_L} - Fe^{-(L-W/2)/\lambda_L}. \quad (2.38)$$

At the left end of the MR film $T_f(-W/2) = T_L(-W/2)$, and substituting these conditions into (2.35) and (2.36), one obtains

$$D[1 - e^{-2(L-W/2)/\lambda_L}] + F[1 - e^{-(L-W/2)/\lambda_L}] = 2A \cosh\left(\frac{W}{2\lambda_f}\right) + C. \quad (2.39)$$

Equations (2.38) and (2.39) are solved for the unknowns D and E in terms of A, C and F:

$$D = \frac{2A \cosh(W/2\lambda_f) + C - F(1 - e^{-(L-W/2)/\lambda_L})}{1 - e^{-2(L-W/2)/\lambda_L}}. \quad (2.40)$$

Continuity of heat flow establishes a third boundary condition, namely,

$$E = -\frac{e^{-2(L-W/2)/\lambda_L}}{1 - e^{-2(L-W/2)/\lambda_L}}\left[2A\cosh\left(\frac{W}{2\lambda_f}\right) + C - F[1 - e^{-(L-W/2)/\lambda_L}]\right] - Fe^{-(L-W/2)/\lambda_L}, \tag{2.41}$$

$$\kappa_L t_L h_L \frac{dT_L(-W/2)}{dx} = \kappa_f t_f h \frac{dT_f(-W/2)}{dx}. \tag{2.42}$$

From this relation, one solves for A in terms of C and F (both known):

$$A = \frac{1}{2}\left[\frac{\coth[(L-W/2)/\lambda_L][F(1 - e^{-(L-W/2)\lambda_L}) - C] - Fe^{-(L-W/2)/\lambda_L}}{\coth[(L-W/2)/\lambda_L]\cosh(W/2\lambda_f) + [(\kappa_f t_f h \lambda_L)/(\kappa_L t_L h_L \lambda_f)]\sinh(W/2\lambda_f)}\right] \tag{2.43}$$

Gibbons (1999) defines several limiting cases in his analysis. If there are no leads, $L = W/2$ and (2.43) reduces to $A = -C/[2\cosh(W/2\lambda_f)]$ since $\coth(0) \to \infty$. Recall that film heating is described with $C = \lambda_f^2 \rho_f J_f^2 / \kappa_f$ and lead heating with $F = \lambda_L^2 \rho_L J_L^2 / \kappa_L$: if $F = 0$ and $L \to \infty$, then (2.43) reduces to

$$A = -\frac{C}{2\left[\cosh(W/2\lambda_f) + (\kappa_f t_f h \lambda_L / \kappa_L t_L h_L \lambda_f)\sinh(W/2\lambda_f)\right]}, \tag{2.44}$$

since $\coth(\infty) \to 1.0$. The form given in (2.44) is very close to that given in Guo and Ju (1997) except for the factor $t_f h/(t_L h_L)$ in the denominator. The final expression for sensor temperature may be written in a more covenient form as

$$T(x, I) = T_0(I)\left[1 - \eta(I)\cosh\left(\frac{x}{\lambda_{mr}(I)}\right)\right] \tag{2.45}$$

where $T_0(I) \equiv C = \lambda_f^2 \rho_0 I^2/[\kappa_f (t_f h)^2]$ (in kelvins or degrees Celsius) and $\eta(I)$ is given by

$$\eta(I) = \left[\cosh\left(\frac{W}{2\lambda_{mr}(I)}\right) + \sinh\left(\frac{W}{2\lambda_{mr}(I)}\right)\frac{\lambda_L \kappa_{mr} t_{mr}}{\lambda_{mr}(I)\kappa_L t_L}\right]^{-1}. \tag{2.46}$$

Figure 2.20 shows temperature profiles using (2.45) and $h_L = h$ in the composite MR/Ta/SAL film described earlier ($t_f = 270$ Å and $\rho_f = 46.4$ µΩ-cm for the parallel resistor); lead thickness is 0.1, 1.0, and 10 times t_f, $\kappa_L = \kappa_f = 35$ W/m-K, $\kappa_0 = 1.0$ W/m-K, $I = 10$ mA, and all other parameters are unchanged from the previous simplified computation. The characteristic decay lengths are λ_f of 0.21 µm and λ_L of 0.06, 0.19, and 0.6 µm. Note the increase in film temperature as leads go

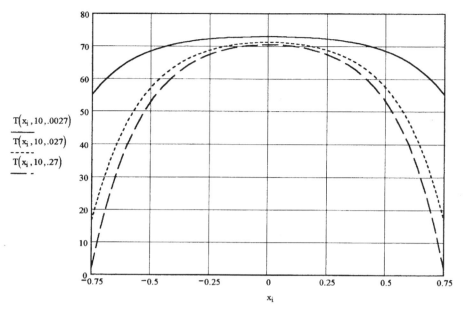

FIGURE 2.20 Temperature profiles across the track width of an MR sensor for $t_L/t_{MR} = 0.1, 1.0, 10$. Bias current = 10 mA.

from thick to thin; as $t_L \to 0$, the film temperature approaches the constant $\lambda_f^2 \rho_0 J^2 / \kappa_f$ along its length. With narrow tracks, thick leads can significantly reduce the average temperature of the sensor. The average film temperature, found by integrating (2.45) over the track width W, is

$$T_{\text{avg}}(I) = T_0(I)\left[1 - \frac{\eta(I)\sinh(W/2\lambda_f)}{W/2\lambda_f}\right]. \tag{2.47}$$

It is interesting to compare the simple estimate of heat rise (2.30), which does not include heat loss through the leads, with the more comprehensive (2.47) for $t_L \to 0$. With $h = h_L = 1.0\,\mu\text{m}$, $t_L = t_f/100$, and $I = 10$ mA, $\Delta T = 79.9$ K [from (2.30)] and $T_{\text{avg}} = 79.7$ K. Plots of ΔT and T_{avg} versus current area are shown in Fig. 2.21. It appears that the simple approach allows quick estimates of the average temperature while the transmission line analysis shows interesting details about the temperature profile along the element and leads.

Current density scales with $1/h^2$ while temperature rise and power density grow somewhat more rapidly because resistance increases with temperature. Figure 2.22 shows temperature rise versus current for different values of stripe height h, and it is clear that small stripe heights are highly vulnerable to thermal damage. The thermal

THERMAL ANALYSIS OF MR HEADS   57

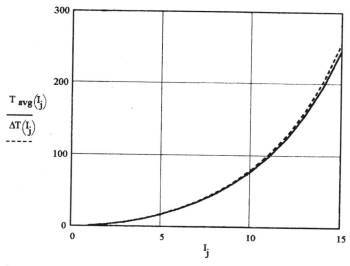

FIGURE 2.21 Comparison of the temperature rise for one- and two-dimensional analyses with no heat loss through the leads.

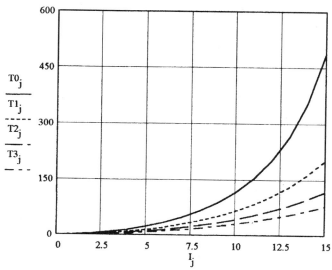

FIGURE 2.22 Temperature rise T_0, T_1, T_2, and T_3 versus bias current for stripe heights h of 0.8, 1.0, 1.2 and 1.4 µm, respectively.

resistance may be defined for the transmission line equations by normalizing the average temperature rise with the total power (P) dissipated at $T = T_{\text{avg}}$:

$$P = I^2 R(T_{\text{avg}}) = I^2 R_0 (1 + \alpha T_{\text{avg}}), \quad \text{where } R_0 = \frac{\rho_0 W}{t_f h}, \quad (2.48)$$

$$R_\tau = \frac{T_{\text{avg}}}{P} = \frac{g/[Wh(1 + \alpha T_{\text{avg}})]}{(2\kappa_0 - \alpha \rho_0 J^2 g t_f)} \left[1 - \frac{\eta(I) \sinh(W/2\lambda_f)}{W/2\lambda_f} \right]. \quad (2.49)$$

This expression is almost independent of temperature and current density because the factors containing α nearly cancel each other. Practically speaking, the prefactor in (2.49) reduces to the relation presented earlier in (2.28), namely $R_\tau \to g/2Wh\kappa_0$. Current and temperature rise are limited by the melting point T_m of the sensor element. For lack of better information regarding very thin films, the melting point of bulk Permalloy ($T_m \approx 1450°\text{C}$) will be used for estimating the melting current I_m for AMR and GMR devices. From (2.28) and (2.30), the temperature rise is equated with the melting point and the relation is solved for the current. This gives the result

$$I_m \simeq \left[\frac{T_m \kappa_0 (g_1 + g_2) t_f h^2}{g_1 g_2 \rho_0 (1 + \alpha T_m)} \right]^{1/2} \quad (2.50)$$

which is linear in stripe height h and independent of track width W. With $\Delta T = T_m = 1450°\text{C}$, gap $\kappa_0 = 1.0$ W/m-K, $g_1 = g_2 = 0.10$ μm, $t_f = 0.0125$ μm, $\rho_0 = 32$ μΩ-cm, and $\alpha = 2 \times 10^{-3}$/K, $I_m \simeq 17.0\, h$ (in milliamperes with h in micrometers). In Chapters 5, 6, and 8, AMR and GMR devices are analyzed with an eye to short- and long-term behavior relative to sense/bias currents. As will be seen, the interplay among device resistance, sense/bias current, power dissipation, thermal resistance, operating temperature, and sensor lifetime is a crucial design and analysis consideration.

REFERENCES

Anderson, R. L., Bajorek, C. H., and Thompson, D. A., *AIP Conf. Proc.*, **10**, 1445 (1972).

Baibich, M. N., Brote, J. M., Fert, A., Nguyen Van Dau, F., Petroff, F., Etienne, P., Creuzet, G., Friederich, A., and Chazelas, J., *Phys. Rev. Lett.*, **61**, 2472 (1988).

Bhattacharyya, M. K. and Simmons, R. F., *IEEE Trans. Magn.*, **MAG-30**, 291 (1994).

Binasch, G., Grunberg, P., Saurenbach, F., and Zinn, W., *Phys. Rev. B*, **39**, 4828 (1989).

Bozorth, R. M., *Ferromagnetism*, Van Nostrand, Princeton, NJ, 1951.

Bozorth, R. M. and Dillinger, J. F., *Physics*, **6**, 285 (1935).

Chaung, Y-S. and Huang, H. L., *J. Appl. Phys.*, **47**, 1775 (1976).

Chen, M., Gharsallah, N., Gorman, G., and Latimer, J., *J. Appl. Phys.*, **69**, 5631 (1991).

REFERENCES

Coehoorn, R., Kools, J. C. S., Rijks, Th. G. S. M., and Lenssen, K-M. H., *Philips J. Res.*, **51**, 93 (1998).

Cullity, B. D., *Introduction to Magnetic Materials*, Addison-Wesley, Reading, MA, 1972.

D'Heurle, F. M. and Ho, P. S., "Electromigration in Thin Films," Chap. 8 in *Thin Films: Interdiffusion and Reactions*, Poate, J. M., Tu, K. N., and Mayer, J. W. (Eds.), Wiley-Interscience, New York, 1978.

Dieny, B., *J. Magn. Magn. Mater.*, **136**, 335 (1994).

Dieny, B., Speriosu, V. S., Parkin, S. S. P., Gurney, B. A., Wilhoit, D. R., and Mauri, D., *Phys. Rev. B*, **43**, 1297 (1991a).

Dieny, B. Speriosu, V. S., Metin, S., Parkin, S. S. P., Gurney, B. A., Baumgart, P., and Wilhoit, D. R., *J. Appl. Phys.*, **69**, 4774 (1991b).

Dieny, B., Humbert, P., Speriosu, V. S., Metin, S., Gurney, B. A., Baumgart, P., and Lefakis, H., *Phys. Rev. B*, **45**, 806 (1992).

Dieny, B., Peireira, L. G., Taylor, R. H., and Yamamoto, S. Y., *IEEE Trans. Magn.*, **MAG-34**, 942 (1998).

Feller, W., *An Introduction to Probability Theory and Its Applications*, Vol. I, 2nd ed. Wiley, New York, 1957.

Frait, Z., *Phys. Status Solidi*, **2**, 1417 (1962).

Fuchs, K., *Proc. Cambridge Phil. Soc.*, **34**, 100 (1938).

Fuke, H., Saito, K., Kamiguchi, Y., Iwasaki, H., and Sahashi, M., *J. Appl. Phys.*, **81**, 4004 (1997).

Gibbons, M., private communication, 1999.

Gill, H. S., U.S. Patent 5,828,531, Oct. 27, 1998.

Glocker, D. A. and Shah, S. I. (Eds.), *Handbook of Thin Film Process Technology*, IoP Publ., Bristol, 1995.

Grünberg, P., U.S. Patent 4,949,039, Aug. 14, 1990.

Guo, Y., and Ju, K., *IEEE Trans. Magn.*, **33**, 2917 (1997).

Hanazono, M., Narishige, S., Hara, S., Mitsuoka, K., Kawakami, K., Sugita, Y., Kuwatsuka, S., Kobayshi, T., and Ohura, M., *J. Appl. Phys.*, **61**, 4157 (1987).

Hannon, D. M., Krounbi, M., and Christner, J., *IEEE Trans. Magn.*, **MAG-30**, 298 (1994).

Hasegawa, H., *J. Appl. Phys.*, **79**, 6376 (1996).

Hathaway, K. B., Fert, A., and Bruno, P., Chap. 2 in *Ultrathin Magnetic Structures II*, Heinrich, B. and Bland, J. A. C. (Eds.), Springer-Verlag, Berlin, 1994.

Heim, D. E., Fontana, Jr., R. E., Tsang, C., Speriosu, V. S., Gurney, B. A., and Williams, M. L., *IEEE Trans. Magn.*, **MAG-30**, 316 (1994).

Huai, Y. and Macchioni, C., unpublished internal data, Read-Rite, Milpitas, 1999.

Hunt, R. P., *IEEE Trans. Magn.*, **MAG-7**, 150 (1971).

Ishiwata, N., Ishi, T., Matsutera, H., and Yamada, K., *IEEE Trans. Magn.*, **MAG-32**, 38 (1996).

Jander, A., Indeck, R. S., Brug, J. A., and Nickel, J. H., *IEEE Trans. Magn.*, **MAG-32**, 3392 (1996).

Kelsall, G. A., *Physics*, **5**, 169 (1934).

Kittel, C., *Introduction to Solid State Physics*, 3rd ed., Wiley, New York, 1966.

Klokholm, E. and Aboaf, J. A., *J. Appl. Phys.*, **52**, 2472 (1981).

Klokholm, E. and Krongelb, S., *Proc. Symp. Mag. Mater. Process. and Devices, Electrochem. Soc.*, **90-8**, 125 (1990).

Kools, J. C. S., *IEEE Trans. Magn.*, **MAG-32**, 3165 (1996).

Kools, J. C. S., Rijks, Th. G. S. M., De Veirman, A. E. M., and Coehoorn, R., *IEEE Trans. Magn.*, **MAG-31**, 3918 (1995).

Krounbi, M. T., Voegeli, O., and Wang, P.-K., U.S. Patent 5,005,096, Apr. 2, 1991a.

Krounbi, M. T., Voegeli, O., and Wang, P.-K., U.S. Patent 5,018,037, May 21, 1991b.

Leal, J. L. and Kryder, M. H., *IEEE Trans. Magn.*, **32**, 4642 (1996).

Lederman, M., *IEEE Trans. Magn.*, **MAG-35**, 794 (1999).

Lide, D. R. (Ed.), *Handbook of Chemistry and Physics*, 79th ed., CRC Press, Boca Raton, FL, 1998.

Lin, T., Tsang, C., Fontana, R. E., and Howard, J. K., *IEEE Trans. Magn.*, **MAG-31**, 2585 (1995).

Markham, D. and Smith, N., *IEEE Trans. Magn.*, **MAG-24**, 2606 (1988).

Maruyama, T., Yamada, K., Tatsumi, T., and Urai, H., *IEEE Trans. Magn.*, **MAG-24**, 2402 (1988).

Mayadas, A. F., Janak, J. F., and Gangulee, A., *J. Appl. Phys.*, **45**, 2780 (1974).

McGuire, T. R. and Potter, R. I., *IEEE Trans. Magn.*, **MAG-11**, 1018 (1975).

McMichael, R. D., Chen, P. J., and Egelhoff, Jr., W. F., *IEEE Trans. Magn.*, **MAG-33**, 3589 (1997).

Mott, N. F. and Jones, H., *Theory of Properties of Metals and Alloys*, Dover, N.Y., 1958.

Néel, L., *Comptes Rendus*, **255**, 1676 (1962).

Nogués, J., and Schuller, I. K., *J. Magn. Magn. Mater.*, 203 (1999).

Parkin, S. S. P., *Phys. Rev. Lett.*, **71**, 1641 (1993).

Parkin, S. S. P., More, N., and Roche, K. P., *Phys. Rev. Lett.*, **64**, 2304 (1990).

Rijks, Th. G. S. M., PhD thesis, Eindhoven Univ. of Tech., 1996.

Shelledy, F. B., and Brock, G. W., *IEEE Trans. Magn.*, **MAG-11**, 1206 (1975).

Shiroishi, Y., Shiiki, K., Yuitoo, I., Tanabe, H., Fujiwara, H., and Kudo, M., *IEEE Trans. Magn.*, **MAG-20**, 485 (1984).

Simmons, R.., Davidson, R., and Gill, H. S., *IEEE Trans. Magn.*, **MAG-25**, 3200 (1989).

Slonczewski, J. C., *J. Appl. Phys.*, **37**, 1268 (1966a).

Slonczewski, J. C., *IBM J. Res. Devel.*, **10**, 377 (1966b).

Smith, N., *IEEE Trans. Magn.*, **MAG-23**, 259 (1987).

Sondheimer, E. H., *Adv. Phys.*, **1**, 1 (1952).

Speriosu, V. S., private communication.

Stoner, E. C., and Wohlfarth, E. P., *Phil. Trans. Roy. Soc.*, **A-240**, 599 (1948).

Tatsumi, T., Tsukamoto, Y., Yamada, K., Motomura, Y., and Aoyama, M., *J. Appl. Phys.*, **69**, 4671 (1991).

Tebble, R. S. and Craik, D. J., *Magnetic Materials* Wiley-Interscience, New York, 1969.

Thompson, W. (Lord Kelvin), *Proc. Roy. Soc. (London)*, **8**, 546 (1857).

Tsang, C., Fontana, Jr., R. E., Lin, T., Heim, D. E., Gurney, B. A., and Williams, M. L., *IBM J. Res. Devel.*, **42**, 103 (1998).

Wang, P-K., Ewasko, R., and Anderson, R., *Dig. Intermag. Conf.*, Paper no. BA-03 (1992).
White, R. L., *IEEE Trans. Magn.*, **MAG-28**, 2482 (1992).
Xiao, M., Devasahayam, A. J., and Kryder, M. H., *IEEE Trans. Magn.*, **MAG-34**, 1495 (1998).
Yamada, K. Maruyama, T., Ohmukai, T., and Urai, H., *J. Appl. Phys.*, **63**, 4023 (1988).
Yoda, H., Iwasaki, H., Kobayshi, T., Tsutai, A., and Sahashi, M., *IEEE Trans. Magn.*, **MAG-32**, 3363 (1996).
Yuan, S. W. , "Micromagnetics of Domains and Walls in Ferromagnetic Soft Materials," Ph.D. thesis, University of California, San Diego, 1992.

3

COUPLED MAGNETIC FILMS

It is experimentally known that the magnitude of spontaneous magnetization in ferromagnetic films is essentially constant throughout a specimen but that the direction of M_S varies from one magnetic domain to another, and that energy is stored in domain walls between the uniformly magnetized regions. High uniaxial anisotropy or strong magnetic fields exert torque on the magnetization in the various domains and uniformly align M_S within the specimen. As a specimen becomes small, it is energetically favorable for the magnetization to align with the dominant anisotropy direction throughout the volume because the energy spent in making a domain wall outweighs the reduction in energy stored in the demagnetization field of the specimen; this is the single-domain state. When two magnetic structures are in close proximity, their interaction alters the distribution of energy within this new system such that magnetic flux couples between the structures, and the magnetization directions within each are mutually interdependent. The problem of multiple-film systems without an applied field was studied many years ago and is given thorough treatment in Slonczewski (1996a,b). Pohm et al. (1984) analyze a system of two magnetic films with current flowing in each film; the films are separated by a nonconducting gap. The width of a domain wall and the energy stored there depend on important details in both of these two-film magnetic systems.

This chapter examines the magnetization distribution in each layer of two ferromagnetic films that are coupled through a nonmagnetic gap layer. As Slonczewski et al. (1988) showed, lamination of ferromagnetic layers leads to elimination of closure domains that would otherwise form in isolated films. In regions along film edges, the magnetization rotates smoothly and creates flux closures across the spacer gap and through the adjacent magnetic film; they called this new flux closure concept the "edge-curling wall" and showed convincing experimental evidence of its existence. The theory of laminated films can be applied to the design and analysis of MR heads, and the task for this chapter is to translate the work of Slonczewski et al. into equations covering the propagation of magnetic flux in two-film structures and to develop the closed-form analytical tools that will carry the reader through much of this book.

MAGNETIZATION DISTRIBUTION IN IDENTICAL FILMS

It seems best to define a simple multiple-film system, solve the equations describing the energy of that system, and then gradually add more complexity such that the reader follows the development from simple to complex. Therefore, the simple multiple-film system becomes two identical ferromagnetic films on both sides of a nonmagnetic film carrying a uniform current. Each magnetic film has uniaxial magnetic anisotropy aligned with the direction of current flow, and the field H_a applied to the magnetization in each film rotates M_S away from the respective easy axes in proportion to the strength of H_a. This simple three-layer geometry is shown in Fig. 3.1.

The magnetic flux couples between the magnetic films through the nonmagnetic layer; thus the stray field (which would otherwise spread around the upper and lower films) is confined to the region *between* the ferromagnetic films. This configuration is the essential idea underlying the work of Slonczewski et al. (1988), in whose system the magnetic films have identical properties. The next level of complexity is to generalize the system to films with different properties, then include exchange

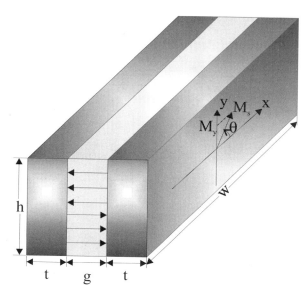

FIGURE 3.1 Simplified geometry of a three-layer MR device: W = track width, h = stripe height, t = film thickness, and g = nonmagnetic gap thickness.

energy, and finally allow current to flow in all three films. In this manner, the describing equations move from a simple, translucent form to a more complex but still identifiable shape. The total energy scales with film area, so this factor is not included in all that follows. In general, there are four terms: anisotropy energy E_K, magnetostatic potential energy E_P, stray field or demagnetizing energy E_D, and exchange energy E_{ex}. (*Note:* The *stray field* is called the *demagnetizing field* H_D by many workers in magnetism, and in this chapter the energy associated with the stray or demagnetizing field is called the demagnetizing energy E_D and use of the term *stray field energy* E_S is avoided. Some readers prefer stray field in place of demagnetizing field and wish to limit the term *demagnetizing* to that flux that passes through a magnetized body in a direction opposite to the magnetization. In this book, usage of the subscript S is limited to connote *stress* or *saturation*.) Ignoring exchange energy for the moment, the total energy of two films (per unit area) can be written as

$$E_T = E_K + E_P + E_D = 2K_U t \sin^2 \theta - 2H_a M_S t \sin \theta + \frac{1}{8\pi} g H_D^2(y) \qquad \text{ergs/cm}^2. \tag{3.1}$$

The anisotropy energy density is given by K_U (in ergs/cubic centimeter), and the applied field from the uniform current sheet is H_a (in oersteds). The stray field H_D, which is confined to the gap (g) between films, arises from changes in the y-component of magnetization ($M_y = M_S \sin \theta$) from both films; in regions where a film is saturated ($M_y = M_S$) the divergence of $M_y = 0$, so $H_D \simeq 0$ there. In general, H_D can be written as

$$H_D(y) = -4\pi t \frac{\partial M_y}{\partial y} = -4\pi t M_S \frac{\partial \sin \theta(y)}{\partial y}, \tag{3.2}$$

so the expression for the energy of this system (with $K_U = M_S H_K / 2$ from Chapter 2) becomes

$$E_T(y) = M_S H_K t \left[\sin^2 \theta - \frac{2H_a}{H_K} \sin \theta + \frac{2\pi M_S t g}{H_K} \left(\frac{\partial \sin \theta}{\partial y} \right)^2 \right]. \tag{3.3}$$

Characteristic Length for Flux Propagation

The expression in (3.3) is integrated along the y-direction over the height of the magnetic stripe to find the total energy:

$$E_{\text{total}} = M_S H_K t \int_{-h/2}^{h/2} dy \left[\sin^2 \theta - \frac{2H_a}{H_K} \sin \theta + \frac{2\pi M_S t g}{H_K} \left(\frac{\partial \sin \theta}{\partial y} \right)^2 \right]. \tag{3.4}$$

This is a variational calculus problem that is solved using Euler's equation; see, for example, Arfken (1968, p. 617). After minimizing the integrand and eliminating extraneous factors, what remains is a second-order linear differential equation,

$$\frac{d^2 \sin\theta}{dy^2} - \frac{1}{\lambda^2}\sin\theta = -\frac{H_a}{\lambda^2 H_K}, \tag{3.5}$$

where $\lambda^2 = 2\pi M_S g t/H_K = \mu t g/2$, since the permeability μ along the hard axis is $4\pi M_S/H_K$. The parameter λ has the dimension of length (in centimeters), so it is called the "characteristic length" for flux propagation. Recalling that $M_y = M_S \sin\theta$, the solution to (3.5) is

$$M_y(y) = A\,\cosh\!\left(\frac{y}{\lambda}\right) + B\,\sinh\!\left(\frac{y}{\lambda}\right) + \frac{M_S H_a}{H_K}. \tag{3.6}$$

The coefficient B is zero because of symmetry requirements at the top and bottom of the stripe, and the coefficient A is found by applying the boundary conditions $M_y(y = \pm h/2) = 0$, for which

$$A = -\frac{M_S H_a}{H_K}\frac{1}{\cosh(h/2\lambda)}. \tag{3.7}$$

The desired solution for the film magnetization is

$$M_y(y) = \frac{M_S H_a}{H_K}\left[1 - \frac{\cosh(y/\lambda)}{\cosh(h/2\lambda)}\right]. \tag{3.8}$$

This is the form given in Tsang (1984), Pohm et al. (1984), and Heim (1986); it is valid when $M_y < M_S$. When a film saturates in the transverse (y) direction, (3.8) must be altered; however, an analysis of saturated films will be deferred until films of different magnetic properties and thicknesses have been discussed. The magnetization distribution given by (3.8) is illustrated in Fig. 3.2; for computational purposes the parameters are for Permalloy films with $M_S = 800$ emu/cm^3, $H_K = 4.0$ Oe, $t = 150$ Å, $\mu = 2500$, $H_a = 5.0$ Oe, and the stripe height $h = 1.0$ μm. With different nonmagnetic gap (g) values of 50, 100 and 200 Å, the characteristic length λ is about 3060, 4330, and 6120 Å, respectively. The magnetization is aligned with the x-direction at the top and bottom edges of the film ($M_x = M_S$; $M_y = 0$) and gradually rotates into the hard axis (y-direction) to a maximum deflection at the film center. Because the magnetization is a constant-magnitude vector, the x- and y- components of M are related by the equation $M_x^2 + M_y^2 = M_S^2$. The three curves show the increasing interaction with reductions in λ.

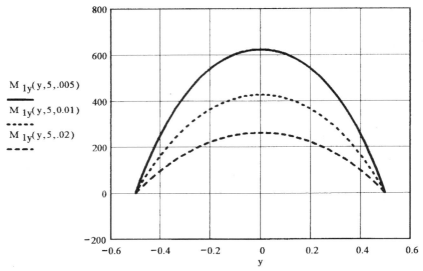

FIGURE 3.2 Magnetization of a Permalloy film versus stripe position for nonmagnetic gaps g of 50, 100, and 200 Å. The characteristic flux decay lengths λ are 3060, 4330, and 6120 Å, respectively.

MAGNETIZATION DISTRIBUTION IN NONIDENTICAL FILMS

Instead of minimizing the total energy of two dissimilar films, the distribution of flux between films of different thicknesses and saturation values can be treated using Ampere's circuital law along with the equations of continuity for B and H at boundaries. Figure 3.3 defines the path of the magnetic circuit. Following Ampere's

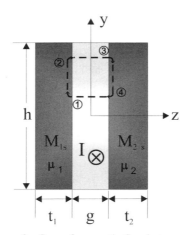

FIGURE 3.3 Pathway for flow of magnetic flux between two magnetic films.

circuital law, the closed line integral along the indicated path is

$$\oint H\, dl = gH_z(y) + H_{1y}(y)\, dy - gH_z(y+dy) - H_{2y}(y)\, dy = \frac{I}{h} dy, \quad (3.9)$$

where $I/2h$ (in amperes per meter) is the applied field H_a in meter-kilogram-second (mks) units. In the cgs system, $H_a = 2\pi I/h$ (in oersteds, with I in milliamperes and h in micrometers). Equation (3.9) can be rewritten as

$$-g\left[\frac{dH_z(y+dy) - H_z(y)}{dy}\right] + H_{1y}(y) - H_{2y}(y) = \frac{I}{h} \quad (3.10)$$

or finally as

$$-g\frac{dH_z(y)}{dy} + H_{1y}(y) - H_{2y}(y) = \frac{I}{h}. \quad (3.11)$$

The relationship between H_{1y} and H_{2y} is found with Gauss's law and substituting $B_1 = \mu_1 H_1$ and $B_2 = \mu_2 H_2$:

$$\int_A B\, dA = B_1 t_1\, dx + B_2 t_2\, dx = \mu_1 H_1 t_1\, dx + \mu_2 H_2 t_2\, dx = 0, \quad (3.12)$$

from which $H_2 = -(\mu_1 t_1/\mu_2 t_2)$. The stray field is confined to the gap and arises from variations in B ($=H + 4\pi M$) within the films. Applying Gauss's law to the gap and film 1,

$$-H_z(y)\, dy\, dx - B_{1y}(y)t_1\, dx + B_{1y}(y+dy)t_1\, dx = 0, \quad (3.13)$$

and with a rearrangement of terms, this is seen to be the gradient of B along y, so

$$H_z(y) = t_1 \frac{dB_{1y}}{dy} = \mu_1 t_1 \frac{dH_{1y}}{dy}. \quad (3.14)$$

This relation and that given above for H_2, are substituted in (3.11),

$$-g \frac{d}{dy}\left(t_1 \frac{dB_{1y}}{dy}\right) + \left(1 + \frac{\mu_1 t_1}{\mu_2 t_2}\right) H_{1y} = \frac{I}{h}, \quad (3.15)$$

and by substituting $B_{1y} = \mu_1 H_{1y}$ and rearranging the terms in (3.15), the final differential equation becomes

$$\frac{d^2 H_{1y}}{dy^2} - \left|\frac{\mu_2 t_2 + \mu_1 t_1}{\mu_1 t_1 \mu_2 t_2 g}\right| H_{1y} = -\frac{I}{\mu_1 t_1 g h}. \quad (3.16)$$

Comparing (3.16) with (3.5), they are of the same form and have similar solutions. The factor between vertical bars is equivalent to but more general than $1/\lambda^2$ in (3.5); thus when $\mu_1 = \mu_2 = \mu$ and $t_1 = t_2 = t$, the characteristic length for two different films reduces to that of identical films:

$$\lambda^2 = \frac{\mu_1 t_1 \mu_2 t_2 g}{\mu_1 t_1 + \mu_2 t_2} = \frac{\mu t g}{2}. \tag{3.17}$$

The solutions to (3.16) for films 1 and 2 are similar to (3.6) and (3.8):

$$H_{1y}(y) = A_1 \cosh\left(\frac{y}{\lambda}\right) + B_1 \sinh\left(\frac{y}{\lambda}\right) + \frac{\lambda^2 I}{\mu_1 t_1 g h}, \tag{3.18}$$

$$H_{2y}(y) = A_2 \cosh\left(\frac{y}{\lambda}\right) + B_2 \sinh\left(\frac{y}{\lambda}\right) + \frac{\lambda^2 I}{\mu_2 t_2 g h}. \tag{3.19}$$

Applying boundary conditions ($H = 0$) to the top and bottom edges of each film, the final solutions become

$$H_{1y}(y) = \frac{\lambda^2 I}{\mu_1 t_1 g h}\left[1 - \frac{\cosh(y/\lambda)}{\cosh(h/2\lambda)}\right], \tag{3.20}$$

$$H_{2y}(y) = \frac{\lambda^2 I}{\mu_2 t_2 g h}\left[1 - \frac{\cosh(y/\lambda)}{\cosh(h/2\lambda)}\right]. \tag{3.21}$$

The magnetization $M_y = \chi H_y = [(\mu - 1)/4\pi]H_y$ and, for a permeability much greater than unity, M_y is well approximated by $M_y = (\mu/4\pi)H_y$, so (3.20) and (3.21) become (by recalling the relation $I/h = 2H_a$)

$$M_{1y} = \frac{2M_{1S}H_a}{H_{1K}(1 + \mu_1 t_1/\mu_2 t_2)}\left[1 - \frac{\cosh(y/\lambda)}{\cosh(h/2\lambda)}\right], \tag{3.22}$$

$$M_{2y} = \frac{2M_{2S}H_a}{H_{2K}(1 + \mu_2 t_2/\mu_1 t_1)}\left[1 - \frac{\cosh(y/\lambda)}{\cosh(h/2\lambda)}\right]. \tag{3.23}$$

These relations collapse to (3.8) for identical films, as they should upon substitution of the film properties.

INFLUENCE OF EXCHANGE ENERGY IN COUPLED FILMS

A quantum mechanical exchange interaction lines up spin magnetic moments in adjacent atoms i and j, and this interaction is expressed in terms of the exchange energy,

$$E_{ij} = -2JS^2 \cos\theta_{ij} \quad \text{ergs,} \qquad (3.24)$$

where J is the exchange integral (positive for ferromagnetic materials) and S is the spin quantum number ($S = 1$ for iron). Many useful references exist for a review of magnetism; see, for example, Chikazumi and Charap (1964) or Soohoo (1965). When the magnetization changes direction, the energy is raised for $J > 0$; thus parallel alignment ($\theta_{ij} = 0$) between spins is the lowest energy. For $J < 0$, antiparallel alignment ($\theta_{ij} = \pi$) is the lowest energy. The exchange integral is related to the Curie temperature, and the values quoted by Chikazumi and Charap (1964), taken from Weiss (1948), are $J = 0.54 k_B T_C$ for a simple cubic (sc) lattice with $S = \frac{1}{2}$, $J = 0.34 k_B T_C$ for body-centered cubic (bcc) lattice with $S = \frac{1}{2}$, and $J = 0.15 k_B T_C$ for bcc lattice with $S = 1$. The Curie temperature T_C is 1043, 853, and 1403 K for Fe, Ni$_{82}$Fe$_{18}$, and Co, respectively, and Boltzmann's constant $k_B = 1.381 \times 10^{-16}$ erg/K.

The exchange constant A (in ergs per centimeter; sometimes called the exchange stiffness) is related to J and the lattice constant a of the material:

$$A = \begin{cases} \dfrac{JS^2}{a} & \text{sc lattice,} \\ \dfrac{2JS^2}{a} & \text{bcc lattice,} \\ \dfrac{4JS^2}{a} & \text{fcc lattice.} \end{cases} \qquad (3.25)$$

For materials of interest here (see Chapter 2) A does not vary significantly from 1×10^{-6} erg/cm. For bcc iron with $a = 2.86$ Å (see Bozorth, p. 864, 1951), the relation above gives $A = 1.5 \times 10^{-6}$ erg/cm. Slonczewski (1966a,b) includes exchange energy in his analysis of laminated films, but he excludes an applied field, while Pohm et al. (1984) treat a system including exchange energy with current flowing in two films separated by a nonconducting gap. Both analyses allow penetration of a small component of field into the film normal to its plane (i.e., the z-direction in Fig. 3.3), so tilting of M_S by H_Z stores some exchange energy. The applied field H_y within each film plane (the x–y plane in Fig. 3.1) rotates spins against exchange (as well as anisotropy forces), so there is additional exchange energy in equilibrium with the torque on M_S from H_y. The energy associated with each of these exchange interactions will now be accounted for, and the influence of this extra energy on the magnetization distribution will be estimated. This problem is related to a standard Néel domain wall problem in thin films or Bloch wall in thick films.

The exchange energy density is written as (see Slonczewski, 1966a)

$$E_{ex}(y,z) = \frac{A}{M_S^2}(\nabla M)^2 = \frac{A}{M_S^2}\left[\left(\frac{\partial M_y}{\partial y}\right)^2 + \left(\frac{\partial M_z}{\partial z}\right)^2\right], \qquad (3.26)$$

and this expression is often simplified by having M_S rotate linearly in a distance (D) to a new direction. The correct solution to (3.26), given in Landau and Lifshitz (1960), will be examined shortly. In the simple view, a reversal in direction (π radians) over a distance D yields an exchange energy density

$$E_{ex} = A\left(\frac{\pi}{D}\right)^2 \qquad \text{ergs/cm}^3, \qquad (3.27)$$

and D depends on the total energy of the "domain wall" between regions of uniform M_S. In thin films, it is energetically favorable for the magnetization to rotate within the film plane; this is the "Néel" type of wall, and the wall energy (per unit area) is composed of exchange, anisotropy, and magnetostatic contributions. That is (reading from Soohoo, 1965, p. 44),

$$E_N = A\left(\frac{\pi}{D}\right)^2 D + \frac{K}{2}D + \frac{\pi t D}{t+D}M_S^2 \qquad \text{ergs/cm}^2. \qquad (3.28)$$

The third term in (3.28) can be identified with the demagnetizing energy of a Néel wall. The energy E_N is minimized with respect to D to find the equilibrium wall thickness. Taking the derivative of (3.28) with respect to D and equating the result to zero, one finds

$$\frac{A\pi^2}{D^2} - \frac{2\pi t^2}{(t+D)^2}M_S^2 = \frac{K}{2}, \qquad (3.29)$$

which must be solved for D. If the film is very thick relative to D, the anisotropy term can be neglected and the approximate solution to (3.29) is

$$D \simeq \pi\sqrt{\frac{2A}{2\pi M_S^2}} \qquad \text{cm.} \qquad (3.30)$$

If $t \simeq D$, the solution for D becomes

$$D \simeq \pi\sqrt{\frac{2A}{2K + 2\pi M_S^2}}, \qquad (3.31)$$

and if t is small relative to D, then the magnetostatic energy of the wall is neglected and the solution for D becomes

$$D \simeq \pi \sqrt{\frac{2A}{K}}. \qquad (3.32)$$

In each case, the wall length D arises from solving an energy minimization problem, and D plays the same role as the characteristic length λ found earlier for coupled films (in which exchange was ignored). In Permalloy films with induced anisotropy, K is roughly 1500 ergs/cm^3, A is about 1×10^{-6} erg/cm, and M_S is 800 emu/cm^3, so the "characteristic" wall length D is about 220 Å according to (3.31). With the geometry and applied field direction given in Fig. 3.1, the magnetization can rotate at most by $\pi/2$ rad (at which point the film is saturated along the hard direction), thus the characteristic wall length is reduced by half to 110 Å.

More accurate estimates of the exchange energy and characteristic wall length are needed. These estimates must include the exchange arising from nonuniform magnetization within the plane as well as the extra exchange energy that arises from tilting the magnetization normal to the film plane. In Slonczewski (1966a), slight tilting of the magnetization toward the inner surfaces of coupled films costs a small amount of exchange energy, displaces the pole density (given by $-\mathbf{\nabla} \cdot \mathbf{M}$) to the inner surfaces, and reduces the demagnetization or stray field energy substantially. This is written

$$\mathbf{\nabla} \cdot \mathbf{M} = \frac{\partial M_z(x,z)}{\partial z} + \frac{dM_y(y)}{dy} = 0, \qquad (3.33)$$

and, with $z = 0$ at the outer surface of film 1 in Fig. 3.3, (3.33) is integrated to find

$$M_z(x,z) = -z \frac{dM_y(y)}{dy}. \qquad (3.34)$$

In this manner, the magnetic field energy is confined to the nonmagnetic gap and the system energy is reduced over that calculated by forcing $M_Z = 0$ everywhere. Slonczewski solves this problem for a system of identical films with no applied field; however, his result is applicable to more complex systems. He reduces the two-dimensional problem to one dimension along z and finds that the tilting of M_S along z is proportional to $\partial M_y/\partial y$ and follows the relation

$$M_z(z) = \frac{\partial M_y}{\partial y} \left[-z + \frac{\lambda_e (e^{z/\lambda_e} - e^{(t-z)/\lambda_e})}{1 + e^{t/\lambda_e}} \right]. \qquad (3.35)$$

A new characteristic length λ_e, identified with the penetration distance of H_Z into the film, is

$$\lambda_e = \sqrt{\frac{A}{2\pi M_S^2}} \quad \text{cm}. \qquad (3.36)$$

For Permalloy, with $A = 1 \times 10^{-6}$ erg/cm and $M_S = 800$ emu/cm^3, the penetration distance allowed by exchange rotations is about 50 Å, or approximately half of the wall thickness found in (3.30). The field associated with M_Z is

$$H_z(z) = -4\pi \left(M_z + z \frac{\partial M_y}{\partial y} \right)$$

$$= -4\pi \frac{\partial M_y}{\partial y} \frac{\lambda_e [e^{z/\lambda_e} - e^{(t-z)/\lambda_e}]}{1 + e^{t/\lambda_e}}. \quad (3.37)$$

These relations (from Slonczewski, 1966b) are shown in the schematic plots of Figs. 3.4a,b. The scales are normalized, but the magnitudes of H_Z and M_Z are estimated by working directly with (3.22), (3.35), and (3.37). At $y = \pm h/2$, H_Z would be a few percent of $4\pi M_S$ for unsaturated Permalloy ($t = g = 100$ Å), and M_Z would be a few percent of M_S at the inner faces of each film, with both quantities depending linearly on H_a/H_k.

Characteristic Length with Exchange Energy Included

The portion of exchange energy associated with tilting M_S out of the film plane and that arising from rotation of M_S within the plane each depend on $\partial M_y/\partial y$. Following (3.26) and operating on (3.35), the resulting energy at the film midplane ($z = t/2$) is

$$E_{\text{ex-tilt}} = \frac{2A}{M_S^2} \left(\frac{\partial M_z}{\partial z} \right)^2 = \frac{2A}{M_S^2} \left(\frac{\partial M_y}{\partial y} \right)^2 \left[1 - \frac{1}{\cosh(t/2\lambda_e)} \right]^2, \quad (3.38)$$

and the exchange energy from rotation in the plane is

$$E_{\text{ex-rotate}} = \frac{2A}{M_S^2} \left(\frac{\partial M_y}{\partial y} \right)^2. \quad (3.39)$$

Including exchange thus shifts the energy budget by the amounts shown in (3.38) and (3.39), and these influences can be accounted for by extending the concept of characteristic length to allow for all energy terms in the total equation. This is seen by returning to the original equation (3.3) for system energy density, expanding it to include (3.38) and (3.39), and writing

$$\frac{E_T(y)}{Wht} = \frac{M_S H_K}{2} \sin^2 \theta - M_S H_a \sin \theta$$
$$+ \left[2\pi M_S^2 tg + 2A \left(1 - \frac{1}{\cosh(t/2\lambda_e)} \right)^2 + 2A \right] \left(\frac{\partial \sin \theta}{\partial y} \right)^2 \quad \text{ergs/cm}^3. \quad (3.40)$$

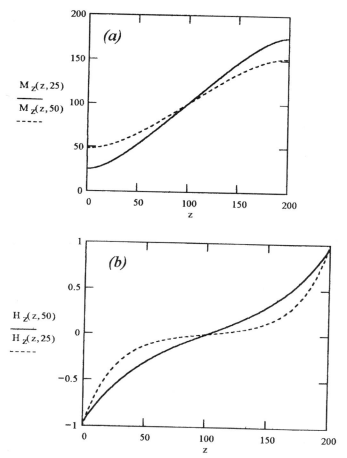

FIGURE 3.4 Schematic plots of (a) the magnetization M_z tilted within the film plane and (b) the associated field H_z through the film thickness z for two different characteristic exchange lengths λ_e of 25 and 50 Å.

Recalling that $M_y = M_S \sin\theta$ and minimizing the energy as before, (3.40) becomes a second-order differential equation of exactly the same form as (3.5) whose solution is the same as (3.6). The characteristic length has grown to include the additional energy terms, so for identical films

$$\lambda^2 = \frac{2\pi M_S t g}{H_K} + \frac{2A}{M_S H_K}\left(1 - \frac{1}{\cosh(t/2\lambda_e)}\right)^2 + \frac{2A}{M_S H_K}. \quad (3.41)$$

The first term is a measure of the stray field or demagnetizing energy, the second is related to the exchange energy for tilting, and the third corresponds to the exchange energy of rotation of M_S in the film plane. In a system of identical films with

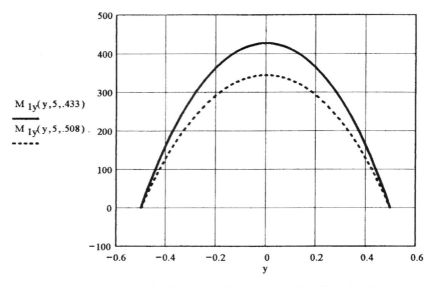

FIGURE 3.5 Magnetization distribution for identical Permalloy films in which exchange energy is ignored (solid line) and included (dotted line). The characteristic lengths are 4330 and 5080 Å, respectively.

$M_S = 800$ emu/cm^3, $t = 150$ Å, $g = 100$ Å, $H_K = 4.0$ Oe, $A = 1 \times 10^{-6}$ erg/cm, and $\lambda_e = 50$ Å computed from (3.36), the characteristic length would be 5080 Å with exchange included and 4330 Å ignoring exchange contributions. The magnetization distributions for $H_a = 5$ Oe are calculated with (3.8) and plotted in Fig. 3.5. In the example given above, the maximum y-component of magnetization is 427 emu/cm^3 ignoring exchange and drops to 344 emu/cm^3, or about 19%, with exchange forces included. For films with different properties, (3.41) is easily rewritten to a more general expression:

$$\lambda^2 = \frac{\mu_1 t_1 \mu_2 t_2 g}{\mu_1 t_1 + \mu_2 t_2} + \frac{A_1}{M_{1S}H_{1K}}\left(1 - \frac{1}{\cosh(t_1/2\lambda_{1e})}\right)^2 + \frac{A_1}{M_{1S}H_{1K}} + \frac{A_2}{M_{2S}H_{2K}}\left(1 - \frac{1}{\cosh(t_2/2\lambda_{2e})}\right)^2 + \frac{A_2}{M_{2S}H_{2K}}. \quad (3.42)$$

SATURATION IN COUPLED FILMS

The relations for magnetization distributions in coupled films [(3.22) and (3.23)] are linear in the applied field H_a until M_y in either film is aligned with its hard axis. At this point the central portion ($y = 0$) of the film with smallest $M_S t$ enters saturation along y; with increasing H_a the saturated zone spreads toward the upper and lower extremes of that film. The differential equation for $M_y = M_S \sin\theta$, written in the form of (3.5), is solved with new boundary conditions describing film saturation. Figure 3.6 describes the coordinates for the saturated and unsaturated regions of film 1 in which $M_{1S}t_1 < M_{2S}t_2$. As before, the applied field arises from current flow in a nonmagnetic conducting sheet between magnetic films. Without injection of signal flux from a magnetic storage disk, the magnetization profile is symmetric about $y = 0$ and the locations of saturation are symmetrically placed about the origin. In the saturated region $(-y_s \leq y \leq +y_s)$ $\sin\theta_1 = 1.0$, and in the unsaturated regions $(-h/2 \leq y < -y_s$ and $+y_s < y \leq +h/2)$, the solution to the differential equation is

$$\sin\theta_1 = A\,\cosh\left(\frac{y}{\lambda}\right) + B\,\sinh\left(\frac{y}{\lambda}\right) + \frac{H_a}{H_{1K}}. \tag{3.43}$$

At the boundary $y = h/2$, the magnetization lies along the x-direction, so $\sin\theta_1 = 0$. At the location of saturation $y = y_s$, the divergence of M_1 is continuous and equal to zero ($\mathbf{\nabla}\cdot\mathbf{M}_1 = \mathbf{\nabla}\cdot\mathbf{M}_{1S} = 0$); thus (3.43) becomes

$$\frac{\partial \sin\theta_1}{\partial y} = 0 = \frac{A}{\lambda}\sinh\left(\frac{y_s}{\lambda}\right) + \frac{B}{\lambda}\cosh\left(\frac{y_s}{\lambda}\right). \tag{3.44}$$

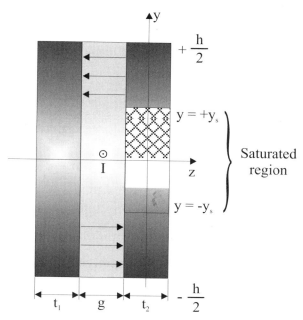

FIGURE 3.6 Saturated region of a thin film is depicted by crosshatches, and the shaded regions remain unsaturated.

At this boundary, $\sin\theta_1 = 1.0$ and is also continuous, so a third equation is defined:

$$\sin\theta_1 = 1.0 = A\,\cosh\left(\frac{y_s}{\lambda}\right) + B\,\sinh\left(\frac{y_s}{\lambda}\right) + \frac{H_a}{H_{1K}}. \tag{3.45}$$

These three equations are solved for the three unknowns, A, B, and y_s; (3.44) is used to find B in terms of A, and the result for B is substituted into (3.43) and (3.45). With a little algebra one finds

$$A = \frac{-H_a}{H_{1K}}\left[\frac{\cosh(y_s/\lambda)}{\cosh(h/2\lambda - y_s/\lambda)}\right], \tag{3.46}$$

$$B = -\frac{\sinh(y_s/\lambda)}{\cosh(y_s/\lambda)} A.$$

The boundary condition at $y = -h/2$ changes to $\sin\theta_1 = \Delta m$ when signal flux is injected, and the solutions for A, B, and the lower saturation point $y_s = y_{sn}$ are altered. It is straightforward to show that $-H_a/H_{1K}$ is replaced with $\Delta m - H_a/H_{1K}$ in (3.46). The locations for film saturation y_s and y_{sn} are taken from (3.45) and (3.46),

$$y_s = \frac{h}{2} - \lambda\,\cosh^{-1}\left(\frac{H_a/H_{1K}}{H_a/H_{1K} - 1}\right), \tag{3.47a}$$

$$y_{sn} = -\frac{h}{2} + \lambda\,\cosh^{-1}\left(\frac{H_a/H_{1K} - \Delta m}{H_a/H_{1K} - 1}\right). \tag{3.47b}$$

The threshold field for saturation H_{S0} [obtained from (3.47) with $y_s = 0$] becomes

$$H_{S0} = H_{1K}\frac{\cosh(h/2\lambda)}{\cosh(h/2\lambda) - 1}. \tag{3.48}$$

The completed solutions to (3.43) for the magnetization profiles $m_p(y \geq 0)$ and $m_n(y \leq 0)$ are

$$m_p = \sin\theta_1 = \frac{H_a}{H_{1K}}\left[1 - \frac{\cosh(|y|/\lambda - y_s/\lambda)}{\cosh(h/2\lambda - y_s/\lambda)}\right], \tag{3.49a}$$

$$m_n = \frac{H_a}{H_{1K}}\left[1 - \frac{\cosh(|y|/\lambda - |y_{sn}|/\lambda)}{\cosh(h/2\lambda - |y_{sn}|/\lambda)}\right] + \Delta m. \tag{3.49b}$$

When $\Delta m = 0$, the solution is symmetric in y about the origin, which is centered in the film plane. Figure 3.7a shows the magnetization profiles $(m_n + m_p)$ for several levels of applied field greater than $H_{S0} = 12$ Oe and $\Delta m = 0$; the parameter values for these curves are $h = 1.0$ μm, $\lambda = 0.52$ μm, $H_{1K} = 4.0$ Oe, and H_a/H_{S0} values of 1, 2, and 4, respectively. Figure 3.7b shows profiles with signal modulation $\Delta m = 0$, ± 0.2 and $H_a/H_{S0} = 4$, which saturates much of the central region of the film.

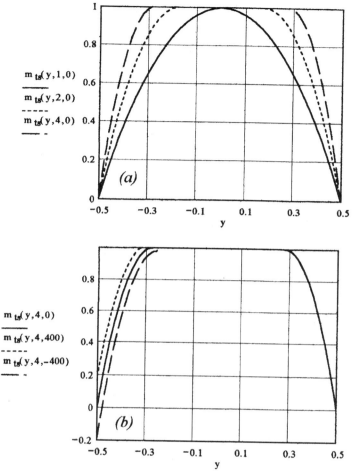

FIGURE 3.7 (a) Magnetization profiles of a thin film in which $H_a/H_{S0} = 1, 2,$ and 4. (b) Saturation profiles for $H_a/H_{S0} = 4$ with signal modulation $\Delta m = 0, \pm 0.2$.

Figure 3.8 is a plot of the saturated locations $|y_{sn}|$ as a function of applied field with $\Delta m = 0, \pm 0.2$; signal flux modulates y_{sn} roughly ± 200 Å around the bias level. Saturation approaches the asymptotes $\pm h/2$ slowly; at $H = 250$ Oe, the saturated region has spread to $y \simeq \pm 0.40$ μm and the remaining 0.1 μm of stripe at the bottom region is sensitive to signal flux.

The conditions of continuity require that $M_{1y}t_1 = -M_{2y}t_2$ in the saturated region of film 1, and magnetic flux (which is often called the "bias flux") couples between films only in the unsaturated regions at the upper and lower ends of the gap. By way of illustration, assume identical Permalloy properties in both films with thicknesses $t_1 = 100$ Å, $g = 100$ Å, and $t_2 = 150$ Å; λ is calculated with (3.36) and (3.42) using $M_S = 800$ emu/cm^3, $H_K = 4.0$ Oe, and $A = 1 \times 10^{-6}$ erg/cm in both films. The

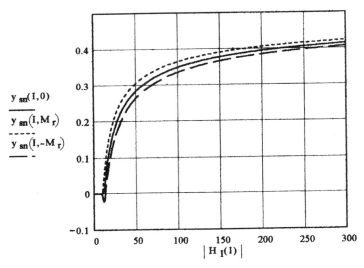

FIGURE 3.8 Location of film saturation as a function of applied field with signal modulation $\Delta m = 0, \pm 0.2$.

magnetization profiles, for which $\lambda \simeq 4400$ Å, are shown in Fig. 3.9a,b, where film 1 (smallest moment–thickness product) is saturated and film 2 is not.

SATURATION DEPENDENCE ON STRIPE HEIGHT

The threshold field for stripe saturation H_{S0} increases rapidly as the stripe height h reduces because the stray field between coupled films is inversely proportional to h. On the other hand, with a fixed current the applied field reduces with increasing h, thus one can define a stripe height at which the saturation current is minimized. The applied field from a current sheet, defined with (3.9), is

$$H_a = \frac{2\pi I}{h} \quad \text{Oe}, \tag{3.50}$$

where I is in milliamperes and h in micrometers, and the minimum current I_{S0} is found by equating the applied field to the threshold field (3.48) for stripe saturation and solving for the current. That is,

$$I_{S0} = \frac{hH_K}{2\pi}\left[\frac{\cosh(h/2\lambda)}{\cosh(h/2\lambda)-1}\right] \tag{3.51}$$

SATURATION DEPENDENCE ON STRIPE HEIGHT 79

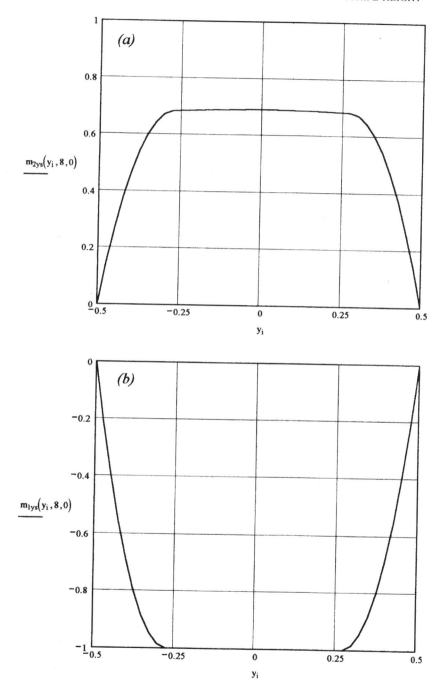

FIGURE 3.9 Magnetization profiles for dissimilar Permalloy films: (*a*) $t_2 = 150$ Å and (*b*) $t_1 = 100$ Å. Spacing between films is $g = 100$ Å.

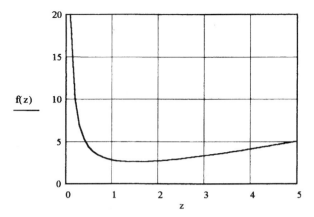

FIGURE 3.10 Normalized relation $f(z) = z \cosh(z)/[\cosh(z) - 1]$ for film saturation depencence on stripe height.

This relation is of the form $f(z) = z \cosh(z)/[\cosh(z) - 1]$, which is plotted in Fig. 3.10. The stripe height for minimum saturation current is obtained by minimizing (3.51) with respect to h:

$$\frac{\partial I_{S0}}{\partial h} = \frac{-\cosh^2(z) + \cosh(z) + z \sinh(z)}{[\cosh(z) - 1]^2} = 0, \quad \text{where } z = \frac{h}{2\lambda}. \tag{3.52}$$

The root is found numerically at $z_0 \simeq 1.5056$ and the minimum stripe height is $h_0 = 2\lambda z_0 \simeq 3.0112\lambda$. In the coupled film example given above, $\lambda = 0.52$ μm, so the minimum current $I_{S0} = 2.56$ mA for the threshold of film saturation at a stripe of 1.6 μm. It is normally impractical to design MR heads for operation at $h = h_0$ because the shape anisotropy ($h < W$) of the sensor [see (2.9)] is chosen to aid the induced anisotropy. The ratio of stripe height to track width is nominally set at $h/W \sim 0.7$ for this reason, and narrow tracks force stripes to be smaller than h_0 in normal practice. Xiao et al. (1998) show the influence of shape anisotropy on MR transfer curves; with $h > W$ the easy axis rotates and aligns in the transverse (y) direction. Controlling stripe height to tight tolerances is a major processing consideration, and lack of stripe control is often first on the list in failure analyses for root causes.

REFERENCES

Arfken, G., *Mathematical Methods for Physicists*, Academic, New York, 1968.

Bozorth, R. M., *Ferromagnetism*, Van Nostrand Co., Princeton, 1951.

Chikazumi, S. and Charap, S., *Physics of Magnetism*, Wiley, New York, 1964. Reprinted by Kreiger, Malabar, FL, 1978.

Heim, D. E., *J. Appl. Phys.*, **59**, 864 (1986).

Landau, L. D. and Lifshitz, E. M., *Electrodynamics of Continuous Media*, Pergamon, Oxford, 1960.

Pohm, A. V., Comstock, C. S., and Pearey, L., *IEEE Trans. Magn.*, **MAG-20**, 863 (1984).

Slonczewski, J. C., *J. Appl. Phys.*, **37**, 1268 (1966a).

Slonczewski, J. C., *IBM J. Res. Devel.*, **10**, 377 (1966b).

Slonczewski, J. C., Petek, B., and Argyle, B. E., *IEEE Trans. Magn.*, **24**, 2045 (1988).

Soohoo, R. F. *Magnetic Thin Films*, Harper and Row, New York, 1965.

Tsang, C., *J. Appl. Phys.*, **55**, 2226 (1984).

Weiss, P. R., *Phys. Rev.*, **74**, 1493 (1948).

Xiao, M., Devasahayam, A. J., and Kryder, M. H., *IEEE Trans. Magn.*, **MAG-34**, 1495 (1998).

4

SENSE ELEMENT BIASING, SHIELDING, AND STABILIZATION

In this chapter, concepts and tools are developed for designing and analyzing bias fields for the active region of MR sensors, shielding the sensor for improved pulse resolution, and stabilizing domain structures with exchange or hard-bias layers at the extreme ends of sensors. Anisotropic and giant MR spin valve sensors are *active* elements because an electric current is required to produce a signal. Current flows through magnetic as well as nonmagnetic films, and the applied magnetic field associated with these current sheets rotates (biases) the magnetization in adjacent magnetic layers to a steady-state angle (the "bias angle") relative to the direction of current flow. For high-density digital recording, Potter (1974) analyzed a sensor placed between very thick layers of magnetic films to shield away approaching magnetic transitions in the recording medium. Shielding improves the resolution of individual transitions and attenuates the signal flux from the written medium, but shields also magnetically interact with the energized sensor and provide an additional source of sensor bias; the magnitude and direction of shield bias depend on the sensor position within the shields and distance between shields. Output signal linearity is a sensitive function of bias level, so a good understanding of bias and its control is required by individuals who design, process, or test MR heads.

Because the readback signal corresponds to rotation of the magnetization vector in a sense layer, it is imperative that discontinuous switching of the magnetization is avoided. These "Barkhausen noises," arising from domain wall formation and movement, are deleterious to signal recovery. Tsang and Decker (1981) traced the origin of Barkhausen noises in MR sensors to transitions in wall states and annihilation of domain walls. Later, Tsang and Decker (1982) showed that domain walls in MR sensors were created primarily by *longitudinal* demagnetizing fields, and this insight stimulated additional work in domain suppression and stabilization of film magnetization. Tsang and Fontana (1982) published on longitudinal biasing with exchange coupling between a NiFe MR layer and FeMn pads at each end of the sensor; they established the minimum longitudinal bias field for stable operation at various track widths and element heights. Cain and Kryder (1989) discussed TbCo exchange bias applied at an angle throughout the NiFe sensor to provide domain stabilization in the longitudinal direction and magnetization biasing in the transverse direction. Lin et al. (1995) studied exchange biasing with NiFe films and concluded that FeMn, NiMn, and NiO AFM layers were appropriate for sensor stabilization at areal densities in use at that time, and Liao et al. (1993) brought attention to limitations of exchange biasing for narrow track widths and short stripe heights appropriate at higher areal densities.

Volume production of AMR sensors ramped up in the early 1990s, and this activity led to increased experience with processing and stabilization issues associated with AFM exchange biasing materials. Hannon et al. (1994) discussed the first- and second-generation (1991 and 1993) MR heads produced by IBM. First-generation heads used patterned FeMn exchange layers for longitudinal biasing of NiFe MR films, and second-generation heads were stabilized with hard magnetic material deposited contiguous to the MR, spacer, and soft adjacent layers. This is now called a *contiguous-junction* or (*CJ*) hard-bias design. Hard-bias designs offer improved stabilization fields at narrow track widths and short stripe heights; since 1993 the literature reflects increased interest in permanent-magnet longitudinal biasing of AMR and GMR elements. For example, see Ishikawa et al. (1993), Liao et al. (1994), Guo et al. (1994), Kikuchi et al. (1994), Champion and Bertram (1996), Shen et al. (1996), Xiao et al. (1998), and Mitsumata et al. (1998). Effective stabilization with CJ hard-bias designs requires careful attention to details of the hard-bias material, junction geometry, and mechanical lapping and polishing processes used to make functional devices. Perhaps the most definitive statement regarding hard-bias CJ deposition is found in Paranjpe et al. (1999), where they call attention to magnetic instabilities arising from overspray of hard-bias magnetic material and from the junction slope at the sense element.

FIELD OF A CONDUCTING RIBBON

In hard disk drives, MR sense elements are typically films about 50–150 Å thick, perhaps 1 μm or so wide, and 1 μm or less high, and the current is about 5–10 mA. As a homework problem in field analysis, this configuration could be described as a bus bar or very thin ribbon of rectangular cross section with uniform current density (see Fig. 4.1). The field of this simple geometry can be derived from the vector magnetic potential A_z of a current sheet with uniform density $K_z = I/h$ (in amperes per meter). The analysis for a *long* conducting ribbon is taken directly from Weber (1965), who uses mks units. The MR sensors are not long, but the analysis nevertheless gives useful insight. Panofsky and Phillips (1962) give a most

SENSE ELEMENT BIASING, SHIELDING, AND STABILIZATION

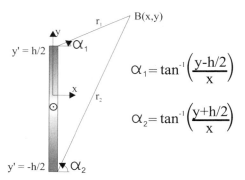

FIGURE 4.1 Geometry for calculating the magnetic field a thin conducting ribbon with uniform current density.

useful discussion of the limits of two-dimensional methods and the neglect of end effects:

$$A_z = \frac{\mu}{4\pi} \int_{-L}^{L} \int_{-h/2}^{h/2} \frac{K_z}{r} \, dy' \, dz' \quad \text{Wb/m}, \quad (4.1)$$

where $r = [x^2 + (y - y')^2 + (z - z')^2]^{1/2}$. The integrand is an even function, so the integral in z' yields

$$A_z = \frac{\mu I/h}{4\pi} \int_{-h/2}^{h/2} [2 \log_e(L + \sqrt{L^2 + R^2}) - 2 \log_e R] \, dy' \quad (4.2)$$

where $R = [x^2 + (y - y')^2]^{1/2}$ and z' is evaluated between 0 and $+L$. The first term in (4.2) is nearly constant for large L and disappears upon differentiation. Therefore, integration in y' of the surviving term $-2 \log R = -\log R^2$ gives

$$A_z = \frac{\mu I/h}{2\pi} \left\{ \left(y - \frac{h}{2}\right) \log_e \frac{r_1}{h/2} - \left(y + \frac{h}{2}\right) \log_e \frac{r_2}{h/2} \right.$$
$$\left. + x \left[\tan^{-1}\left(\frac{y - h/2}{x}\right) - \tan^{-1}\left(\frac{y + h/2}{x}\right) \right] \right\}, \quad (4.3)$$

where $r_1 = [x^2 + (y - \frac{1}{2}h)^2]^{1/2}$ and $r_2 = [x^2 + (y + \frac{1}{2}h)^2]^{1/2}$. The flux density B is obtained directly from the negative gradient of the potential (strictly speaking, $B = \nabla \times A_z$):

$$B_x(x, y) = -\frac{\partial A_z}{\partial y} = \frac{\mu I/h}{2\pi} \log_e\left(\frac{r_1}{r_2}\right) \quad (4.4a)$$

$$B_y(x, y) = -\frac{\partial A_z}{\partial x} = \frac{\mu I/h}{2\pi} \left[\tan^{-1}\left(\frac{y + h/2}{x}\right) - \tan^{-1}\left(\frac{y - h/2}{x}\right) \right] \quad \text{Wb/m}^2. \quad (4.4b)$$

FIELD OF A CONDUCTING RIBBON 85

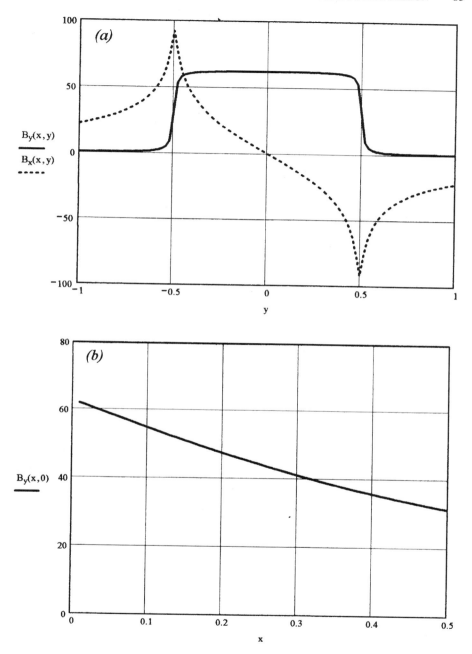

FIGURE 4.2 (*a*) The x and y field components of a very thin conducting ribbon of height $h = 1.0$ μm as a function of y-position along the stripe at a distance $x = 0.01$ μm. (*b*) The y-component of the field at $y = 0$ plotted as a function of distance x from the thin conducting ribbon.

In a nonmagnetic medium, the permeability in mks units is $\mu = 4\pi \cdot 10^{-7}$ H/m (1.0 H = 1.0 W/A), and a flux density of 1.0 W/m^2 = 1.0 T = 10^4 G. For a thin sheet, $t \ll h$ and for small x, $B_y(x, 0)$ can be written

$$B_y(x, 0) = \frac{\mu I/h}{2\pi}\left[\tan^{-1}\left(\frac{h}{2x}\right) - \tan^{-1}\left(\frac{-h}{2x}\right)\right] \simeq \frac{\mu I/h}{2\pi}\left[\pi - \frac{4x}{h}\right], \quad (4.5)$$

so at the surface of the sheet, $4x/h = 2t/h \to 0$ and $B_y \simeq \mu I/2h$ or $H = B/\mu = I/2h$. This result is obtainable directly from Ampere's circuital law, but Ampere's law does not reveal the details contained in (4.4a,b) near the conductor edges. At the upper and lower extremes ($y = \pm\frac{1}{2}h$), for example, the field along y drops to one-half the value at $y = 0$. It is common to express the field of a current sheet as $H = I/2h$ and ignore the edge effects. In the mks or International System (SI) of units, a current of 10 mA in a sheet with $h = 1.0$ µm produces a field of 5.0 kA/m near the surface and away from an edge; in cgs units $H = (4\pi \cdot 10^{-3})I/2h = 62.8$ Oe. It is also common to find I/h expressed in units of milliamperes per micrometer, which is identical to kiloamperes per meter. Figures 4.2a,b show field plots according to (4.4a,b) along the y-direction for $x = 0.01$ µm, $h = 1.0$ µm, and $I = 10$ mA.

FIELD OF A CURRENT SHEET BETWEEN SHIELDS

The magnetic field in the space around a conducting sheet is substantially altered when it is placed between two slabs of high-permeability magnetic shield material. This problem in field analysis can be solved by using $\nabla \times H = J = I/ht$ (in amperes per square meter), integrating for H, applying the necessary boundary conditions at the slabs (shields) and conductor surfaces, and finding the coefficients satisfying these conditions. Figure 4.3 defines idealized geometry for this problem: The film is thin, very high along the y-direction, and long in the z-direction (into the plane of the figure), so end effects are ignored. This leaves only the x-direction, which is confined to the small gap region between shields.

The current density is uniform in region 2 and aligned in the positive z-direction, so the curl of H becomes

$$\nabla \times H = \frac{\partial H_y}{\partial x} - \frac{\partial H_x}{\partial y} = \begin{cases} 0 & \text{(region 1)} \quad (4.6a) \\ J & \text{(region 2)} \quad (4.6b) \\ 0 & \text{(region 3).} \quad (4.6c) \end{cases}$$

The permeability of both shields is very high; thus the tangential field H_y disappears at $x = +g_2$ and $x = -g_1$. Inside the conductor, Ampere's law shows that $|H_y(x, y)|$

FIELD OF A CURRENT SHEET BETWEEN SHIELDS

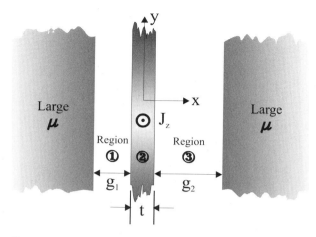

FIGURE 4.3 Geometry for calculating the field of a thin current sheet between two thick magnetic shields.

increases linearly with $|x|$ and reaches a maximum value at the surfaces $x = \pm\frac{1}{2}t$. These observations suggest the following three equations for regions 1, 2, and 3:

$$H_{y1} = A_1(x + g_1), \quad (4.7a)$$

$$H_{y2} = A_2 x + B_2, \quad (4.7b)$$

$$H_{y3} = A_3(x - g_2). \quad (4.7c)$$

Symmetry along the y-direction suggests an additional three equations:

$$H_{x1} = C_1 y, \quad (4.8a)$$

$$H_{x2} = C_2 y, \quad (4.8b)$$

$$H_{x3} = C_3 y. \quad (4.8c)$$

Continuity in H_x requires $C_1 = C_2 = C_3$, and applying (4.6a,b,c) to (4.7a,b,c) and (4.8a,b,c) yields the relations $A_1 = C_1$, $A_2 - C_2 = J$, and $A_3 = C_3$. This terminology is cleaned up by writing $A = A_1 = A_3 = C_1 = C_2 = C_3$ and $A_2 = A + J$. At the boundary $x = \frac{1}{2}t$ between regions 2 and 3, one obtains the relation

$$A\left(\tfrac{1}{2}t - g_2\right) = \tfrac{1}{2}(A + J)t + B_2, \quad (4.9)$$

and at the boundary $x = -\frac{1}{2}t$ between regions 1 and 2 one writes a second equation,

$$A\left(-\tfrac{1}{2}t + g_1\right) = (A + J)\left(-\tfrac{1}{2}t\right) + B_2. \quad (4.10)$$

The coefficient B_2 is eliminated by subtracting (4.10) from (4.9) and solving the result for A; thus

$$A = \frac{-Jt}{g_1 + g_2}, \quad (4.11)$$

and B_2 is found by adding (4.9) and (4.10) and substituting (4.11) for A:

$$B_2 = \frac{-Jt(g_1 - g_2)}{2(g_1 + g_2)}. \quad (4.12)$$

The field in each region can now be written as

$$H_{y1} = -Jt\left(\frac{x + g_1}{g_1 + g_2}\right), \quad (4.13a)$$

$$H_{y2} = J\left[\frac{(g_1 + g_2 - t)x + (g_2 - g_1)t/2}{g_1 + g_2}\right], \quad (4.13b)$$

$$H_{y3} = -Jt\left(\frac{x - g_2}{g_1 + g_2}\right). \quad (4.13c)$$

In the limit as $t/(g_1 + g_2) \to 0$, it is useful to define $H_0 = \frac{1}{2}Jt = I/2h$ (A/m) $= 2\pi I/h$ (Oe) with I (mA) and h (µm) and rewrite (4.13a,b,c) as

$$H_{y1} = -2H_0\left(\frac{x + g_1}{g_1 + g_2}\right), \quad (4.14a)$$

$$H_{y2} = H_0\left[\left(1 - \frac{t}{g_1 + g_2}\right)\frac{2x}{t} + \frac{g_2 - g_1}{g_1 + g_2}\right] \simeq H_0\left(\frac{2x}{t} + \frac{g_2 - g_1}{g_1 + g_2}\right), \quad (4.14b)$$

$$H_{y3} = -2H_0\left(\frac{x - g_2}{g_1 + g_2}\right). \quad (4.14c)$$

The second term in (4.14b) directly reveals the influence of shields on the field at the conductor surface; for symmetrical placement of the conductor ($g_1 = g_2$) the magnetic "images" in the shields cancel, but the field can be increased or decreased depending on the asymmetrical location of the conductor. As $g_1 \to 0$, $H_{y2}(x = \frac{1}{2}t) \to 2H_0$, and as $g_2 \to 0$, $H_{y2}(x = \frac{1}{2}t) \to -2H_0$, so the conductor location can be used to enhance or degrade the bias field on an adjacent magnetic film in a shielded sensor. Indeed, in the earlier period of MR head design, Shelledy and Brock (1975) discussed "self-biased" heads and showed that ratios g_2/g_1 greater

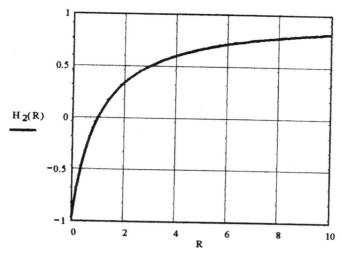

FIGURE 4.4 Normalized field at the centerline of a thin conducting sheet plotted as a function of the gap ratio $R = g_2/g_1$. If the sheet is magnetic, it will magnetize upward or downward depending on its position between the shields.

than 4 could achieve useful bias levels. Figure 4.4 plots (4.14b) normalized by H_0 for $x = 0$ (conductor centerline) as a function of the ratio $R = g_2/g_1$.

ONE MAGNETIC FILM BETWEEN SHIELDS

Coupling between magnetic films was analyzed in Chapter 3, and some of the concepts introduced there may be combined with ideas discussed in this present and other chapters. If a single thin anisotropic film is magnetized by an adjacent current sheet, the demagnetization energy forms a substantial portion of the energy budget, so the magnetization level along the hard direction grows approximately as $M_y = M_s H_a/(H_k + H_D)$, where the demagnetizing field H_D might be as large as 100 Oe and the anisotropy field H_k is normally about 4 Oe. (Recall a discussion of the Stoner–Wohlfarth model in Chapter 2.) In Chapter 3, it was shown that flux coupling between magnetic films substantially reduces the demagnetizing energy; thus the film magnetization increases more easily with applied field as $M_y \propto M_s H_a/H_k$. This is also true in the situation of a magnetic film between shields. The geometry is essentially the same as that given in Fig. 4.3 with the conducting layer replaced by a magnetized film of thickness t. As before, flux goes from the film along the $\pm x$-directions to the shields, and end effects of a finite height film are ignored. This simplification does no serious damage to the accuracy of the analysis except for very large gaps between film and shields.

The analysis proceeds with $\nabla \cdot B = \nabla \cdot (H + 4\pi M) = 0$, and Gauss's law is used to find the relationship between stray (demagnetizing) fields in each region:

$$\int H \, dA = -4\pi \int \nabla \cdot M \, dV,$$

$$(H_{D3} - H_{D1}) \, dy \, dz = -4\pi t \, dy \, dz \frac{\partial M_y}{\partial y}, \qquad (4.15)$$

$$H_{D3} - H_{D1} = -4\pi t \frac{\partial M_y}{\partial y}.$$

The integral along the path between shields provides another relationship between the stray (demagnetizing) fields:

$$\int H_D \, dl = H_{D1} g_1 + H_{D3} g_2 = 0,$$

$$H_{D1} = -\frac{g_2}{g_1} H_{D3}, \qquad H_{D3} = -\frac{g_1}{g_2} H_{D1}. \qquad (4.16)$$

The results in (4.16) are substituted into (4.15) to find the demagnetizing fields with shields,

$$H_{D1} = \frac{g_2}{g_1 + g_2} \left(4\pi t \frac{\partial M_y}{\partial y} \right), \qquad (4.17\text{a})$$

$$H_{D3} = -\frac{g_1}{g_1 + g_2} \left(4\pi t \frac{\partial M_y}{\partial y} \right). \qquad (4.17\text{b})$$

Characteristic Length for a Shielded Film

The demagnetizing energy (per unit area) for the field in a gap is $E_D = (1/8\pi) g H_D^2$, so the total energy/per square centimeter between shields is

$$E_D = \frac{1}{2} g_1 \left(\frac{4\pi g_2 t}{g_1 + g_2} \frac{\partial M_y}{\partial y} \right)^2 + \frac{1}{2} g_2 \left(\frac{4\pi g_1 t}{g_1 + g_2} \frac{\partial M_y}{\partial y} \right)^2$$

$$= \frac{1}{2} \left(\frac{4\pi g_1 g_2 t M_s}{g_1 + g_2} \frac{\partial \sin \theta}{\partial y} \right)^2. \qquad (4.18)$$

The energy equation is then written and solved as before [see Eq. (3.1)–(3.8)]. The characteristic length λ now includes the parallel leakage paths g_1 and g_2 to the shields:

$$\lambda = \sqrt{\frac{4\pi M_s t g_1 g_2}{(g_1 + g_2) H_k}} = \sqrt{\frac{\mu t g_1 g_2}{g_1 + g_2}}. \qquad (4.19)$$

It is useful to compare the magnetizing behavior of an isolated thin magnetic film (the uniform rotation Stoner–Wohlfarth model) to that of a film between shields. The solution to (2.10) for $\alpha = \frac{1}{2}\pi$ (i.e., $H = H_y$; $H_x = 0$) is $M_y = M_s \sin\theta = M_s H_y / (H_{ku} + H_D)$ with $M_y \leq M_s$; that is, the magnetization initially grows linearly with applied field with a slope $\chi = dM/dH = M_s/(H_{ku} + H_D)$, where H_{ku} is the uniaxial anisotropy field (about 4 Oe) and H_D is the transverse demagnetizing field $|H_D| = N_y M_s$, which depends on film geometry. For Permalloy with $h = 0.8$ μm, $W = 1.25$ μm, $t = 125$ Å, and $M_s = 800$ emu/cm^3, $H_D \simeq 8tWM_s/[h(W^2 + h^2)^{1/2}] = 84$ Oe using Yuan's (1992) derivation of the demagnetizing field for single-domain films. The magnetizing slope of an isolated single-domain film is thus $\chi \simeq 800/(4 + 84) = 9.1$, which is dominated by the demagnetizing field. Placing the film between shields provides two magnetic pathways for flux closure, and the demagnetizing energy can be reduced if the nonmagnetic gaps are not too large. With shields, the magnetization distribution becomes

$$M_y(y) = \frac{M_s H_a}{H_k}\left[1 - \frac{\cosh(y/\lambda)}{\cosh(h/2\lambda)}\right], \tag{4.20}$$

which is the result for identical coupled films as expressed by Equation (3.8), but the characteristic length is given by (4.19) in the present situation. Reading directly from (4.19) and (4.20), the magnetizing slope at $y = 0$ is

$$\frac{dM}{dH} = \frac{M_s}{H_k}\left[1 - \frac{1}{\cosh(h/2\lambda)}\right], \tag{4.21}$$

which is about $800/12.9 H_k \simeq 15.7$ with $\lambda = 1.22$ μm for $g_1 = g_2 = 0.08$ μm. In other words, through a reduction in demagnetizing energy (or the "effective" H_D), shields improve the sensitivity (at the film center) by the factor $15.7/9.1$, or about 1.73, over the unshielded example.

One cannot speak of the "demagnetizing field" in the simple manner as was done with the isolated, uniform film model because the flux splits according to (4.17), the demagnetizing fields (in general) are different on opposite sides of the film, and in any case, H_D is highly nonuniform along the film height. Taking the derivative of (4.20) and substituting the result into (4.17) gives

$$H_{D3} = \frac{-g_1}{g_1 + g_2}\left(4\pi t \frac{\partial M_y}{\partial y}\right) = \frac{4\pi g_1 t M_s H_a}{(g_1 + g_2)H_k}\left[\frac{1}{\lambda}\frac{\sinh(y/\lambda)}{\cosh(h/2\lambda)}\right], \tag{4.22}$$

which scales linearly with H_a until the film reaches saturation at $y = 0$. For the symmetric case $g_1 = g_2 = g$, (4.22) reduces to

$$H_{D3} = H_a \frac{\lambda}{g}\frac{\sinh(y/\lambda)}{\cosh(h/2\lambda)}. \tag{4.23}$$

92 SENSE ELEMENT BIASING, SHIELDING, AND STABILIZATION

TWO MAGNETIC FILMS BETWEEN SHIELDS

With two magnetized films between shields, there are three regions for splitting up the total demagnetizing energy. The geometry is sketched in Fig. 4.5, and the reasoning used for one film and two regions is applied here. Using $\nabla \cdot B = 0$ and Gauss's law for each film, the demagnetizing fields for each region are

$$H_{D1} - H_{D2} = -4\pi t_1 \frac{\partial M_{y1}}{\partial y} = \rho_1, \qquad (4.24a)$$

$$H_{D3} - H_{D2} = -4\pi t_2 \frac{\partial M_{y2}}{\partial y} = \rho_2, \qquad (4.24b)$$

and the line integral of H_D between shields gives

$$H_{D1}g_1 + H_{D2}g + H_{D3}g_2 = 0. \qquad (4.25)$$

Solving the simultaneous equations (4.24a,b) and (4.25) for the demagnetizing fields yields

$$H_{D1} = \left(\frac{g_1}{G} - 1\right)\rho_1 - \left(\frac{g_2}{G}\right)\rho_2, \qquad (4.26a)$$

$$H_{D2} = \left(\frac{g_1}{G}\right)\rho_1 - \left(\frac{g_2}{G}\right)\rho_2, \qquad (4.26b)$$

$$H_{D3} = \left(\frac{g_1}{G}\right)\rho_1 + \left(1 - \frac{g_2}{G}\right)\rho_2, \qquad (4.26c)$$

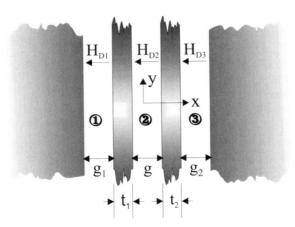

FIGURE 4.5 Geometry for computing the magnetic fields arising from two thin conducting magnetic fields placed in the gap between thick magnetic shields.

where $G = g_1 + g + g_2$, which is nearly equal to the shield-to-shield distance if the magnetic films are very thin. The demagnetization energy per square centimeter $E_D = (\frac{1}{2}) \int H_D^2 \, dx$ is integrated along the gaps; thus

$$E_D = \tfrac{1}{2}g_1 H_{D1}^2 + \tfrac{1}{2}g H_{D2}^2 + \tfrac{1}{2}g_2 H_{D3}^2. \tag{4.27}$$

In the symmetric case where $g_1 = g_2 \gg g$, (4.26a,b,c) reduce to

$$H_{D1} \simeq -\tfrac{1}{2}(\rho_1 + \rho_2), \tag{4.28a}$$

$$H_{D2} \simeq \tfrac{1}{2}(\rho_1 - \rho_2), \tag{4.28b}$$

$$H_{D3} \simeq -\tfrac{1}{2}(\rho_1 + \rho_2), \tag{4.28c}$$

and (4.27) becomes

$$E_D = \tfrac{1}{8} G(\rho_1 + \rho_2) - \tfrac{1}{2} g \rho_1 \rho_2. \tag{4.29}$$

Minimizing E_D with respect to ρ_2 yields the useful insight (already discussed in Chapter 3)

$$\rho_2 = -\frac{g_1 - g + g_2}{G}\rho_1 \simeq -\rho_1 \tag{4.30}$$

for very small g_2. In other words, if a two-magnetic-film system is dominated by demagnetization energy (i.e., anisotropy, exchange, and magnetostatic energies are small), flux closure ($\rho_1 = -\rho_2$) between films is the preferred state. Close examination of (4.26a,b,c) shows the demagnetizing fields in regions 1 and 3 (H_{D1} and H_{D3}) are small relative to H_{D2} for tightly coupled films.

LOSS OF SIGNAL FLUX TO THE SHIELDS

The signal flux entering the MR sensor [$\phi_s = 4\pi t W H_y$ (medium); see (1.21)] leaks to the shields as it travels up the element. Thompson (1974) analyzes flux leakage using a transmission line method and shows, for boundary conditions $\phi(y = 0) = \phi_s$ and $\phi(y = h) = 0$, that the flux decay is

$$\phi(y) = \phi_s \frac{\sinh[(h-y)/\lambda]}{\sinh(h/\lambda)}, \tag{4.31}$$

where λ is the characteristic length derived in (4.19) with $g_1 = g_2 = g$ and h is the stripe height. Flux decay is plotted in Fig. 4.6 for several values of λ with $h = 1.0$ and $\phi_s = 1.0$; note that decay is approximately linear for $\lambda/h > 1$ and becomes increasingly exponential for $\lambda/h < 1$. Typical values for AMR elements might be $h = 0.8$ μm, $\mu = 2000$, $g = 0.08$ μm and $t = 0.01$ μm, so the characteristic length

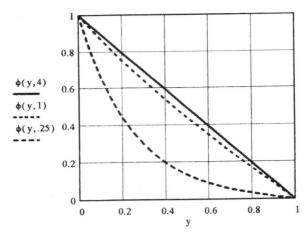

FIGURE 4.6 Decay of magnetic flux injected into a thin magnetic film at $y = 0$ computed as a function of position y along the film, which is placed between two magnetic shields. The flux decay length λ equals 0.25, 1.0, and 4 μm for a stripe whose height $h = 1.0$ μm.

would be 0.89 μm and $h/\lambda \simeq 1$. Flux decay would be nearly linear in this example, so the average signal flux along the sensor would be about $0.5\phi_s$. In other words, a shielded MR sensor has a maximum flux efficiency of about 50% for short stripe heights.

SHIELD DIMENSIONS AND SENSOR FIELD WITH UNIFORM EXCITATION

The height, width, and thickness of magnetic shields affect the response of an MR element to external fields. Magnetoresistive heads are tested in quasi-static uniform fields and dynamically tested on rotating magnetic disks where the signal field is highly nonuniform. Feng et al. (1991) experimentally studied readback waveform distortions induced by uniform external fields applied to shielded AMR heads. They found that shortening the shields along the sensor height (y-direction) reduced the sensitivity along this axis and shield symmetry along the x-direction reduced the sensitivity to cross-track uniform fields. One would expect a uniform field to magnetize the shields substantially, whereas a nonuniform field from a written transition would have very limited influence on the shield magnetization. The results of uniform and nonuniform field testing are well correlated, but the specific details of shield geometry must be understood to make absolute comparisons between MR heads with different designs. This problem is analyzed by Smith (1989), Koehler (1994), Yuan et al. (1995), and Suzuki and Ishikawa (1998). Smith (1989) showed, for example, that in uniform field testing the head amplitude for 175-μm-high shields was 67% of that for shields 390 μm high, with all other variables held constant. He analyzed the uniform field response of two identical shields and showed the total field at the sensor location is the sum of the applied (uniform) field and the field from the magnetized shields. Long and thin shields have small demagnetizing factors for the long dimension, so they magnetize easily in uniform fields, and the head amplitude is thus greater when compared to short and thick shields.

Yuan et al. (1995) analyzed two shields under uniform external field excitation using a Fourier expansion of Laplace's equation (in two dimensions). The general profile is of the form

$$H_y(x, y) = \sum_{n=1}^{\infty} C_n \cos\left[\frac{(2n-1)\pi x}{G_{ss}}\right] \exp\left[-\frac{(2n-1)\pi y}{G_{ss}}\right], \quad (4.32)$$

however, a comparison between a simple exponential and the first 20 terms in (4.32) shows that the field $H_y(0, y) = C \exp(-\pi y/G_{ss})$ adequately describes the field profile in the sensor region between shields. Yuan et al. (1995) approximated the nonuniform field excitation from a recording medium with a line source at a spacing d from the gap centerline ($x = 0$). They found, in agreement with Koehler (1994), that the field profile from nonuniform excitation agrees almost perfectly with a simple exponential function $H_y(0, y) = D \exp(-\pi y/G_{ss})$. The *shapes* of field profiles arising from uniform and nonuniform excitations thus share in essential features, but nonuniform excitation does not magnetize the shields to as great an extent as uniform excitation. Yuan et al. (1995) found a closed-form expression describing field amplitude at $y = 0$ arising from uniform excitation of two shields whose thicknesses and heights are (t_1, h_1) and (t_2, h_2), respectively. The function for a shield is

$$F(t, h) = \left(t + \frac{G_{ss}}{2}\right)\sqrt{1 + \left(\frac{h}{t + G_{ss}/2}\right)^2 + 1} - \frac{G_{ss}}{2}\sqrt{1 + \left(\frac{h}{G_{ss}/2}\right)^2 + 1}, \quad (4.33)$$

and the field for two shields is

$$H_y(y = 0) = \frac{H_0}{2\sqrt{2}t_1} F(t_1, h_1) + \frac{H_0}{2\sqrt{2}t_2} F(t_2, h_2). \quad (4.34)$$

As a specific example, assume $G_{ss} = 0.20$ μm, $t_1 = 2.5$ μm, $h_1 = 100$ μm, $t_2 = 3$ μm, and $h_2 = 60$ μm; the field at the bottom of the sensor is increased by the factor $H_y/H_0 \simeq 3.22$. Therefore, a highly nonuniform field amplitude [see Eq. (1.18)] of 300 Oe, for example, at the bottom of the sensor would produce a signal equivalent to a uniform field excitation of about 93 Oe. The relationship given in (4.34) is plotted in Fig. 4.7 for identical shields and several values of thickness with $G_{ss} = 0.20$ μm.

96   SENSE ELEMENT BIASING, SHIELDING, AND STABILIZATION

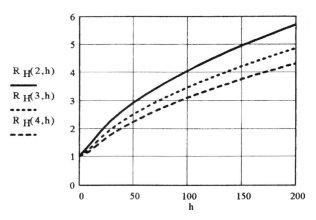

FIGURE 4.7   Ratio $R = H(y=0)/H_0$ of the field along the centerline and at the bottom of two identical magnetic shields to the strength of a uniform external magnetic field H_0. Plots are given as a function of shield height h (μm) for shield thickness t equal to 2, 3, and 4 μm. The shield-to-shield gap $G_{ss} = 0.20$ μm.

SENSOR STABILIZATION WITH EXCHANGE PINNING

In their studies on the origin of Barkhausen noise in MR sensors, Tsang and Decker (1981, 1982) observed the behavior of domain wall formation and movement in unshielded Permalloy samples. They found that longitudinal demagnetizing energy was the primary factor in creating buckling domain patterns when magnetization reversals were along the traverse direction; cycling the magnetization with a transverse field caused "systematic creation, intensification and annihilation of buckling domain structures." Tsang and Fontana (1982) applied a uniform bias field in the longitudinal direction (H_{Bx}) to films with various stripe heights (h) and aspect ratios (W/h) and determined the bias required to suppress buckling domains and achieve quiet MR responses. Their work, summarized in Fig. 4.8, shows experimental values of H_{Bx} for Permalloy films with 400-Å films of Permalloy; the line plots are calculated using a theoretical estimate for the demagnetizing field H_{Dx} of flat ellipsoidal films (see Chikazumi and Charap, 1978, p. 21) given by the relation

$$H_{Dx} = \pi^2 M_s \frac{t}{W}\left[1 - \frac{W-h}{4W} - \frac{3}{16}\left(\frac{W-h}{W}\right)^2\right] \quad \text{Oe} \quad (4.35)$$

In Chapter 2, it was shown that a uniform bias field along the sensor x-direction reduces the sensitivity along the y-direction according to the relation $M_y = M_s H_y/(H_{Ku} + H_D + H_x)$. The loss of sensitivity from an x-directed field can arise from an external field or from uniform exchange pinning along the sensor. Tsang (1989) studied sensor stabilization with patterned exchange biasing applied only in

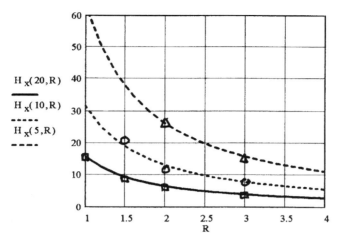

FIGURE 4.8 Comparison of experimental stabilization fields for MR films with theoretical demagnetizing fields of flat ellipsoidal films. (Tsang and Fontana, 1982).

the tail regions on either side of the active MR sensing region. His approach preserved the MR sensitivity along the y-direction and produced stable films in the x-direction at the sensor ends, where the demagnetization fields are strongest. Permalloy films were deposited to 400 Å thickness, shaped to 5 × 100-μm rectangles, then 500 Å of FeMn was deposited and shaped to patterns 70, 50, and 30 μm long on each side of a 10-μm-wide MR sensing region. The devices were heated to 240°C then cooled in a longitudinal external field to establish an exchange biasing field of about 24 Oe [a result compatible with Eq. (2.18) and the discussion of AFM materials in Chapter 2]. The normalized MR responses were compared with control devices having no exchange biasing, and Tsang found quiet, stable transfer curves for each of the exchange pattern widths, no loss of signal for the stabilized devices, and good agreement with the theoretical results of Anderson et al. (1972). Tsang's results are given in Fig. 4.9.

Tsang and Lee (1982) studied the temperature dependence of exchange anisotropy in Permalloy–FeMn and Cu–FeMn–Permalloy systems and found significant differences in their $H_{ex}(T)$ behaviors. The drop in exchange field with temperature was nearly linear for the NiFe(400 Å)–FeMn(500 Å) system and quite nonlinear for the Cu(1000 Å)–FeMn(100 Å)–NiFe(400 Å) system. In the NiFe–FeMn system, $H_{ex} \simeq 25$ Oe ($t = 400$ Å) $T = 25°C$, and the blocking temperature T_B (see Chapter 2) was about 150°C, thus devices operating at 100°C would have perhaps 10 Oe of pinning field. For a much thinner film ($t = 150$ Å), pinning would increase to about 27 Oe [see Eq. (2.18)]; this level of stabilization might be sufficient for a device 5 μm high and 10 μm wide [see Eq. (4.35) and Fig. 4.9], but this is unacceptable for high areal densities. The Cu–FeMn–NiFe system produced higher exchange ($H_{ex} \simeq 50$ Oe at $T = 25°C$) with $T_B \simeq 165°C$, but the thick Cu layer (providing an fcc substrate for the FeMn layer) would form a galvanic couple with NiFe and

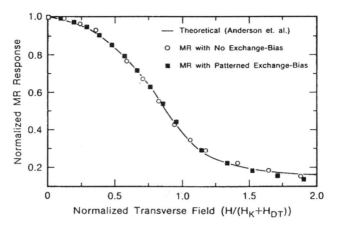

FIGURE 4.9 Theoretical and experimental MR transfer curves. Reprinted with permission from C. Tsang, *IEEE Trans. Magn.*, **MAG-25**, 3692. Copyright 1989 by IEEE.

create reliability issues. Lin et al. (1995) found improved exchange fields and blocking temperatures with NiFe–NiMn systems and the necessity of annealing at 320°C for 30 h to achieve $H_{ex} \simeq 273$ Oe and $T_B \simeq 425°C$ for NiFe films at 150 Å thickness.

Liao et al. (1993) modeled the biasing characteristics of unshielded MR/SAL sensors with exchange pinning at the end regions and compared model results with experimental elements. The sensors were 300-Å Permalloy 50 μm long, 3 μm high, with an active track region 5 μm wide. Pinning at about 40 Oe was achieved with FeMn patterns on each side of the active region. They found that the MR bias level at 10 mA bias current was influenced by the demagnetizing field operating in the active region. The demagnetizing fields at the center of the element depended on the aspect ratio (W/h) in the manner derived by Yuan (1992), namely

$$H_{Dx} = \frac{8M_s t}{\sqrt{W^2 + h^2}} \frac{h}{W} (1 - \cos\theta), \tag{4.36a}$$

$$H_{Dy} = \frac{-8M_s t}{\sqrt{W^2 + h^2}} \frac{W}{h} \sin\theta, \tag{4.36b}$$

where θ is the bias angle. They showed that the simple moment–ratio formula for MR/SAL biasing,

$$\sin\theta = \frac{(M_s t)_{SAL}}{(M_s t)_{MR}}, \tag{4.37}$$

overestimates the MR biasing level because the x-axis demagnetizing field is neglected in the soft adjacent and MR layers. Their model and experimental results are shown in Figs. 4.10a,b.

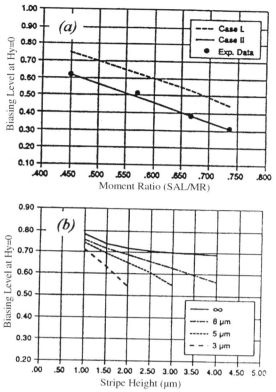

FIGURE 4.10 (a) Modeled and experimental results of the biasing level, $\cos^2 \theta$ at $H_y = 0$. (b) Effect of the stripe height on the biasing level with track width as a parameter. Reprinted with permission from S. H. Liao, S. W. Yuan, and H. N. Bertram, *IEEE Trans. Magn.*, **MAG-29**. Copyright 1993 by IEEE.

SENSOR STABILIZATION WITH PERMANENT MAGNET (HARD-BIAS) PINNING

With increasing areal density, the critical dimensions of sensor elements become smaller and thus the minimum longitudinal field necessary for domain stabilization increases; according to (4.36a), the horizontal demagnetizing field increases by $2^{1/2}$ if h and W each reduce by a factor of 2. The first-generation AMR heads in IBM disk drives (described by Hannon et al., 1994) operated at 135 Mbits/in.^{2}, had a read width W of about 6 μm, MR thickness of 300 Å, and presumed height h of perhaps 4 μm ($W/h \simeq 1.5$). With these values, and the studies of Tsang and Fontana (1982), the required stabilization field would be approximately 36 Oe, and patterned exchange pinning with FeMn would likely provide marginal stabilization over a large population of production devices. Second-generation IBM disk drives were introduced about one year later at 160 Mbits/in.^{2} and the design abandoned exchange pinning in favor of permanent magnet (PM, or "hard-bias") stabilization.

The sensor width was reduced to about 4 μm with a presumed height of 2.5–3 μm, and the thickness was reduced to about 240 Å, so the required stabilization field calculated with (4.35) would be about 44 Oe. At operating temperatures in the range of 50–75°C, patterned exchange stabilization with FeMn would drop to unacceptable levels. It was well established (see Ishikawa et al., 1993) that a trade-off existed between PM hard biasing, which could effectively stabilize narrow tracks with a loss of signal sensitivity, and patterned exchange biasing, which suffered from insufficient stabilization with no loss in sensitivity. The simulation results of Ishikawa et al. (1993) showed that the resistance change ΔR of an *unshielded* MR film with patterned exchange layers was about twice that of a film with PM biasing; the effective anisotropy field along the easy axis was increased by the stray field from two hard-bias layers placed symmetrically at each side of the sensor.

Field of a Uniformly Magnetized Slab

Insight into hard-bias stabilization is gained by examining the magnetic field surrounding a PM thin film. Three-dimensional, finite-element micromagnetic computations are required to address many details of permanent magnets joined contiguously with AMR or GMR films for stabilization, but closed-form analytic techniques are justified where the needs for rapid computation and semiquantitative insight outweigh numerical accuracy. The two-dimensional field around a uniformly magnetized slab is derivable from the vector potential $\mathbf{A}$ (see Williams, 1984, p. 180):

$$\mathbf{A} = \iint_S \frac{\mathbf{M} \times \hat{\mathbf{n}}}{r} \, dS, \tag{4.38}$$

where $r = [(x-x')^2 + (y-y')^2 + (z-z')^2]^{1/2}$ is the distance from a source point to a field point, dS is an element of surface area ($dS = dx' \, dz'$), $\mathbf{M} = M_r \hat{\mathbf{i}}$ is the magnetization along the x-direction and $\hat{\mathbf{n}}$ is a unit vector in the outward normal direction on each surface. Figure 4.11 is a sketch of one hard-bias magnet in the x–z plane; the magnet is long (high) in the x–y plane, and ignoring end-effects (arising from finite stripe height along y) *will overestimate the unshielded fields by roughly a factor of 2*. (Shielding attenuates leakage fields to an extent that the practical consequences of overestimating the *unshielded* field will be small.) The solution to this problem is similar to that of the thin conducting ribbon [Eq. (4.3)] except the magnet cross section is thick with four vertices at $(x', z') = \pm L, \pm T$. The external field of the magnet (where $\mu = 1.0$) is $\mathbf{H} = \mathbf{B} = \nabla \times \mathbf{A}$; omitting details of the derivation, one obtains

$$H_x(x,z) = 2M_r \left[\tan^{-1}\left(\frac{x+L}{z+T}\right) - \tan^{-1}\left(\frac{x+L}{z-T}\right) \right.$$
$$\left. + \tan^{-1}\left(\frac{x-L}{z-T}\right) - \tan^{-1}\left(\frac{x-L}{z+T}\right) \right], \tag{4.39}$$

$$H_z(x,z) = M_r \left[\log_e\left(\frac{(x-L)^2 + (z+T)^2}{(x-L)^2 + (z-T)^2}\right) - \log_e\left(\frac{(x+L)^2 + (z+T)^2}{(x+L)^2 + (z-T)^2}\right) \right]. \tag{4.40}$$

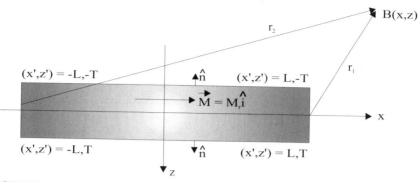

FIGURE 4.11 Sketch of a hard-bias magnet in the x–y plane. The magnet is very long in the x–y plane, which goes into the plane of the page.

The field components $H_x(x, z, L, T)$ and $H_z(x, z, L, T)$ are plotted in Figs. 4.12a,b for $M_r = 500$ emu/cm^3, $L = 10$ μm, and T values of 0.01, 0.02 and 0.05 μm (PM thickness $\delta = 2T$). The maximum value of $H_x(x = L, z = 0, L, T) = 2\pi M_r \simeq 3140$ G is nearly independent of thickness when $L \gg T$, and H_x falls off rapidly with distance along $x \geq L$.

Along any plane $z = T + \Delta$, the flux reverses direction near the PM edges at $x = \pm L$; this is shown in Fig. 4.13 for $T = 0.02$ μm, $L = 10$ μm, and several values of distance Δ above the magnet. The reversal in field direction is extremely important when depositing soft magnetic films over magnets or depositing magnets onto soft magnetic films because domain walls are formed along the regions of field reversal. Champion and Bertram (1996) modeled the hysteresis and baseline shift of MR/SAL heads stabilized with PMs and found that careful attention must be given to the overlap region between layers of soft magnetic films and the PM material. Poor stability resulted when the overlap angle of the junction was small (about 10°), and good performance was predicted for junction angles of 30° or more. Shen et al. (1996) compared MR head designs with abutted (CJ) and overlaid PMs and found that CJ devices had less hysteresis in the transfer curves from reverse field sweeping. Magnetic tests showed loss of sensitivity and narrowing of the effective track width with CJ designs, and overlaid magnet designs showed higher sensitivity, broader magnetic read width, and more side reading. Zhu and O'Connor (1996) define a *stability coefficient* (SC) for SAL-biased MR heads that matches the magnetic moment densities of the PM and sensor films; SC = $[M_r\delta(\text{PM})]/[M_s t(\text{MR})/\sqrt{2}]$, and SC values of 1.5–2.0 are usually required for acceptable stability. Xiao et al. (1998) addressed processing details of CJ-stabilized MR heads and found the *unshielded* longitudinal bias field decreased with increasing track width. They optimized the properties of CoCrPt permanent-magnet material to achieve $M_r = 703$ emu/cm^3, $H_c \simeq 1500$ to $H_c \simeq 1200$ Oe (dependent on thickness δ values of 100–500 Å, with a Cr underlayer of about 50 Å), and $S^* = 0.95$. Paranjpe et al. (1999) showed the desirability of collimated deposition of hard-bias material over a two-layer photo-resist stencil to avoid overspray of the hard-bias material on the sense layers. They

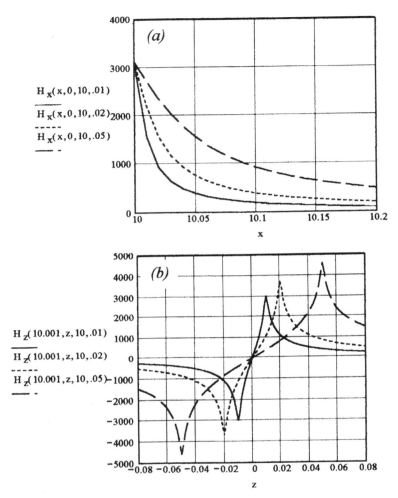

FIGURE 4.12 Field components: (a) $H_x(x, z, L, T)$ and (b) $H_z(x, z, L, T)$: $2L$ is the full width and $2T$ is the full thickness of the magnet.

also showed that the width of the transition between hard-bias and MR material can be significantly reduced (from 0.5 µm down to 0.15 µm) with good collimation of the ion beam. Each of these steps produces improved magnetic stability of the MR sensor. Examples of good and poor hard-bias junctions are sketched in Figs. 4.14a,b.

The field of two magnets is found by linear superposition of the fields from isolated sources translated along x by the distances $x_1 = L + W/2$ and $x_2 = -(L + W/2)$, where W is the physical width of the sensor element. The geometry of CJ hard-bias pinning (with an idealized 90° junction angle) is shown in Fig. 4.15; angles other than 90° can be simulated with a superposition of several very thin permanent-magnet films stacked to form a staircase arrangement at each side of the MR/SAL element. The PMs are defined using Equation (4.40) with x replaced by

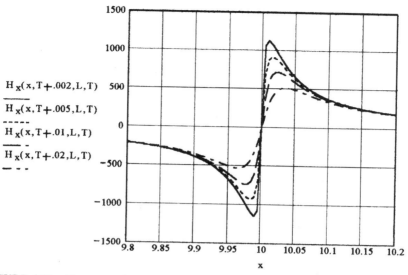

FIGURE 4.13 Flux reversal near the edge of a PM for various planes above the top surface of the magnet.

$x - x_1$ and $x - x_2$ for $H_{1x}(x, z)$ and $H_{2x}(x, z)$, respectively. Placing a PM-stabilized sensor between shields substantially reduces the leakage field along the sensor. Yuan et al. (1995) showed that field attenuation by shields follows a simple exponential function

$$H(x) \simeq H_0 \exp\left(-\frac{\pi x}{G_{ss}}\right), \qquad (4.41)$$

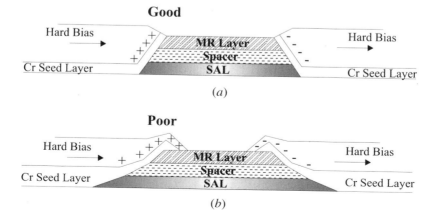

FIGURE 4.14 Sketches of (*a*) good and (*b*) poor CJ geometry of a hard-bias MR sensor.

104 SENSE ELEMENT BIASING, SHIELDING, AND STABILIZATION

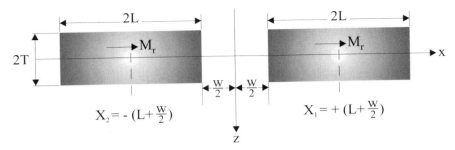

FIGURE 4.15 Geometry of superposition of two PMs representing an idealized CJ hard-bias pinning configuration.

where H_0 is the field at the edge of a magnet and x is the distance away from the edge. Using this understanding here, the estimated leakage field along the shielded sensor element is

$$H_{ts}(x, z) \simeq H_{1x}(x, z) \exp\left[\frac{\pi}{G_{ss}}\left(x - \frac{W}{2}\right)\right] + H_{2x}(x, z) \exp\left[-\frac{\pi}{G_{ss}}\left(x + \frac{W}{2}\right)\right]. \quad (4.42)$$

Figure 4.16 shows the estimated leakage field in the sensor region; for this calculation, $M_r = 500$ emu/cm^3, PM thickness $\delta = 2T = 400$ Å, PM length = $2L = 20$ μm, sensor width $W = 2.2$ μm, and shield–shield gap $G_{ss} = 0.20$ μm. The graph limits are set to $x = \pm 1.0$ μm way from an edge) to expand the resolution of the field along the sensor.

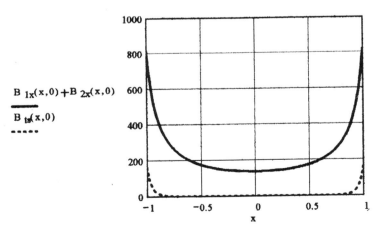

FIGURE 4.16 Plots of the estimated hard-bias leakage field with and without shields on a sensor.

Magnetic Width of a Hard-Bias Stabilized Sensor

The shields attenuate the field to nearly zero throughout most of the sensor region, and the effective magnetic width of the sensor is reduced somewhat along each edge. The amount of reduction can be estimated by selecting a leakage field threshold above which the element sensitivity would be nil [e.g., use Eq. (4.35) and let $H_x \sim H_{Dx}$]. Reading directly from (4.35) and (4.41) and solving $H(\Delta x) \sim H_{Dx}$ for the distance from an edge, one obtains

$$\Delta x \simeq \frac{G_{ss}}{\pi} \log_e \left(\frac{H_0}{H_{Dx}}\right) \simeq \frac{G_{ss}}{\pi} \log_e \left(\frac{2M_r}{\pi M_s t / W}\right) \tag{4.43}$$

under the assumption that $W \simeq h$. Under this reasoning, the magnetic width ($W_M = W - 2\Delta x$) is dominated by G_{ss} and is only *logarithmically sensitive* to the PM and MR film properties. It is useful at this point to plug in some numbers, establish ballpark values of the track width reduction $2\Delta x$, and compare the results with experiment; let $M_r = 400$ emu/cm^3, $\delta = 400$ Å, $M_s = 800$ emu/cm^3, $t = 150$ Å, $h \simeq 1.0$ µm, the physical sensor width W be 1.95–3.0 µm, and G_{ss} be 0.15–0.20 µm. Reading directly from (4.43), the lower and upper values of $2\Delta x$ are 0.36 µm for the thin-gap, narrow-track case, and 0.53 µm for the thick-gap, wide-track case. In other words, the magnetic width is about 0.5 µm narrower than the physical width. Figure 4.17 shows that experiment compares favorably with this

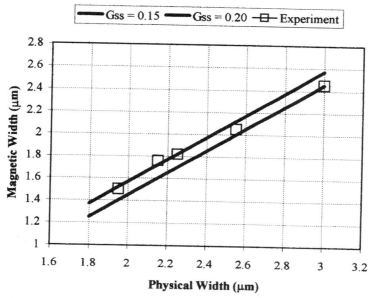

FIGURE 4.17 Comparision of experimental and theoretical magnetic width as a function of the physical width of the MR sensor.

simple analysis; each plotted point is the mean of 54–98 devices and the error bars are at the $\pm 1\ \sigma$ level.

Many workers in the magnetic recording industry refer to the "width" of an MR sensor without qualification regarding whether the term refers to the physical (geometrical) or the magnetic width. Measurement of the magnetic width is straightforward and unambiguous, whereas the physical width requires high-resolution electron microscopy for its measurement and there remains a basic uncertainty regarding the precise location of the junction between the sense element and the hard-bias material. For these reasons, the magnetic width has become the preferred term, and if the word "width" is used without qualification, it is best to interpret its meaning as the "magnetic width."

REFERENCES

Anderson, R. L., Bajorek, C. M., and Thompson, D. A., *AIP Conf. Proc.*, **10**, 1445 (1972).

Cain, W. C. and Kryder, M. H., *IEEE Trans. Magn.*, **MAG-25**, 2787 (1989).

Champion, E. and Bertram, H. N., *IEEE Trans. Magn.*, **MAG-32**, 13 (1996).

Chikazumi, S. and Charap, S., *Physics of Magnetism*, Kreiger, Huntington, N.Y., 1978.

Feng, J. S., Tippner, J., Kinney, B. G., Lee, J. H., Smith, R. L., and Chue, C., *IEEE Trans. Magn.*, **MAG-27**, 4701 (1991).

Guo, Y., Zhu, J.-G., and Liao, S. H., *IEEE Trans. Magn.*, **MAG-30**, 3861 (1994).

Hannon, D. M., Krounbi, M., and Christner, J., *IEEE Trans. Magn.*, **MAG-30**, 298 (1994).

Ishikawa, C., Suzuki, K., Koyama, N., Yoshida, K., Sugita, Y., Shinagawa, K., Nakatani, Y., and Hayashi, N., *J. Appl. Phys.*, **74**, 5666 (1993).

Kikuchi, K., Kobayashi, T., Kawai, T., and Sakama, A., *IEEE Trans. Magn.*, **MAG-30**, 3858 (1994).

Koehler, T. R., *IEEE Trans. Magn.*, **MAG-30**, 3867 (1994).

Liao, S. H., Yuan, S. W., and Bertram., H. N., *IEEE Trans. Magn.*, **MAG-29**, 3873 (1993).

Liao, S. H., Torng, T., and Kobayashi, T., *IEEE Trans. Magn.*, **MAG-30**, 3855 (1994).

Lin, T., Tsang, C., Fontana, R. E., and Howard, J. K., *IEEE Trans. Magn.*, **MAG-31**, 2585 (1995).

Mitsumata, C., Kikuchi, K., and Kobayashi, T., *IEEE Trans. Magn.*, **MAG-34**, 1453 (1998).

Panofsky, W. K. H. and Phillips, M., *Classical Electricity and Magnetism*, Addison-Wesley, Reading, MA, 1962.

Paranjpe, A. P., Kools, J. C. S., Schwartz, P. V., Song, K., Bergner, B., Heimanson, D. H., and Van Ysseldyk, R. W., Paper GB-01, *Mag. And Mag. Mater. Conference*, San Jose, 1999.

Potter, R. I., *IEEE Trans. Magn.*, **MAG-10**, 502 (1974).

Shelledy, F. B. and Brock, G. W., *IEEE Trans. Magn.*, **MAG-11**, 1206 (1975).

Shen, Y., Chen., W., Jensen, W., Ravipati, D., Retort, V., Rottmayer, R., Rudy, S., Tan, S, and Yuan, S., *IEEE Trans. Magn.*, **MAG-32**, 19 (1996).

Smith, R. L., Digest #EQ-11, Intermag Conference (1989).

Suzuki, Y. and Ishikawa, C., *IEEE Trans. Magn.*, **MAG-34**, 1513 (1998).

Thompson, D. A., *AIP Conf. Proc.* **24**, 528 (1974).
Tsang, C., *IEEE Trans. Magn.*, **25**, 3692 (1989).
Tsang, C. and Decker, S. K., *J. Appl. Phys.*, **52**, 2465 (1981).
Tsang, C. and Decker, S. K., *J. Appl. Phys.*, **53**, 2602 (1982).
Tsang, C. and Lee, K., *J. Appl. Phys.*, **53**, 2605 (1982).
Tsang, C. and Fontana, Jr., R. E., *IEEE Trans. Magn.*, **MAG-18**, 1149 (1982).
Weber, E., *Electromagnetic Theory*, Section 15, Dover, New York, 1965.
Williams, E. M., *The Physics and Technology of Xerographic Processes* Wiley, New York, 1984. Reprinted by Kreiger (1993).
Xiao, M., Devasahayan, A. J., and Kryder, M. H., *IEEE Trans. Magn.*, **MAG-34**, 1495 (1998).
Yuan, S. W., "Micromagnetics of Domains and Walls in Ferromagnetic Soft Materials," Ph.D. thesis, University of California, San Diego, 1992.
Yuan, S. W., Kobayashi, T., and Liao, S. H., *IEEE Trans. Magn.*, **MAG-31**, 2627 (1995).
Zhu, J-G. and O'Connor, D. J., *IEEE Trans. Magn.*, **MAG-32**, 3401 (1996).

5

AMR HEAD DESIGN AND ANALYSIS

The previous four chapters introduced system requirements under which recording heads must operate, presented motivation for improving read signal output, reviewed material properties of thin films, and developed mathematical tools for designing and analyzing AMR and GMR sensors. It is now appropriate to apply these tools to practical design problems that the digital magnetic recording industry has addressed and solved in a variety of ways. Out of many ideas in the literature, this chapter discusses AMR design concepts that found their way into successful hard disk drive products. Potter (1974) introduced the shielded AMR head for improving pulse resolution, Thompson (1974) discussed sensor biasing techniques for shielded elements, Shelledy and Brock (1975) compared self-biased (i.e., shield-biased) and shunt-biased elements, O'Connor, et al. (1985) analyzed shunt-biased tape heads, and Smith (1987) analyzed MR biasing with a soft adjacent layer (SAL). Smith et al. (1992) introduced and analyzed the unshielded dual-stripe or dual-sensor MR (DSMR) head, and Bhattacharyya and Simmons (1994) compared shielded DSMR, SAL, and barber-pole MR heads. Anthony et al. (1994) and Hsu et al. (1995) gave experimental results for shielded DSMR heads, and Guo, et al. (1996) and Zhu and O'Connor (1996) compared performance and process tolerances of shielded DSMR and SAL MR heads. These approaches to AMR design are found in disk or tape drive products, and the signal/biasing considerations of each has advantages and disadvantages over competing designs.

The details behind signal amplitude and appropriate biasing of MR sensors are buried, onionlike, in many layers beneath Equation (1.5), which is repeated here for convenience and emphasis of its importance:

$$\text{Signal} = I \ \Delta R(H) = I \frac{\Delta \rho(H)}{t} \frac{W}{h} \quad \text{V.} \quad (5.1)$$

In multifilm designs, the total current divides among the various conductors, and the current I flowing through the MR layer is found with Ohm's law, assuming the layers are of a thickness where this law is valid. In the simple uniform rotation (Stoner–Wohlfarth) view, the AMR effect $\Delta\rho(H)$ is quadratic in nature, so finding a transfer curve operating point that maximizes signal and minimizes signal nonlinearities is roughly equivalent to finding bias and input excitation fields that meet predetermined criteria for the output signal voltage. Chapters 3 and 4 treated highly *nonuniform* magnetization profiles that arise from magnetic coupling between films, so transfer curves of coupled devices will be more complicated than simple parabolas. Sensor resistance increases as MR thickness t is reduced, and thus I^2R heating and its dissipation must be addressed in the design process. Choice of MR track width W and stripe height h each influence sensor resistance as well as its sensitivity to signal fields (through film demagnetization effects). This chapter begins with an analysis of self-biased AMR heads, moves in gradual complexity to shunt biasing, and then analyzes DSMR- and SAL-biased AMR heads. Because commercial success and volume production go hand in hand (145 million disk drives, or over 700 million MR heads shipped in 1998), the final task of this chapter is to examine critical design and process control parameters of SAL-biased heads using Monte Carlo analytical techniques. Specific design objectives must be chosen to pin down the head geometry: An areal density of 3 Gbits/in.^{2}, for example, translates to 13 ktpi and 230 kbpi for a bit cell ratio $R_{bc} = 18$ (see Chapter 1 scaling rules), and the physical sensor width [see (1.4)] is $W = 0.64/\text{tpi} = 49$ µin. $= 1.25$ µm. The necessary pulsewidth [estimated with (1.12) and channel density $U \simeq 2.3$] becomes PW50 $\simeq 10.0$ µin., or 0.25 µm, which constrains the shield-to-shield gap (G_{ss}), film thickness (t), magnetic spacing (d), transition length (πa), and medium thickness (δ) [see (1.10a)]. With $d + a = 0.10$ µm, a medium thickness $\delta = 0.015$ µm, an element $t = 0.0125$ µm, and $G_{ss} \simeq 0.215$ µm, the required PW50 can be supported. These parameters will serve as a common base for the case studies that follow.

BIASING THE AMR HEAD FOR SIGNAL LINEARITY

Self-Biasing (Shield Biasing)

Perhaps the simplest technique for biasing an MR element is to place it between shields in an off-center position; this is called self-biasing or shield biasing. Much can be learned by applying the design and analysis tools of Chapters 3 and 4 to derive the output signal of a self-biased AMR head. Reading directly from Eq. (4.14b) and Fig. 4.3, the field applied to the center ($x = 0$) of a current sheet between shields is

$$H_a = H_{y2} = H_0\left(\frac{g_2 - g_1}{g_1 + g_2}\right), \qquad (5.2)$$

where $H_0 = 2\pi I/h$ (in oersteds), with I in milliamperes and h in micrometers. In Fig. 4.3, positive current flows out of the page toward the reader, so a sensor close to the left shield ($g_2 > g_1$) would experience a positive (upward) bias field. The element is a film of Permalloy ($Ni_{82}Fe_{18}$) with $M_s \simeq 800$ emu/cm^3, $H_k \simeq 4$ Oe, and exchange stiffness $A = 1 \times 10^{-6}$ erg/cm; thickness t is chosen based on signal requirements at a selected areal density, film resistivity depends on thickness ($\rho \simeq 32$ μΩ-cm at $t \simeq 125$ Å, for example), and the maximum AMR effect $\Delta\rho_0 = 0.64$ μΩ-cm is constant at room temperature.

Magnetization Bias Profile The Permalloy sensor magnetizes according to the analysis worked out in Chapter 4. Substituting (5.2) into (4.20), the magnetization profile becomes

$$M_y(y) = \frac{M_s}{H_k} \frac{2\pi I}{h} \left[\frac{g_2 - g_1}{g_1 + g_2}\right] \left[1 - \frac{\cosh(y/\lambda)}{\cosh(h/2\lambda)}\right] \quad (5.3)$$

because the demagnetizing flux (also called the "fringing" flux) is shunted by the two shields. Recall [from (4.19)] the characteristic length for flux decay to the shields is $\lambda = [\mu t g_1 g_2/(g_1 + g_2)]^{1/2} \simeq 1.08$ μm for $g_1 = 0.05$ mm, $g_2 = 0.15$ mm, $\mu = 2500$, and $t = 125$ Å. Including exchange energy in the MR layer [see (3.42)] slightly increases the characteristic length ($\lambda \simeq 1.11$ μm). At $I = 10$ mA and $h = 0.80$ mm, the magnetization profiles for $g_2/g_1 = 2, 3, 4$ are plotted in Fig. 5.1; the characteristic lengths are 1.18, 1.08, and 1.0 μm, and the corresponding

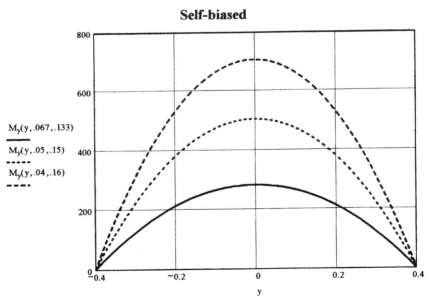

FIGURE 5.1 Magnetization profiles M_y (emu/cm^3) vs. y (μm) for gap ratios $g_2/g_1 = 2, 3, 4$.

applied fields H_a are 25.9, 39.3, and 47.1 Oe, respectively. Small shifts in sensor location obviously have substantial impact on sensor biasing.

Signal flux from the recording medium enters the MR sensor and leaks away to the shields as it travels up the element. In unsaturated high-permeability sensors, the induction can be written as $B = H + 4\pi M = 4\pi M[\mu/(\mu - 1)] \simeq 4\pi M$ [see Potter (1974) and Bertram (1994, p.177)] and the signal magnetization in the sensor follows the signal flux through the relation

$$\Delta M_y^{\text{signal}}(y) = \frac{\Phi^{\text{signal}}(y)}{4\pi W t}. \tag{5.4}$$

At the transition center, signal flux (in cgs units of maxwells) entering the sensor is [see (1.7)]

$$\phi(y_e) = \frac{2}{\pi} 4\pi M_r \, \delta W \frac{y_e}{g} \left[f\left(\frac{g+t/2}{y_e}\right) - f\left(\frac{t/2}{y_e}\right) + f\left(\frac{-g-t/2}{y_e}\right) - f\left(\frac{-t/2}{y_e}\right) \right]$$
$$= \frac{16 M_r \, \delta W \, y_e}{g} \left[f\left(\frac{g+t/2}{y_e}\right) - f\left(\frac{t/2}{y_e}\right) \right]. \tag{5.5}$$

The effective readback spacing between the medium and transducer is $y_e = d + a + \frac{1}{2}\delta$ and the function $f(z) = z \tan^{-1}(z) - 0.5 \ln(1 + z^2)$. The trade-off between output signal amplitude and pulse distortion is controlled by setting the medium $M_r \delta$ for a given spacing. Normally $M_r \delta$ will be about $0.25 M_s t$ or $0.5 M_s t$ for sensor designs having one or two magnetic layers, respectively. Reading from (4.31), (5.4), and (5.5), the signal magnetization along the sensor height is

$$\Delta M_y^{\text{signal}}(y) \simeq \frac{\phi(y_e)}{4\pi W t} \frac{\sinh[(h/2 - y)/\lambda]}{\sinh(h/\lambda)}, \tag{5.6}$$

and the magnetization profile is the algebraic sum of the bias and signal levels. This is shown in Fig. 5.2 using (5.3) and (5.6) with the previously defined device parameters and with $I = 10$ mA, $M_r = 270$ emu/cm^3, $\delta = 150$ Å, $M_s = 800$ emu/cm^3, and $d + a = 1000$ Å.

The Bias Angle The voltage changes associated with variations in M_y depend on resistance changes ΔR *averaged* over the height h. The phenomenological expression for the AMR effect is discussed in Chapter 2 and the parabolic AMR transfer curve is written

$$\Delta \rho(H) = \Delta \rho_0 [1 - \langle \sin \theta^2(H) \rangle] = \Delta \rho_0 \left[1 - \left(\frac{\langle [M_y^{\text{bias}}(H) + \Delta M_y^{\text{signal}}]^2 \rangle}{M_s^2} \right) \right], \tag{5.7}$$

where the brackets $\langle \cdots \rangle$ indicate an average value over the stripe height h, and $\Delta \rho_0$ is the maximum change in resistivity. If $M_y(y)$ (bias plus signal) *is not saturated*

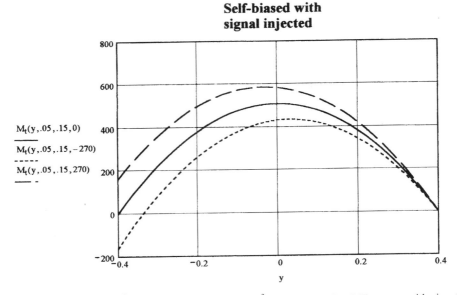

FIGURE 5.2 Magnetization profiles M_y (emu/cm^3) vs. y (μm) of an MR sensor with signal injected from a medium with $M_r = 0, -270, +270$ emu/cm^3. Medium thickness $\delta = 150$ Å.

anywhere, the average total magnetization level over the stripe is found by integrating (5.3) plus (5.6) and dividing by h. The result is $m_t[H(I), M_r \delta] = m_y^{\text{bias}}[H(I)] + m_y^{\text{signal}}(M_r \delta)$, where the average bias is expressed as

$$m_y^{\text{bias}}(I) = \sin\theta_{\text{bias}} = \frac{\langle M_y^{\text{bias}}(I)\rangle}{M_s} = \frac{2\pi I}{H_k h}\left(\frac{g_2 - g_1}{g_1 + g_2}\right)\left[1 - \frac{\tanh(h/2\lambda)}{h/2\lambda}\right] \quad (5.8)$$

and θ_{bias} is the "bias angle." The average signal magnetization is written

$$\Delta m_y^{\text{signal}}(M_r \delta) = \frac{\langle \Delta M_y^{\text{signal}}\rangle}{M_s} = \frac{\phi(y_e)}{4\pi M_s Wt}\frac{\lambda[\cosh(h/\lambda) - 1]}{h \sinh(h/\lambda)} = \frac{\phi(y_e)}{4\pi M_s Wt}\frac{\tanh(h/2\lambda)}{(h/\lambda)}. \quad (5.9)$$

If h is small with respect to λ, then $\tanh(h/2\lambda)/(h/\lambda) \to \frac{1}{2}$, and the average signal magnetization becomes $\phi(y_e)/8\pi M_s Wt$. The sensor voltage is found by direct substitution of (5.7), (5.8), and (5.9) into (5.1).

Saturation Current for the MR Layer With increasing bias current, the MR layer begins to saturate at the stripe center and the linear analysis given by (5.7) must be replaced with the analysis of coupled film saturation given in Chapter 3. The self-biased film enters saturation at the threshold field H_{S0} defined by (3.48), and the magnetization profile $m_y(y) = M_y(y)/M_s$ is altered according to (3.49a,b). The bias current at onset of film saturation is found by equating (3.48) with (5.2); solving for $I = I_{S0}$, one obtains

$$I_{S0} = \frac{h}{2\pi}\left(\frac{g_1+g_2}{g_2-g_1}\right)H_k\frac{\cosh(h/2\lambda)}{\cosh(h/2\lambda)-1} \quad \text{mA}. \quad (5.10)$$

Profiles for I values of 10, 15.9, and 30 mA [H = 39, 62.4, 118 Oe, computed with (5.2) and θ_{bias} = 23.3°, 38.9°, 45.4°] are shown in Fig. 5.3; the saturation threshold field H_{S0} = 62.4 Oe (at I_{S0} = 15.9 mA) for the selected geometry. The average *total* magnetization $m_{\text{av}}(I, M_r, \delta) = m_y(y, I) + \Delta m_{\text{sig}}(y, M_r, \delta)$ for a partially saturated film is readily calculated by numerical integration of (3.49) plus (5.9) over the range $-\frac{1}{2}h \leq y \leq \frac{1}{2}h$, under the constraint that $m_{\text{av}}(I, M_r, \delta) \leq 1.0$ everywhere along y. A magnetization curve $m_{\text{av}}(I, 0)$ for a numerically averaged nonuniform film is shown in Fig. 5.4. The stripe h is divided into 40 equal increments for the computations. Normalized AMR transfer curves, which are of the form $\cos^2\theta = 1 - m_{\text{av}}^2$, are graphed in Figs. 5.5a,b. The three curves shown in Fig. 5.5a are for zero, positive, and negative signal flux, and parabolic shapes are followed until the film enters saturation at $I = I_{S0}$. For higher currents, the slope gradually reduces toward zero. In Fig. 5.5b, the transfer curves are calculated for constant bias current with variable signal M_r (the curve for I = 2 mA is seriously underbiased). At a given bias current (or field), the transfer curve is modulated up or down by the signal flux; that is, the average resistance is increased or decreased around the bias resistance.

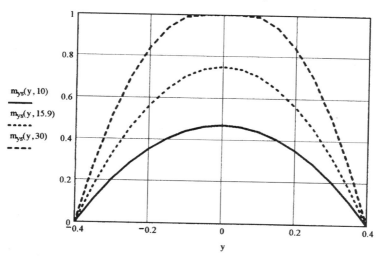

FIGURE 5.3 Normalized bias profiles for an MR sensor at bias currents of 10, 15.9, and 30 mA showing unsaturated and saturated behavior.

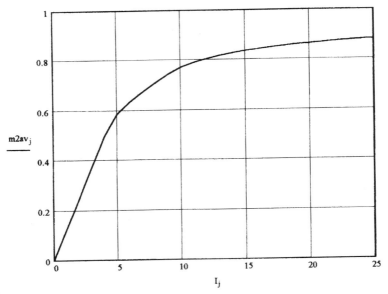

FIGURE 5.4 Average normalized magnetization of an MR film showing onset of saturation at 15.9 mA.

Output Signal and Asymmetry Output voltages for positive and negative input signals are derived directly from the transfer curves $1 - m_{av}^2[H(I), M_r \delta]$, where $m_{av} = m_y(I) + \Delta m(M_r, \delta)$. Under conventional practice, *positive* or *negative* signals V_p or V_n are produced by positive or negative voltage shifts from the quiescent level across the MR element, so in terms of the transfer curve the output signals are written (where, for sake of brevity, $M_r \delta$ is replaced with M_r)

$$V_{\text{signal}}(I, M_r) = I\,\Delta R[1 - m_{\text{avg}}^2(I, M_r)], \qquad (5.11\text{a})$$

$$V_p(I, M_r) = V_{\text{signal}}(I, -M_r) - V_{\text{signal}}(I, 0), \qquad (5.11\text{b})$$

$$V_n(I, M_r) = V_{\text{signal}}(I, 0) - V_{\text{signal}}(I, +M_r), \qquad (5.11\text{c})$$

$$V_{p\text{-}p}(I, M_r) = V_p(I, M_r) + V_n(I, M_r). \qquad (5.11\text{d})$$

Signal amplitude asymmetry is defined as the difference over the sum of outputs:

$$\text{Asymmetry} = \frac{V_p - V_n}{V_p + V_n}. \qquad (5.12)$$

Asymmetry is often expressed as a percent instead of a ratio and is an indicator of proper biasing. For this design, zero asymmetry is obtained at the average bias angle of 38.9° [$\sin^{-1}(m_y^{\text{bias}})$]; this bias angle is close to the value of 45° normally found in the literature and is often recommended as the preferred design point for maximizing

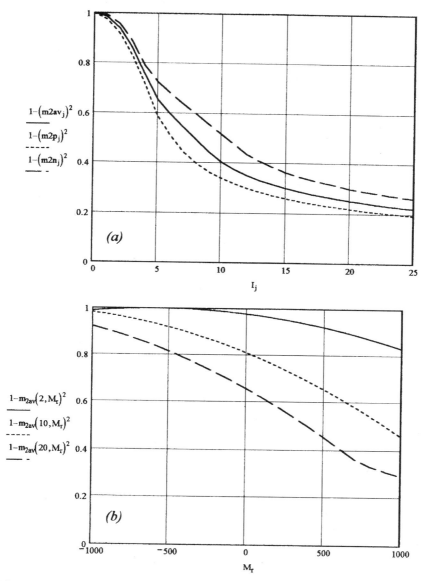

FIGURE 5.5 (*a*) Normalized AMR transfer curves for zero, positive, and negative signal flux plotted as a function of bias current. (*b*) Normalized transfer curves (each at a constant bias current) for variable signal flux.

the dynamic range of an MR head. It would be a matter of choice for the head designer to design for minimum asymmetry or for maximum dynamic range and, more importantly, to understand that trade-offs are involved in such choices.

Plots of positive, negative, and peak-to-peak signal (all in millivolts) and asymmetry versus sense/bias current (in milliamperes) are given in Figs. 5.6a,b; device heating is ignored here, so the signal increases quadratically with I until saturation begins. These calculations show the signal peaks at $I \sim I_{S0}$ and then flattens with increasing current. Generally speaking, MR sensors are operated in the linear region and saturation is avoided. Peak-to-peak signal and signal asymmetry are calculated for $M_r \, \delta = 4 \times 10^{-4}$ (often written as 0.4 memu/cm^2, where memu is shorthand for a milli-emu of magnetic dipole moment). At $t = 125$ Å the sensor resistivity (ρ) is about 32 $\mu\Omega$-cm (see Fig. 2.1a), so the resistance $R = \rho W/(th) = 40 \, \Omega$, $\Delta R_0 = 0.8 \, \Omega$ and the AMR "figure of merit" defined by $\Delta R/R$ is 2.0% at room temperature. In Fig. 5.5a at 15 mA, the operating point at

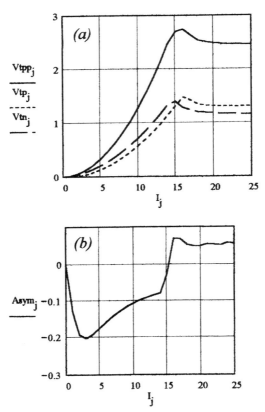

FIGURE 5.6 (a) Plots of positive, negative, and peak-to-peak signal (all in mV) and (b) signal asymmetry versus bias current (mA). Device heating is ignored.

BIASING THE AMR HEAD FOR SIGNAL LINEARITY   117

$M_r \delta = 0$ is $1 - m_{\text{avg}}^2 = 0.61$; a signal of -0.4 memu/cm^2 shifts the curve upwards to 0.72 and $+0.4$ memu/cm^2 shifts the curve downward to 0.49. The peak-to-peak modulation is $0.23 \Delta R = 0.23 \times 0.80\ \Omega = 0.18\ \Omega$, which produces a peak-to-peak voltage of 2.76 mV. Fig. 5.6b shows that signal asymmetry is negative for $I < I_{S0}$ and positive for greater sense/bias currents. Requirements for good signal linearity determine the limits on MR signal asymmetry, which is normally about $\pm 10\%$, so (ignoring device heating for the moment) the acceptable operating current is above 12 mA for the nominal design ($g_2/g_1 = 3.0$ with $h = 0.8\ \mu$m). High negative asymmetry is evidence of an underbiased and overdriven sensor; signal input scales with $M_r \delta$, and thus amplitude and asymmetry each increase monotonically with input level. In the example here, the medium $M_r \delta$ is high and could be reduced to improve asymmetry at the expense of signal at operating currents below I_{S0}.

Temperature of the MR Layer   The sense/bias current for zero asymmetry is approximately equal to the threshold current for onset of saturation (I_{S0}) defined with (5.10). The thermal analysis in Chapter 2 gives the temperature profile of an MR element and its average value at any bias current; reading from (2.35) and (2.43) for the situation where heating in leads can be ignored, the temperature along the sensor is

$$T(x) = \frac{\lambda_f^2 \rho_f J^2}{\kappa_f}\left[1 - \frac{\cosh(x/2\lambda_f)}{\cosh(W/2\lambda_f) + (\kappa_f t_f h \lambda_L / \kappa_L t_L h_L \lambda_f)\sinh(W/2\lambda_f)}\right] \quad (5.13)$$

and the symbols are defined in (2.33) and (2.34). The average temperature rise, given by (2.47), is repeated here for convenience:

$$T_{\text{avg}} = \frac{\lambda_f^2 \rho_f J_f^2}{\kappa_f}\left[1 - \frac{\sinh(W/2\lambda_f)}{(W/2\lambda_f)[\cosh(W/2\lambda_f) + (\kappa_f t_f h \lambda_L / \kappa_L t_L h_L \lambda_f)\sinh(W/2\lambda_f)]}\right]. \quad (5.14)$$

For the present example, the temperature rise $\Delta T = T_{\text{avg}}$ and the resultant $\Delta R = \Delta R_0 \exp(\alpha_{\Delta\rho} \Delta T)$ [see (2.24b) and Chapter 2, Table 2.5 for $\Delta\rho_{\text{AMR}}(T)$] are calculated as a function of bias current; the results are plotted in Figs. 5.7a,b.

The current density J_f and device resistance each increase rapidly with short stripe height h, so the thermal situation deteriorates dramatically for reductions in h. With increasing temperature, R increases, ΔR decreases, signal is lost, and thermal degradation processes are accelerated. Severe restrictions are thus necessarily placed on the operating current. With temperature effects included, the peak-to-peak signal versus current is shown in Fig. 5.8a for a nominal stripe height of 0.8 μm; Fig. 5.8b shows the maximum signal over a range of stripe heights. At $h = 0.4$ μm signal is maximum at 6 mA and the sensor melts [see (2.50)] at $I_m = 19.8h = 7.9$ mA, whereas the signal maximizes at 12 mA with $h = 1.0$ μm, and the melting current has increased to 19.8 mA. At $h = 2.0$ μm, signal maximizes at 8 mA and

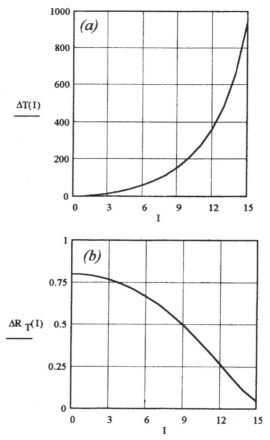

FIGURE 5.7 (*a*) Sensor temperature rise (K) as a function of bias current (mA). (*b*) AMR effect $\Delta R(\Omega)$ as a function of bias current.

$I_m = 39.6$ mA. Figure 5.8*c* plots the sensor element temperature rise at the maximum signal level for stripes between 0.40 and 2.0 μm high; the temperature variation for $h \sim 0.5$ μm is a computational artifact arising from the estimated current for maximum signal.

A simple model of thermal resistance (2.28) provides the result $R_\tau = g_1 g_2/[\kappa_0 W h(g_1 + g_2)]$. With a thermal conductivity $\kappa_0 = 1.0$ W/m-K for the gap material, the design with $g_2/g_1 = 4$ would have a thermal resistance $R_\tau = 3.2 \times 10^4$ K/W and an unacceptable temperature rise of 160 K at 5 mW dissipation (11.4 mA sense/bias); at this temperature rise, the AMR effect is reduced to half of its room temperature value. Meeting a goal of about 50 K temperature rise (or less) requires a gap material like AlN with a thermal conductivity $\kappa_0 \simeq 3$ W/m-K, or the head would have to operate in an underbiased mode at

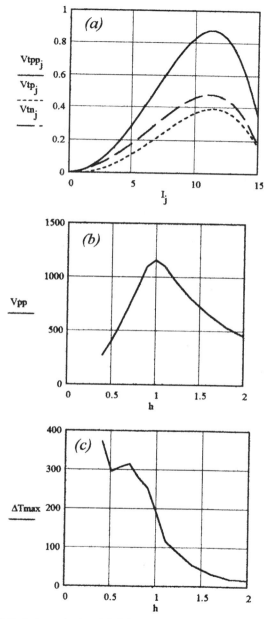

FIGURE 5.8 (*a*) Peak-to-peak signal (mV) versus bias current (mA). (*b*) Maximum peak-to-peak signal as a function of stripe height (μm). (*c*) Sensor temperature rise (K) as a function of stripe height.

$I \simeq 6.5$ mA. It appears that good signal output, low asymmetry, and low sensor heating are incompatible with this type of design at 3 Gbits/in.^{2} Successful high-volume manufacturing of a self-biased head would require very tight controls on stripe height h and deposition thicknesses of the insulating gaps (g_1 and g_2), a material with $\kappa_0 \simeq 3$ W/m-K or greater, and the detection electronics would have to cope with excessive signal asymmetry.

Comparison between Model and Experiment Shelledy and Brock (1975) give information in their paper that allows a comparison of experiment with the model computations arising from (5.1)–(5.11a,b,c,d). Their experimental device is a self-biased AMR tape head; the sense element is NiFe with $H_k = 5$ Oe, $M_s = 800$ emu/cm^3, $t = 600$ Å, $W \simeq 940$ μm (effective), $h = 10$ μm, $g_1 = 0.43$ μm, and $g_2 = 1.07$ μm. The medium is γFe_2O_3 with $4M_r \delta/(d + a) \approx 460$ Oe. At a sense current of 12.5 mA (set by power limits and temperature rise) and a linear density of 1600 transitions/in. (63 flux reversals/mm), the measured signal from the oscilloscope trace was about 12 mV (peak to peak) with the element "badly underbiased." Model calculations give $V_{pp} \simeq 13.5$ mV and a large asymmetry of -103.5%. [Here $V_p = -0.3$ mV and $V_n = 13.8$ mV. Note the definition of V_p in (5.11b): A *negative* V_p implies that the signal flux for $-M_r \delta$ drives the transfer curve over the maximum and down below the zero signal level.] It appears that model and experiment agree within roughly 12.5% for estimating the signal, provided the medium field (460 Oe) is reasonable.

Shunt Biasing

Shunt biasing is accomplished by laminating a thin magnetic film with a nonmagnetic conductor to form a parallel resistance network. The geometry is given in Fig. 5.9, where t and ρ are the MR layer thickness and resistivity and t_s and ρ_s refer to the shunt layer. These films are electrically shorted together and brought to the external circuitry through two leads. Assuming that Ohm's law is valid (see the discussion in Chapter 2 on resistivity of very thin films), the resistance of each layer is

$$R_{MR} = \frac{\rho}{t}\frac{W}{h} \quad \Omega \qquad R_S = \frac{\rho_S}{t_S}\frac{W}{h} \quad \Omega \qquad (5.15)$$

and the total resistance is $R_T = R_{MR}R_S/(R_{MR} + R_S)$. The total current ($I$) supplied through the leads splits according to the relations $I_{MR}/I = R_T/R_{MR}$ and $I_S/I = R_T/R_S$. The shunt current supplies a bias field $H_{y3} = -2H_0(x - g_2)/(g_1 + g_2)$ [see (4.14c)]; x (defined in Fig. 4.3) is the distance between shunt and MR film centerlines [$x = (t_S + t)/2$], $H_0 = 2\pi I_S/h$ in oersteds, with I_S in milliamperes and h in micrometers. The total field acting on the MR layer is the sum of self-biasing

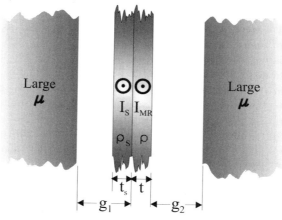

FIGURE 5.9 Geometry for a shunt-biased MR head with shields.

[(5.2) or (4.14b)] and conductor biasing (4.14c) terms (with appropriately modified terminology):

$$H_T(I) = H_{MR}(I) + H_S(I) = \frac{2\pi I_{MR}}{h}\left(\frac{g_2 - g_1}{g_1 + g_2}\right) + \frac{4\pi I_S}{h}\left(\frac{g_2 - (t_S + t)/2}{g_1 + g_2}\right)$$

$$= \frac{2\pi I}{h(g_1 + g_2)}\left[\frac{R_T}{R_{MR}}(g_2 - g_1) + \frac{2R_T}{R_S}\left(g_2 - \frac{t_S + t}{2}\right)\right]. \quad (5.16)$$

With practical choices of shunt resistance and g_2, (5.16) indicates useful bias can be achieved even in the case of symmetrical gaps ($g_1 = g_2$), where self-biasing is eliminated. (As will be seen shortly, device heating would be excessive at currents where signal asymmetry is small, so a shunt-biased design would be used at lower currents and large negative asymmetry.)

Magnetization Bias Profile Demagnetization flux of the magnetic layer goes to the shields, and the magnetization profile is obtained with (4.20) and (5.16). The characteristic length is found with (4.19), as in the example for a self-biased design:

$$M_y(y, I) = \frac{M_s}{H_k}\frac{2\pi I}{h(g_1 + g_2)}\left[\frac{R_T}{R_{MR}}(g_2 - g_1) + \frac{2R_T}{R_S}\left(g_2 - \frac{t_S + t}{2}\right)\right]\left[1 - \frac{\cosh(y/\lambda)}{\cosh(h/2\lambda)}\right]. \quad (5.17)$$

Choice of shunt material, resistivity, and thickness depends on several factors: The shunt material should not galvanically corrode the NiFe layer, the shunt current should be less than the MR layer current, the gap G_{ss} should support the required PW50, and device fabrication should not be highly difficult. Large gap asymmetry is incompatible with adequate insulation thickness between the shunt and shield unless G_{ss} is opened up accordingly. Since PW50 is a fundamental determinant of a recording system, a thick shunt design would not perform well at high linear densities, so design challenges emerge. Gaps should be small and thermal conductivity adequate to dissipate heat effectively.

In the present design example, initial design parameters are chosen and then analyzed to define and quantify the consequences. Choose a minimum insulation thickness $g_1 - t_S = 425$ Å and let the Ta shunt layer t_S, MR layer t, and g_2 equal 375, 125, and 1200 Å, respectively. This gives $r = 1.5$ ($g_1 = 800$ Å) to provide self-biasing and support the required $G_{ss} = 0.215$ μm; the characteristic decay length for this system is $\lambda = 1.23$ μm. Tantalum provides good crystalline [111] texture for the NiFe layer, does not corrode, and, depending on thickness and deposition conditions, has a resistivity of about 150 μΩ-cm. Sensor width $W = 1.25$ μm and height $h = 0.8$ μm are unchanged, so layer resistances $R_S = 62.5$ Ω, $R_{MR} = 40$ Ω, and $R_T = 24.4$ Ω. Total current splits 61 and 39% to the MR and shunt layers, respectively. Output signal (without heating) and asymmetry for $M_r \delta = 0.4$ memu/cm^2 is graphed in Figs. 5.10a,b.

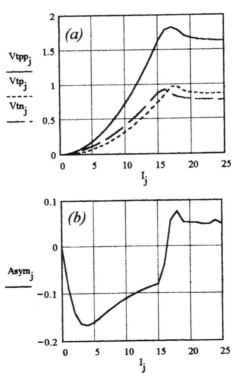

FIGURE 5.10 (a) Output signal (mV) and (b) signal asymmetry as a function of bias current (mA) for a shunt-biased MR head with shields. Heating is ignored.

BIASING THE AMR HEAD FOR SIGNAL LINEARITY

Saturation Current for the MR Layer   The total current for onset of saturation (and zero asymmetry) (I_{Sa0}) is found by substituting (5.16) into (3.48) and solving for I; this yields

$$I_{Sa0} = \frac{h(g_1 + g_2)H_k \cosh\left(\frac{h}{2\lambda}\right)}{2\pi\left[\frac{R_T}{R_{MR}}(g_2 - g_1) + \frac{2R_T}{R_S}\left(g_2 - \frac{t_S + t}{2}\right)\right]\left[\cosh\left(\frac{h}{2\lambda}\right) - 1\right]}. \quad (5.18)$$

For the present example, $I_{Sa0} = 20.3$ mA and the temperature rise would be unacceptable even with improved thermal conductivity. The saturation current depends on stripe height [see (3.52)], and at $h_0 \simeq 3.7$ μm, I_{Sa0} reduces to 8.3 mA, but the required shape anisotropy ($W \simeq 1.4h_0$) precludes such a large stripe height. Figures 5.11a–d show calculations for signal, asymmetry, temperature, and ΔR as a function of bias current; AlN gap material is used with $\kappa_0 = 3.0$ W/m-K, for which the thermal resistance $R_\tau \simeq 1.6 \times 10^4$ K/W. To meet a criterion of 50 K rise, this design would have to operate at 3.1 mW ($I \simeq 11$ mA) where the signal at the head terminals would be 600 μV (peak to peak) and asymmetry would be -10%. It appears shunt biasing suffers from the same problems as self biasing: To obtain low asymmetry, bias currents must be high and overheating produces signal loss arising from the temperature coefficient of the AMR effect. Thermal resistance and current density are inversely proportional to stripe height h, and thus short stripes will not survive routine production tests.

Dual-Sensor MR (DSMR) Biasing

In Chapter 3, the behavior of coupled magnetic films was analyzed, and it was shown that the magnetization vectors in each film rotate in opposite directions such that flux closes between films. Designs based on tight coupling between magnetic films facilitate biasing at reduced levels of bias current; this is a useful feature because as areal density increases, the MR sensor area (Wh) decreases inversely, and power density (in watts/per square centimeter) increases unless the optimal bias current is reduced. The analysis of coupled films is extended here to explore signal and biasing attributes of the DSMR head invented by Voegeli (1975). The geometry of a shielded DSMR arrangement is sketched in Fig. 5.12; signal is *differentially* sensed across the two films, and thus a *three-wire configuration* is necessary. Sense element thermal fluctuations arising from head–medium interactions (thermal asperities) are nearly equal in the two sensors if the gap g between films is small, and thus good immunity to this type of common mode noise is possible with DSMR designs. There will be more discussion on this important subject later in this book. Guo et al. (1996) and Bhattacharyya and Simmons (1994) are useful references for DSMR analysis.

Ideally, both films will have identical geometry and properties, but the analysis of DSMR behavior must anticipate process variations, so the equations of nonidentical coupled films will be used [see (3.22), (3.23), and (3.42).] All gaps between

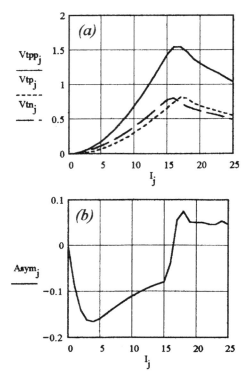

FIGURE 5.11 (a) Signal (mV), (b) asymmetry, (c) temperature rise (K), and (d) ΔR (Ω) as a function of bias current (mA) of a shunt-biased head. Gap thermal conductivity is 3.0 W/m-K.

magnetic layers and shields are insulating, and if the elements are placed symmetrically between shields, the currents $I_1 = I_2 = I$ flow in the same direction to produce equal and opposite bias fields on films 1 and 2. Differences in gap thicknesses ($g_1 \neq g_2$) readily disturb this ideal because self-biasing and conductor biasing polarities are opposed. The bias fields, derived from (4.14a,b,c), are the sum of self-biasing and conductor biasing terms:

$$H_1(I_1, I_2) = \frac{2\pi}{h} \left[\frac{I_1(g_2 + g + t_2 - g_1) - I_2(2g_1 + t_1)}{G_{ss}} \right] \quad (5.19a)$$

$$H_2(I_1, I_2) = \frac{2\pi}{h} \left[\frac{I_2(g_2 - g - t_1 - g_1) + I_1(2g_2 + t_2)}{G_{ss}} \right]. \quad (5.19b)$$

Gap symmetry is defined by the ratio $r = g_2/g_1$, and the ratio of bias magnitudes $|H_1/H_2|$ changes rapidly with r. This behavior is plotted in Fig. 5.13 for selected

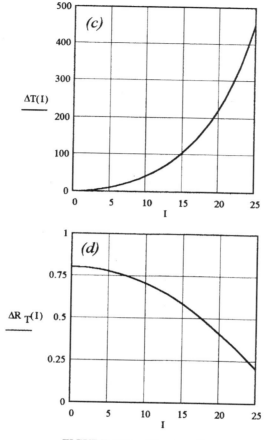

FIGURE 5.11 (*Continued*)

values of $g_1 + g_2 + g + t_1 + t_2 = G_{ss} = 2150\,\text{Å}$, $g = 350\,\text{Å}$, $t_1 = t_2 = 150\,\text{Å}$, and $g_1 + g_2 = 1500\,\text{Å}$. The analytical form of the bias field ratio is

$$\frac{H_1}{H_2} = \frac{Ar - B}{Br - A}, \tag{5.20}$$

where $A = g_1 + g + g_2$, $B = 3(g_1 + g_2) - g$, and equal currents are assumed in (5.19a,b).

Magnetization Bias Profiles With identical sense layers and a small spacer g (much less than g_1 or g_2), flux closure between the magnetic films establishes equal and opposite magnetization profiles with small leakage of bias flux to the shields, and profiles are calculated with (3.22), (3.23), or (3.49) (for saturated films) with

126 AMR HEAD DESIGN AND ANALYSIS

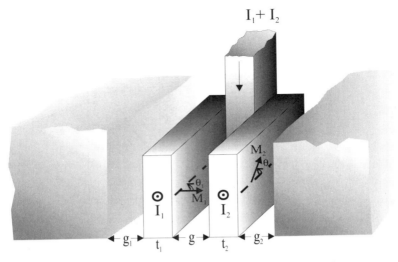

FIGURE 5.12 Geometry of a shielded DSMR head.

appropriate substitution of (5.19a,b) for the applied fields. If $g_1 \neq g_2$, the analysis is more complicated because the magnetization profiles and bias flux decay characteristic lengths of each film are different. Using the analysis in Chapter 3 and drawing from Bhattacharyya and Simmons (1994), the magnetization profiles are

$$M_1(y) = \frac{M_{1s}H_1(I)}{H_{1k}}\left[1 - \frac{\cosh(y/\lambda_1)}{\cosh(h/2\lambda_1)}\right], \qquad (5.21a)$$

$$M_2(y) = \frac{M_{2s}H_2(I)}{H_{2k}}\left[1 - \frac{\cosh(y/\lambda_2)}{\cosh(h/2\lambda_2)}\right], \qquad (5.21b)$$

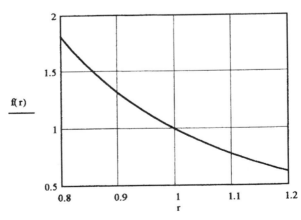

FIGURE 5.13 Ratio of bias field-magnitudes $f(r) = |H_1/H_2|$ as a function of gap asymmetry ratio $r = g_2/g_1$.

where H_1 and H_2 are defined with (5.19a,b) and the bias flux decay lengths (λ_1, λ_2) for each layer are

$$\lambda_1^2 = \frac{\mu_1 t_1 \mu_2 t_2 g}{\mu_1 t_1 + \mu_2 t_2 (1 + g/g_1)}, \tag{5.22a}$$

$$\lambda_2^2 = \frac{\mu_1 t_1 \mu_2 t_2 g}{\mu_1 t_1 (1 + g/g_2) + \mu_2 t_2}. \tag{5.22b}$$

These relationships ignore the influence of exchange energy, but modifications according to (3.41) or (3.42) are easily made. Under most circumstances (for a 3-Gbit/in.^{2} design, for example) the bias flux decay lengths would be nearly equal (within about 2%). The factor $1 + g/g_{1,2}$ is ignored and exchange energy is included in all that follows, and thus $\lambda_1 = \lambda_2 = \lambda$ in further calculations. For the present example, $\lambda = 0.86$ μm including exchange energy.

Because two active sensors are present, the signal flux divides proportionately between them and the appropriate medium $M_r \delta$ for DSMR application would be approximately double that for a single-element design provided g *is small* relative to the transition length. [The dual MR head (DMR) analyzed by Smith et al. (1992) is designed with g *large* with respect to the transition length. The DMR signal analysis is different from the DSMR design, but this subject is not addressed in this book.] Signal magnetization of each element in a DSMR design is derived from (5.6) with modifications for division of signal flux between elements:

$$\Delta M_{1y}^{\text{signal}}(y) \simeq \left(\frac{\phi(y_e) t_1}{4\pi W (t_1 + t_2)^2}\right) \frac{\sinh[(h/2 - y)/\lambda_{\text{sh}}]}{\sinh\left(\frac{h}{\lambda_{\text{sh}}}\right)} \tag{5.23a}$$

$$\Delta M_{2y}^{\text{signal}}(y) \simeq \left(\frac{\phi(y_e) t_2}{4\pi W (t_1 + t_2)^2}\right) \frac{\sinh[(h/2 - y)/\lambda_{\text{sh}}]}{\sinh\left(\frac{h}{\lambda_{\text{sh}}}\right)}. \tag{5.23b}$$

Signal flux flows from the written transition, up the sensors, through the shields, and back to the medium, so the signal flux decay length λ_{sh} is greater than the bias flux decay length λ across the gap g between tightly coupled thin films. In general, the signal flux decay length could be different for each sensor if their properties are significantly different, but with nearly identical films, signal flux decay length is defined by extending the relation (4.19) derived in Chapter 4 to read

$$\lambda_{\text{sh}} = \sqrt{\frac{\mu_1 t_1 g_1 (g + t_2 + g_2)}{G_{\text{ss}}}}. \tag{5.24}$$

Output Signal and Asymmetry The change in resistance of each sensor is found with (5.7) upon substituting the appropriate relations from (5.21a,b) and (5.23a,b).

Normally DSMR heads will operate in the linear region, so the average bias and signal levels over the stripe height can be found analytically [see (5.8) and (5.9) for the single AMR sense layer], and numerical integration is used for saturated films. Extending (5.1), (5.7), and (5.11a,b,c,d) to the DSMR configuration gives an output signal that is written as the difference in signal voltage of the two sensors. That is, with a differential connection, the expression is written

$$V_{\text{signal}} = I_1 \Delta R_1 \left[1 - \left(\frac{\langle [M_{1y}^{\text{bias}} + \Delta M_{1y}^{\text{signal}}]^2 \rangle}{M_{1s}^2} \right) \right]$$
$$- I_1 \Delta R_2 \left[1 - \left(\frac{\langle M_{2y}^{\text{bias}} + \Delta M_{2y}^{\text{signal}}]^2 \rangle}{M_{2s}^2} \right) \right]. \qquad (5.25)$$

The magnetizations in films 1 and 2 are biased in the opposite directions, so the injected signal flux increases the magnetization in one film and reduces it in the other and the output signal is twice the value for a single film.

Guo et al. (1996) define DSMR amplitude asymmetry as

$$A_{\text{DSMR}} = \frac{(A_1 V_{1\text{pp}} - A_2 V_{2\text{pp}})}{V_{1\text{pp}} + V_{2\text{pp}}}, \qquad (5.26)$$

where the terminology A_i and $V_{i\text{pp}}$ refer to the asymmetry [see (5.12)] and peak-to-peak voltage, respectively, of each sensor ($i = 1, 2$). Inspection of (5.26) shows that amplitude asymmetry of a DSMR design should be very low compared to single-element designs because the total asymmetry is the difference between asymmetries of individual elements. A mismatch between film thicknesses of perhaps 10% would still produce acceptable asymmetry. With $t_1 = 150$ Å, $t_2 = 135$ Å, $W = 1.25$ μm, $h = 0.8$ μm, $M_r \delta = 1.2$ memu/cm^2, $d + a = 0.10$ μm, and all other variables unchanged from those stated earlier, the magnetization profiles shown in Figs. 5.14a,b are slightly mismatched. With $t_2 < t_1$ the normalized magnetization in film 1 is determined by the ratio $[M_{2s} t_2]/[M_{1s} t_1]$, and film 2 is limited to $M_2(y) \leq M_{2s}$. The thinner layer (2) enters saturation at about 8 mA when positive signal flux is injected. This behavior is reflected in the output signal of layer 2 as a function of sense current, which is shown in Fig. 5.15a. Total asymmetry, given in Fig. 5.15b, is about -0.6% to -0.9% over most of the unsaturated range, then goes to about -5.7% as element 1 enters saturation. With perfectly symmetric and matched layers, asymmetry is zero at any level of sense/bias current.

Saturation Current The saturation current I_{S0} for either layer of identical films is found with (5.19b) ($I_1 = I_2$) and (3.48) and solving for the current to obtain the

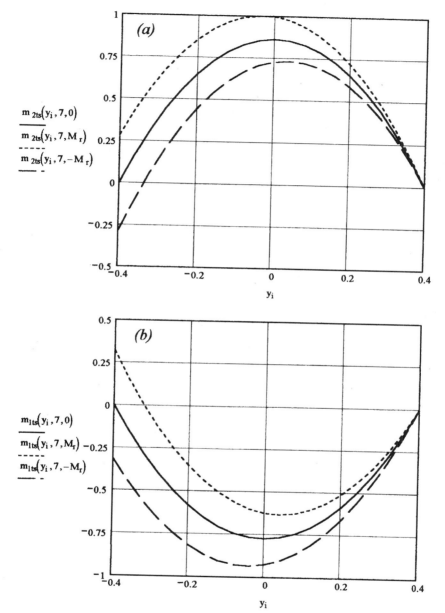

FIGURE 5.14 Normalized bias profiles for slightly mismatched DSMR elements with positive, zero, and negative signals injected: $M_r \delta = 1.2 \, \text{memu/cm}^2$.

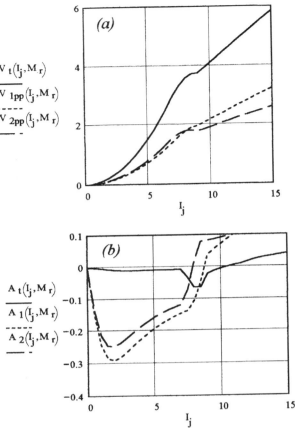

FIGURE 5.15 (*a*) Output signal (mV) and (*b*) asymmetry as a function of sense current (mA) for slightly mismatched DSMR elements. Heating is ignored. Individual sensors and their total are shown.

relation

$$I_{S0} = \frac{hG_{ss}}{2\pi(3g_2 - g_1 - g - t_1 + t_2)} \frac{H_{2k}\cosh(h/2\lambda)}{\cosh(h/2\lambda) - 1}, \quad (5.27)$$

For the design example above $I_{S0} = 9.3$ mA per layer or 18.6 mA total for identical layers. At this current (per layer) the peak-to-peak signal (without heating) is 3.9 mV. Assuming the thermal resistance is 1.25×10^4 K/W [see (2.28) and Chapter 2, Table 2.5 for AlN with $\kappa_0 = 3.0$ W/m-K], the total power dissipated is 6.2 mW with a temperature rise of 77.5 K, and the signal drops to 3.1 mV (a 21% reduction). Signals with and without heating are plotted in Fig. 5.16. Meeting a requirement of 50 K rise (or less) is achievable at 4.0 mW dissipation (7.5 mA and a signal of

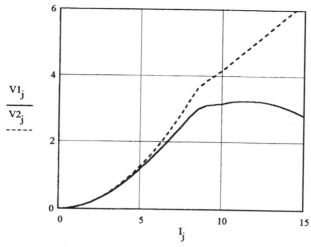

FIGURE 5.16 Output signal (mV, p-p) as a function of bias current (mA) with and without heating.

2.5 mV). Because signal asymmetry is zero and independent of current (with matched elements and symmetric gaps), the DSMR design has advantages for signal detection and processing.

Pulsewidth of a DSMR Head The DSMR elements are separated by a gap g so the readback pulse PW50 is broader than that obtained with single-element designs. This can be understood by examining Fig. 5.17, which shows normalized pulses of layers 1 and 2 and the normalized sum of the individual pulses, all as a function of position along the data track. The individual pulses are calculated with Lorentzian functions (see Chapter 1 and Fig. 1.2). Guo et al. (1996) show that PW50 for a DSMR design follows the relation [see (1.10b)]

$$\text{PW50} \simeq \sqrt{\frac{G_{ss}^2 + (g + 2t)^2}{2} + 4(d + a)(d + a + \delta)}, \qquad (5.28)$$

where $G_{ss} = g_1 + g_2 + g + 2t$. With the present example, PW50 would be approximately 0.27 μm, which, at a linear density of 230 kbpi, would force a channel density $U = \text{PW50}/B \simeq 2.4$ [recall (1.12), where $B = 1/\text{density} \simeq 0.110$ μm]. A single-layer design would produce pulses about 5% slimmer. Very small insulating gaps suffer from shorting between elements, and from elements to shields, so multilayer designs require excellent processing equipment and discipline to avoid yield issues.

Comparison between Model and Experiment Hsu et al. (1995) published experimental results for a 1.0-Gbit/in.² DSMR design and give sufficient information to

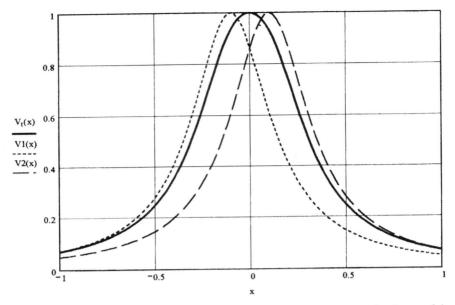

FIGURE 5.17 Normalized signal pulses for layers 1 and 2 and the normalized sum of the individual pulses of a DSMR head as a function of position along the data track.

allow a comparison with model calculations. The sensor layers are Permalloy with $M_s = 800$ emu/cm^3, $H_k = 4$ Oe, $A = 1 \times 10^{-6}$ erg/cm, $t_1 = t_2 = 250$ Å, $W = 2.0$ μm, and $h = 1.8$ μm. Shields are Permalloy and the insulator thicknesses are $g_1 = g_2 = 900$ Å and $g = 500$ Å. Medium $M_r \delta = 1.2$ memu/cm^2, $d = 0.066$ μm, and the calculated transition parameter [see (1.17) and (1.18)] is $a = 0.06$ μm. The experimental signal is 900 μV (peak to peak) at 10 mA per sensor with low asymmetry. (The authors give an "asymmetry ratio" $V_p/V_n = 0.94$ and do not discuss gap asymmetry or mismatched elements.) The stripe height h for various devices is apparently different and the output ranges between 600 and 1000 μV overall. Model calculations from (5.25) and the supporting equations of this chapter give a signal of 982 μV (peak to peak) at 10.0 mA per element; this is above experiment by 9%. The calculated asymmetry is zero because the model elements are identical and symmetrically placed between shields. From (5.28) the estimated PW50 is 0.33 μm, which agrees well with the experimental value of 0.342 μm. As before, if devices are operated well into saturation, the model calculations of this chapter are not expected to agree closely with experiment. Finite-element, micromagnetic numerical calculations would be more appropriate for this regime.

Soft Adjacent-Layer Biasing with Insulating Spacer

Biasing techniques for AMR sensors have evolved from a simple element between shields to a complex arrangement of coupled films and thin dielectric layers between

BIASING THE AMR HEAD FOR SIGNAL LINEARITY

shields. The "induced bias" MR head invented by Beaulieu and Nepela (1975) is similar in many ways to a DSMR head, so much of the preceding analysis is applicable to this discussion. The sense current of an MR film develops a magnetic field that magnetizes an adjacent magnetic film; the flux of the adjacent film couples through the MR layer and magnetizes it in an opposing direction. If the SAL is electrically insulated from the rest of the structure, no current flows through it. The maximum induced bias field on an *unshielded* MR layer is achieved when the SAL saturates, and there can be no additional biasing from current flow. Saturation magnetization and thickness of soft adjacent and MR layers are chosen to saturate the SAL and establish a useful MR bias level. The MR layer is chosen to provide adequate signal and reasonable I^2R Joule heating at the operating current. A bias angle [see (5.34)] of about 45° establishes a useful design center (see Bertram, 1994, p. 175). Using numerical techniques, Smith (1987) analyzed the unshielded SAL-biased MR head with perfect insulation between layers, and Bhattacharyya and Simmons (1994) included shields in their transmission line analysis of the insulated SAL biasing technique.

Figure 5.18 defines the shielded SAL–MR geometry; it is essentially the shielded DSMR head with an isolated SAL (1) coupled magnetically with the MR layer (2) through which a current (I) flows. The equations for a DSMR bias field (5.19a,b) are modified with $I_1 = 0$ and $I_2 = I$ to read

$$H_1(I) = \frac{-2\pi I}{h} \left[\frac{2g_1 + t_1}{G_{ss}} \right], \tag{5.29a}$$

$$H_2(I) = \frac{2\pi I}{h} \left[\frac{g_2 - (g + t_1 + g_1)}{G_{ss}} \right]. \tag{5.29b}$$

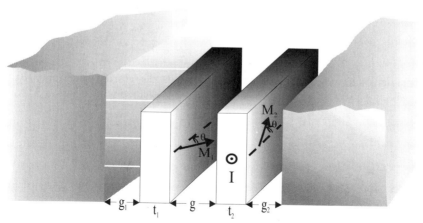

FIGURE 5.18 Geometry of shielded SAL-biased MR head with insulating spacer.

134 AMR HEAD DESIGN AND ANALYSIS

Bias Profiles for SAL and MR Layer Magnetization profiles are calculated from the DSMR relations (5.21a,b) with a modification that includes the influence of $H_2(I)$ on the MR layer (2) as well as the bias flux coupled from the SAL (1). The characteristic flux decay length λ for SAL–MR coupling is found with (3.42), and (5.24) is used for the flux decay length λ_{sh} for MR–shield coupling and its biasing influence on M_2. Saturation of the SAL is forced by setting $M_{2s}t_2 > M_{1s}t_1$ (it is assumed all flux of the SAL couples through the MR layer); (3.49) is used in normalized form to describe the magnetization bias profile for a saturated SAL. That is,

$$m_1^{\text{bias}}(y, I) = \frac{H_1(I)}{H_{1k}}\left[1 - \frac{\cosh[(|y| - |y_s|)/\lambda]}{\cosh[(h/2 - |y_s|)/\lambda]}\right]. \quad (5.30)$$

The MR layer carries the SAL flux and the additional biasing from magnetic images in each shield; under the assumption that the shield image bias flux returns to the shields without disturbing the SAL bias, the normalized profile for film 2 becomes

$$m_2^{\text{bias}}(y, I) = -\frac{M_{1s}t_1}{M_{2s}t_2}m_1^{\text{bias}}(y, I) + \frac{H_2(I)}{H_{2k}}\left[1 - \frac{\cosh(y/\lambda_{sh})}{\cosh(h/2\lambda_{sh})}\right]. \quad (5.31)$$

Saturation Current of the SAL The assumption stated above is compatible with the coupled film analysis in Chapter 3 where flux travels from the MR layer, passes through the SAL normal to its plane (a direction for which the permeability is 1.0), continues to the shield, and travels up the shield over to the MR layer, where flux closure is achieved. The active region at the lower end of a saturated SAL stripe will conduct a portion of the signal flux, and thus the location of SAL saturation is modulated up or down depending on the signal flux polarity. The saturation location y_s is defined by (3.47b) with the provision $H_1(I) \geq H_{1S0}$ [see (3.48)]. Here, Δm is the normalized SAL magnetization at the lower boundary $(-\frac{1}{2}h)$ and is calculated with (5.23a). As shown in Chapter 3, Δm shifts the saturated location to

$$y_S(I) = -\frac{h}{2} + \lambda \cosh^{-1}\left[\frac{|H_1(I)| - \Delta m\, H_{1k}}{|H_1(I)| - H_{1k}}\right]. \quad (5.32)$$

With signal flux typical for a 3.0-Gbit/in.^{2} AMR design, the saturation location shifts about ± 150 Å. Saturation threshold current is defined with (3.48) and (5.29a):

$$I_{1S0} = \frac{H_{1S0}hG_{ss}}{2\pi(2g_1 + t_1)}. \quad (5.33)$$

Profiles of $m_1(y, I)$ and $m_2(y, I)$ for $M_{1s} = M_{2s} = 800$ emu/cm^3, $t_2 = 125$ Å, and $t_1 = 125, 100, 62$ Å are plotted in Figs. 5.19a,b; the other parameters are $g_1 = g_2 = 750$ Å, $g = 75$ Å, $W = 1.25$ μm, $h = 0.80$ μm, $H_{1k} = H_{2k} = 4.0$ Oe, $A_1 = A_2 = 1 \times 10^{-6}$ erg/cm, and $I = 8.0$ mA (with I_{1S0} values of 1.9, 1.8, and 1.6 mA). The normalized profiles for the SAL layer (m_1) are not identical because

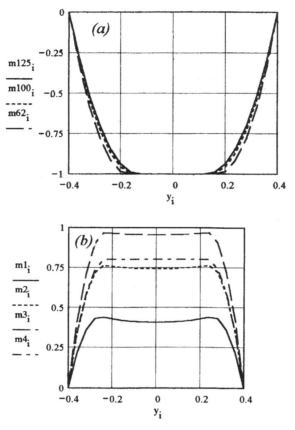

FIGURE 5.19 (a) Normalized magnetization of a 125-Å SAL with the adjacent MR layer thicknesses equal to 125, 100, and 62 Å. (b) Magnetization profiles m_1, m_2, m_3 of the MR layer whose thicknesses are 125, 100, and 62 Å, respectively. The flat magnetization profile $m4$ ($\simeq 0.8$) is calculated with no gap asymmetry and a SAL thickness of 100 Å.

the decay length is proportional to $t_1^{1/2}$; in this example, λ is 0.441, 0.422, and 0.385 μm, including exchange energy. Figure 5.19b reveals the influence of self-biasing on the MR layer; with $g_1 = g_2$, the MR layer is off center by a distance $\frac{1}{2}g = 37.5$ Å toward the right-hand shield, so $H_2(I = 8 \text{ mA}) = -3.0$ Oe, which slightly reduces the positive bias arising from the SAL. The fourth curve of Fig. 5.19b ($m4_i$, which is flat for much of the stripe) is calculated for an MR layer with no gap asymmetry ($g_1 = 713$ Å, $g = 75$ Å, and $g_2 = 788$ Å) and a SAL thickness $t_1 = 100$ Å. The stripe height for minimum I_{S0} [from (3.52)] is $h_0 = 3.0112\lambda = 1.33, 1.27, 1.16$ μm for each thickness. This is nearly equal to the track width ($W = 1.25$ μm) required for this design example, so the necessary shape anisotropy ($h/W \sim 0.7$) forces h to values below h_0. The numerical analysis by Smith (1987) of an unshielded SAL/MR head compares well with the closed-form approach taken here. He tabulates the saturation current for stripes varying from 2.5 to 20 µm for an unshielded SAL/MR design where $t_1 = 340$ Å, $t_2 = 400$ Å, $g = 1000$ Å, with $H_k = 5$ Oe and $M_s = 800$ emu/cm^3 in both films. Using (3.42) to estimate the decay length, one obtains $\lambda = 1.947$ µm, which gives $h_{opt} = 5.86$ µm for his design. Smith's data are plotted in Fig. 5.20, and the analytical estimate of 5.86 µm agrees well with the numerical result.

As in the DSMR analysis, in the active region $-\frac{1}{2}h \leq y \sim -\frac{1}{4}h$ of an otherwise saturated SAL, this layer conducts flux in parallel with the MR layer, and signal flux (5.5) can be split approximately in proportion to the relative thickness (t_1 or t_2 divided by $t_1 + t_2$) of each layer. The signal magnetization of each layer is estimated with (5.23a,b) in the same manner as was done for the DSMR head. Profiles for bias with positive and negative signals for both layers are shown in Figs. 5.21a,b, with $M_r \delta = 0.6$ memu/cm^2, $d + a = 0.10$ µm, $t_1 = 100$ Å, $t_2 = 125$ Å, and $g = 75$ Å.

Bias Angle With the insulated SAL, a signal voltage exists only across the MR layer and signal is computed with (5.1) and (5.7), including appropriate modifications. The average normalized bias is obtained from numerical integration of (5.31) for $m_{2y} = \sin\theta_b$, where θ_b is the "bias angle," which is normally set at approximately 45° to maximize the useful operating range for signal excursions about the bias point:

$$m_{2av}^{bias}(I) = \sin\theta_b = \frac{1}{n+1}\sum_i^n m_2(y_i, I), \quad (5.34)$$

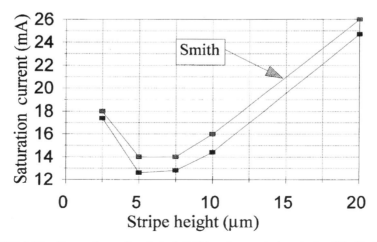

FIGURE 5.20 Comparison of analytical (3.51) and numerical analysis of Smith (1987) for the saturation current of an MR element versus stripe height.

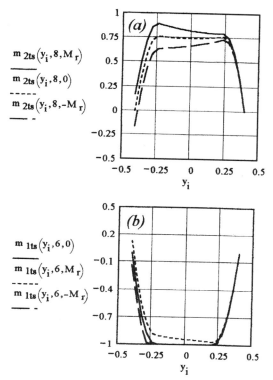

FIGURE 5.21 Normalized bias ± signal profiles for (a) MR layer (125 Å) and (b) SAL (100 Å).

and y_i can be broken into equal increments (about 20–40) along the stripe height h. The average normalized signal magnetization, taken from (5.23b), is found analytically,

$$\Delta m_2^{\text{sig}}(M_r, \delta) = \frac{\phi(y_e)t_2}{4\pi M_{2s}W(t_1+t_2)^2} \frac{\tanh(h/2\lambda_{\text{sh}})}{h/\lambda_{\text{sh}}}$$
$$\simeq \frac{\phi(y_e)t_2}{8\pi M_{2s}W(t_1+t_2)^2} \quad \text{for } h/\lambda_{\text{sh}} \text{ small,} \qquad (5.35)$$

and the total average normalized magnetization is $m_{\text{avg}}(I, M_r, \delta) = m_{2\text{av}}(I) \pm \Delta m_2^{\text{sig}}(M_r, \delta)$. As before, this average cannot exceed saturation of the MR layer. The transfer curve is $\langle \cos^2\theta \rangle = 1 - \langle \sin^2\theta \rangle = 1 - m_{\text{avg}}^2$, which is shown in Figs. 5.22a,b as a function of current with fixed $M_r \delta$ and at fixed current with varying medium $M_r \delta$, respectively. The influence of MR current on self-biasing is seen in the upward curvature of Fig. 5.22a as current increases beyond 7 mA.

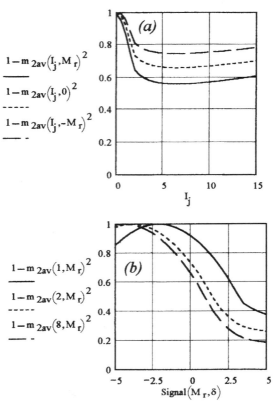

FIGURE 5.22 Normalized transfer curves of an MR element as a function of (*a*) current (mA) with signal $= 0 \pm M_r \delta$ (fixed M_r) and (*b*) $M_r \delta$ for several values of fixed bias current.

Output Signal and Asymmetry With the maximum available AMR effect $\Delta\rho_0 = 0.64$ μΩ-cm, $\Delta R_0 = \Delta\rho_0 W/t_2 h = 0.80$ Ω, and sensor resistance $R_0 = \rho_0 W/(t_2 h) = 40$ Ω at room temperature. The direct current (DC) signal and asymmetry, defined with (5.11a,b,c,d) and (5.12), are plotted in Figs. 5.23*a,b* with and without heating for $M_r \delta = 0.60$ memu/cm^2 and all other parameters unchanged. The unheated signal curve ($V2_j$) is slightly sublinear with increasing current because of self-biasing; with a centered MR element the signal is linear in current above I_{S0} (5.33). Average stripe temperature is calculated with (5.14) (also see the thermal analysis in Chapter 2), and this result is applied to the thermal dependence of the AMR effect; calculations of temperature rise and the AMR effect with current are plotted in Figs. 5.24*a,b*. An advantage of the insulated SAL biasing technique arises from the nearly constant amplitude asymmetry with sense/bias current. A disadvantage of this design is it uses dielectric layers containing materials that may degrade the NiFe MR layer at elevated temperatures, and device function could suffer after long periods of use. A discussion of these reliability issues is deferred to Chapter 8.

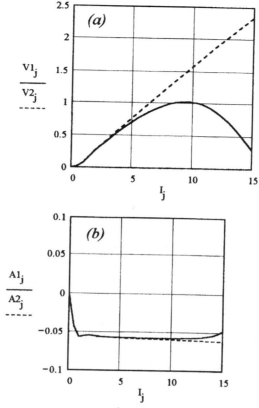

FIGURE 5.23 (*a*) AMR signal (mV) and (*b*) asymmetry with and without heating plotted as a function of bias current (mA).

SAL BIASING WITH A CONDUCTING SPACER

The SAL-biased transducer with a conducting spacer is a practical device compatible with long-term reliability and high-volume manufacturing, and for these reasons it is found in the majority of AMR-based hard disk drives produced since 1991. For long-term reliability, spacer materials are selected for thermal stability at elevated temperatures, and this consideration has led to high-resistivity metallic spacers (e.g., tetragonal phase of tantalum and nonmagnetic NiFeCr) that do not degrade the NiFe sensor layer. The electrical resistivity of useful metallic spacers is in the range of 100–200 $\mu\Omega$-cm in thin films, and thus current is shunted from the MR layer into the spacer and SAL, and the readback signal reduces accordingly. The shunted current generates additional biasing for the MR layer, so it is possible to operate in an overbiased (positive asymmetry) mode. Self, shunt, and SAL bias techniques have been examined individually in the first half of this chapter, and their combined

140 AMR HEAD DESIGN AND ANALYSIS

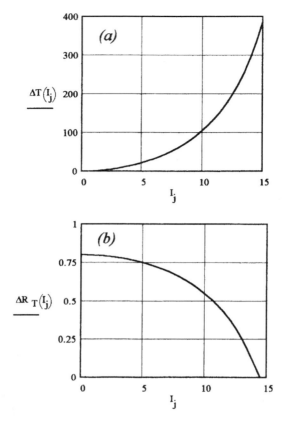

FIGURE 5.24 (*a*) Temperature rise (K) and (*b*) AMR effect $\Delta R(\Omega)$ plotted as a function of bias current (mA).

influence is now studied. The resistivity and AMR effect of the SAL become very important when the SAL is electrically connected to the MR layer because the signal responses of the SAL and MR layer are in opposition and the SAL shunts some of the sense current. This is in sharp contrast to the insulated SAL where the device designer will primarily be concerned with the thickness and M_s of the soft layer.

Bias Fields

The geometry for SAL biasing with a conductive spacer is taken from Fig. 5.18; the spacer g between magnetic films is a nonmagnetic material with resistivity ρ_g of about 100–200 $\mu\Omega$-cm. The resistance of the three conducting layers is described by the relation (2.25b), which is repeated here for convenience:

$$R_t = \frac{R_1 R_2 R_3}{R_1 R_2 + R_1 R_3 + R_2 R_3}, \quad \text{where } R_i = \frac{\rho_i}{t_i}\frac{W}{h}. \tag{5.36}$$

Current flow in any layer is found with the relation $I_i = I_t R_t / R_i$ and the applied fields on the SAL (1) and MR (2) are derived from (4.14a,c). The SAL and MR bias fields, respectively, are

$$H_1(I) = \frac{2\pi I}{h} \left[\frac{(R_t/R_1)(g_2 + t_2 + g - g_1) - (R_t/R_2)(2g_1 + t_1) - (R_t/R_3)(2g_1)}{G_{ss}} \right] \tag{5.37}$$

and

$$H_2(I) = \frac{2\pi I}{h} \left[\frac{(R_t/R_2)(g_2 - g - t_1 - g_1) + (R_t/R_1)(2g_2 + t_2) + (R_t/R_3)(2g_2)}{G_{ss}} \right]. \tag{5.38}$$

The first term in each of these equations describes shield biasing, and the remaining terms describe shunt biasing from adjacent layers. With very large R_1 and R_3 these relations collapse to those of the insulated SAL (5.29a,b), as expected. The magnetization profiles are defined by the same equations used with the insulated SAL, namely (5.30) and (5.31). The characteristic length for flux decay with MR/SAL coupling (λ) is defined with (3.42) for dissimilar films, which includes the exchange energy of each layer, and the decay length for shield biasing (λ_{sh}) of the MR layer, found using (4.19) with appropriate modifications, is

$$\lambda_{sh} = \sqrt{\frac{\mu_2 t_2 g_2 (g + t_1 + g_1)}{G_{ss}}}. \tag{5.39}$$

SAL and MR Layer Saturation

The field H_{1S0} for onset of SAL saturation (without signal) is estimated with (5.32) taking y_s and Δm equal to zero and solving for the field $|H_1(I)| = H_{1S0}$. The result is

$$H_{1S0} = H_{1k} \frac{\cosh(h/2\lambda)}{\cosh(h/2\lambda) - 1}, \tag{5.40}$$

and saturation current I_{1S0}, derived from (5.37) and (5.40), is

$$I_{1S0} = \frac{H_{1S0} h G_{ss}}{2\pi |(R_t/R_1)(g_2 + g + t_2 - g_1) - (R_t/R_2)(2g_1 + t_1) - (R_t/R_3)(2g_1)|}, \tag{5.41}$$

where the magnitude bars ($|\cdots|$) in the denominator define a positive value for I_{1S0}. The SAL magnetizes according to (3.49a,b) where the saturation profile in the lower half of the stripe shifts with signal and the upper half remains fixed. The SAL bias profile is of greater interest because it is the bias flux that couples through the MR layer across the spacer gap g and establishes most of the MR bias level. Signal flux goes from the medium through the SAL and MR layer and flows through the shields back to the medium. The normalized bias magnetization profile of the SAL is calculated with (5.30), and the total magnetization (including signal) is

$$m_{1t}(y, I, M_r, \delta) = m_1^{\text{bias}}(y, I) + \Delta m_1^{\text{signal}}(y, M_r, \delta). \tag{5.42}$$

Signal magnetization for the SAL is calculated with (5.5) and (5.23a) normalized by M_{1s}. Some logical conditions must be placed on m_{1t}: First, when calculating the total magnetization, it can never exceed saturation, so $|m_{1t}| \leq 1.0$, and second, a signal field (flux) of appropriate polarity and magnitude can take saturated regions out of saturation. The average magnetization is found by numerical integration; that is,

$$m_{1\text{av}}(I, M_r) = \frac{1}{n+1} \sum_i^n m_{1t}(y_i, I, M_r). \tag{5.43}$$

The MR layer bias is described by (5.31), where the first term is the SAL demagnetization flux coupled through the MR layer and the second term with $H_2(I)$ includes current bias from the SAL, the spacer, and shield (self) biasing [see (5.38)]. The average bias and bias angle are calculated using (5.34) with appropriate substitutions. As before, the total (bias + signal) normalized magnetization cannot exceed 1.0 (saturation), and a saturated MR layer can be pulled out of saturation by a negative signal field. The field at onset of MR saturation H_{2S0} is estimated with (5.31) at $y = 0$ and $m_1^{\text{bias}}(0, I) = -1$; with no signal present, the result is

$$H_{2S0} = H_{2k}\left(1 - \frac{M_{1s}t_1}{M_{2s}t_2}\right) \cdot \left[\frac{\cosh(h/2\lambda_{\text{sh}})}{\cosh(h/2\lambda_{\text{sh}}) - 1}\right]. \tag{5.44}$$

With a signal present, the total magnetization is the sum of (5.23b) (normalized by M_{2s}) and (5.31),

$$m_{2t}(y, I, M_r, \delta) = m_2^{\text{bias}}(y, I) + \Delta m_2^{\text{signal}}(y, M_r, \delta), \tag{5.45}$$

and the logical conditions placed on the SAL are valid for the MR layer as well. That is, $m_{2t} \leq 1.0$ and a negative signal field can pull a saturated MR film out of saturation. With a positive signal impressed, the saturation field reduces, although a simple closed-form solution for H_{2S0} has not been found for this situation. As an approximation, the bias term $M_{1s}t_1/M_{2s}t_2$ in (5.44) is replaced with $M_{1s}t_1/M_{2s}t_2 + \Delta m_2$, where $\Delta m_2 \simeq 0.25\, \Delta M_2^{\text{sigmal}}(y=0)/M_{2s}$ [using (5.23b) for the signal]. The saturation field with signal then reads

$$H_{2S0}(\text{with signal}) \simeq H_{2k}\left(1 - \frac{M_{1s}t_1}{M_{2s}t_2} - \frac{\Delta m_2}{4}\right)\left[\frac{\cosh(h/2\lambda_{\text{sh}})}{\cosh(h/2\lambda_{\text{sh}}) - 1}\right], \tag{5.46}$$

and this equation becomes a useful design tool for defining the sense/bias current at which signal asymmetry becomes positive (onset of MR saturation). This current is found by equating $H_2(I)$ in (5.38) with (5.46) and solving for I. That is,

$$I_{2S0}(h) = \frac{H_{2S0}(h)hG_{ss}}{2\pi[(R_t/R_1)(2g_2 + t_2) + (R_t/R_2)(g_2 - g - t_1 - g_1) + (R_t/R_3)(2g_2)]}. \quad (5.47)$$

The optimal stripe height $h_0 = 3.0112\lambda_{sh}$ is found by minimizing (5.47) with respect to h, but shape anisotropy requires $h \sim 0.7W$, and thus high track densities preclude increasing h to an optimal value (about 3 µm).

MR Layer Transfer Curves

The MR transfer curve $\langle\cos^2\theta\rangle = 1 - \langle\sin^2\theta\rangle$ is found with the total normalized magnetization averaged along the stripe (y) direction, where

$$\langle\sin\theta\rangle = m_{2av}(I, M_r) = \frac{1}{h}\int_{-h/2}^{h/2} m_{2t}(y, I, M_r)\, dy. \quad (5.48)$$

Figure 5.25 shows the MR transfer curves versus sense/bias current for positive, zero, and negative signals, and Fig. 5.26 shows transfer curves for fixed currents and variable input signals. Saturation of the SAL is apparent in Fig. 5.25 at $I \simeq 2$ mA,

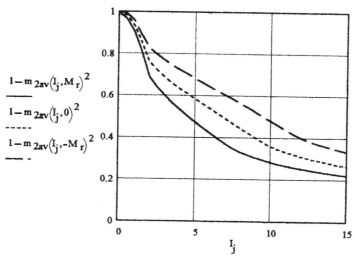

FIGURE 5.25 Normalized transfer curves for SAL-biased AMR head with conductive spacer plotted as a function of bias current for positive, zero, and negative signal flux.

144 AMR HEAD DESIGN AND ANALYSIS

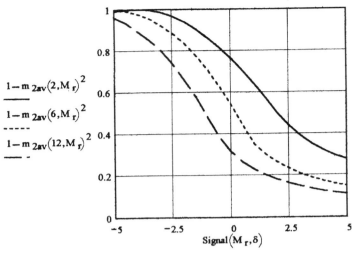

FIGURE 5.26 Transfer curves for fixed current and variable input signals (memu/cm^2).

and MR saturation occurs more gradually at about 8, 10, and 12 mA for positive, zero, and negative signals.

Bias Curves: Signal and Asymmetry Versus Bias Current and MR/SAL Moment Ratio

Readback signal is calculated from the sum of the SAL and MR responses,

$$V_{\text{signal}}(I, M_r, \delta) = I_1 \, \Delta R_1[1 - m_{1\text{av}}^2(I, M_r, \delta)] + I_2 \, \Delta R_2[1 - m_{2\text{av}}^2(I, M_r, \delta)], \quad (5.49)$$

where ΔR_1 and ΔR_2 are defined as $\Delta R_i = (\Delta \rho_i W / h t_i) \exp(\alpha_{\Delta \rho} \, \Delta T)$. The AMR effect of the SAL subtracts from the MR layer output, and thus identical films would produce zero output signal. Figures 5.27a,b show calculated signals (positive, negative, and peak to peak) and signal asymmetry [calculated with (5.12)] as a function of sense/bias current for a 3.0-Gbit/in.^{2} design with a NiFeRh SAL(t_1) = 80 Å, $g = 75$ Å, and NiFe MR(t_2) = 120 Å; the MR/SAL moment ratio, defined as $[M_s t]_{\text{MR}}/[M_s t]_{\text{SAL}}$, is $800 \times 120 / 724 \times 80 = 1.66$ for this example. The SAL properties are taken from Table 2.1, Chapter 2, and the spacer is Ta with a resistivity $\rho = 160 \, \mu\Omega$-cm. Negative peak amplitude is greater than positive amplitude until the MR film enters saturation, at which point the amplitudes are equal [see (5.47)]. With increasing bias current, negative amplitude decreases while positive amplitude continues to increase to a maximum (limited by heating in this example), so asymmetry rapidly changes from negative to positive values. With good design practice, the desired operating current is placed somewhat below the transition between negative and positive asymmetry; the small negative asymmetry

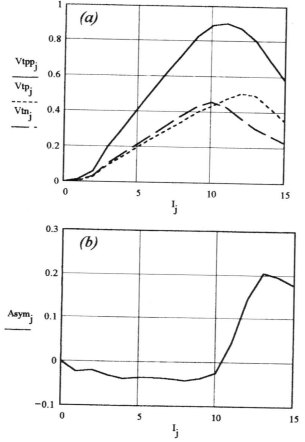

FIGURE 5.27 (*a*) Output signal (mV) and (*b*) asymmetry as a function of bias current (mA). Heating is included in the calculations.

($\sim -4\%$) that exists at this bias level is fairly constant with reasonable variations in operating current.

Plots of readback signal versus bias current are called *bias curves*, and useful diagnostic information about device behavior is revealed by their shapes. The initial behavior is quadratic in bias current and shows buildup of bias from SAL–MR coupling until the SAL is saturated. This is followed by a sublinear rise in signal, then a maximum occurs followed by either an inflection point and nearly constant signal, or by a gradual loss of signal (depending on details of the device design, materials properties, and construction). The signal maximum arises where reduction of the AMR effect with device heating is matched by signal increase with higher current, or by MR layer saturation, or perhaps by both effects. The behavior is quite complicated and is best analyzed with tools such as those developed in this book. For example, Fig. 5.28 shows peak-to-peak signal bias curves for two different

146 AMR HEAD DESIGN AND ANALYSIS

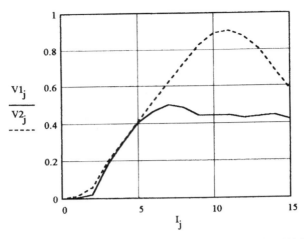

FIGURE 5.28 Comparison of signal bias curves for two devices: V1 has a thick SAL (100 Å) and high-thermal-conductivity insulating gaps; the maximum in signal arises from overbiasing. V2 has a thinner SAL (80 Å) and low-thermal-conductivity insulating gaps; the maximum in signal arises from device heating.

devices: The first curve maximizes at about 7 mA bias current then levels out at higher currents, while the second curve maximizes at 11 mA with a monotonic decrease thereafter. The first device has a thick-SAL (100-Å), high-thermal-conductivity material in the insulating gaps $g_{1,2}$ [$\kappa_0 = 10$ W/m-K; see (2.26) and (2.45)], and the maximum signal is the result of overbiasing. The second head has a thinner SAL (80-Å), low-thermal-conductivity gap material ($\kappa_0 = 1.0$), and the maximum is caused by device heating (ΔT is 128 K at 11 mA and 304 K at 15 mA).

Comparison between Experiments and Calculations

Analytical tools are useful to the extent they are based on sound physics and agree reasonably well with experiment. With the tools developed here, model calculations of bias curves for AMR heads require specification of 31 fundamental independent variables or parameters. These quantities fall into five broad categories:

- Device geometry: seven variables (parameters)
- Magnetic properties: six variables (parameters)
- Electrical properties: eight variables (parameters)
- Thermal properties: six variables (parameters)
- Medium and head–medium interface: three variables, one parameter

The distinction between variables and parameters is justified by observing that a quantity often serves different purposes in mathematical computations. Stripe height (h), for example, influences bias curves and it serves as a parameter when computing signal as a function of bias current. On the other hand, h serves as a variable when computing signal at a fixed bias current as a function of stripe height.

The experimental data base for SAL-biased AMR heads is constructed from dynamic test results at 35 kfci density (16.67 Mbits/sec at 476 in./sec) and 8.0 mA bias current for over 800 devices built from 24 wafers. In this experiment, the following design parameters were varied:

- SAL layer (NiFeRh) thickness $t_1 \simeq 58, 66, 80$ Å
- MR layer (NiFe) thickness $t_2 \simeq 78, 120$ Å
- Shield-to-shield gap $G_{ss} \simeq 0.16, 0.18, 0.20$ μm
- MR track width (magnetic) $W \simeq 1.0, 1.25, 1.50$ μm

The Ta gap between MR and SAL was held constant at 75 Å thickness, and stripe heights were selected by resistance measurements for a nearly constant $h \simeq 0.96$ μm. Heads were flown at a nominal magnetic spacing $d = 650$ Å over a medium with the properties $M_r = 350$ emu/cm^3, $\delta = 200$ Å, $H_c = 2650$ Oe, $S^* = 0.87$, and estimated transition parameter [see (1.18)] $a = 400$ Å. Each experimental data point represents the average of 25–40 devices at a given design point. As an example, the MR/SAL thickness ratio was set at a minimum of 78 Å/80 Å = 0.98 and a maximum of 120 Å/58 Å = 2.07, track widths ranged form 1.0 to 1.5 μm, and the shield-to-shield gap length ranged from 0.16 to 0.20 μm for each thickness ratio. The MR/SAL moment ratio is found by multiplying the thickness ratio by $M_s^{MR}/M_s^{SAL} = 800/724 = 1.11$. Gap asymmetry ($A_g$), defined by the relation

$$A_g = \frac{g_1 + t_1 + g - g_2}{g_1 + t_1 + g + g_2}, \qquad (5.50)$$

was held to 0.06 ± 0.01 throughout the experiment. The insulating material for g_1 and g_2 was sputtered Al$_2$O$_3$ with $\kappa_0 = 1.0$ W/m-K. Scatter diagrams for peak-to-peak signal (calculated) versus signal (experiment) and asymmetry (calculated) versus asymmetry (experiment) are given in Figs. 5.29a,b with lines of slope equal to 1.0 drawn to aid the eye. Correlation coefficients (product moment correlation $R = \text{cov}(x, y)/[(\text{var } x)(\text{var } y)]^{1/2}$) are 0.8398 for the signal and 0.8078 for the asymmetry diagrams, respectively. On average, the calculated signal is greater than experiment and calculated asymmetry is less. In view of the 31 independent variables required to calculate the output of AMR heads and the extreme difficulty (and expense) of accurate measurements and experimentation, the correlations shown above, while not excellent, are acceptable and the essential correctness of the physics of AMR devices can be assessed.

Calculated and experimental bias curves are compared in Fig. 5.30a (peak-to-peak signal) and Fig.5.30b (positive and negative signals) for a device with MR/SAL thickness ratio of 0.98 (78 Å/80 Å.) The experimental results contain wide-band white noise that is not included in the calculations, and thus at low bias

148 AMR HEAD DESIGN AND ANALYSIS

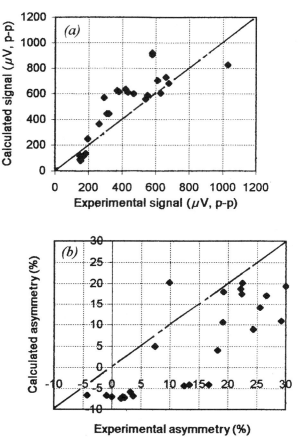

FIGURE 5.29 Comparison of calculated and experimental output data (800 devices from 24 wafers). (*a*) Signal (μV, p-p) scatterplot of calculation versus experiment; correlation coefficient is 0.8398. (*b*) Asymmetry (%) scatterplot of calculation versus experiment; correlation coefficient is 0.8078.

currents the calculated signal is below the noise floor of the actual device and the experiments do not reveal important details of device behavior at low levels of signal. (The test equipment uses peak detection circuitry, so low signal levels are not discriminated from the peak value of noise spikes.) For example, from $I = 0$ to $I = 2$ mA in Fig. 5.30*a*, the calculations (Vtpp) show a *negative* peak-to-peak signal followed by a reversal in polarity between 2 and 3 mA, whereas the experimental result (Evpp) shows a noise floor rising from about 0.07 mV (p-p) up to 0.15 mV (p-p). The calculated behavior can be understood by examining the definitions of positive and negative signals given in (5.11b,c), looking at a transfer curve at low bias current (see, e.g., Fig. 5.22*b*), and seeing that negative input flux drives the

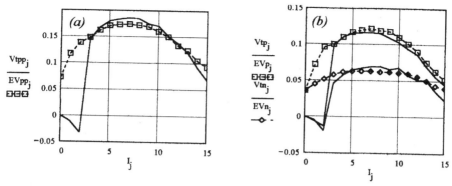

FIGURE 5.30 Comparison of calculated and experimental bias curves for a SAL-biased AMR head with conductive gap with MR = 78 Å and SAL = 80 Å: (*a*) peak-to-peak signal (mV) and (*b*) positive and negative signals (mV, 0-p).

signal voltage over the top of the transfer curve and down to a level *below* the bias point at zero flux. At bias currents above 3 mA, calculation and experiment are in good agreement; the loss in signal above 7 mA arises from device heating and magnetic saturation. At 10 mA the calculated negative voltage (Vtn) is 65 µV, the experimental result (EVn) is 60 µV, and the results for positive signal (Vtp and EVp) are 102 and 110 µV, respectively.

At a thickness ratio of 2.07 (120 Å/58 Å) the results of calculation and experiment are also in reasonable agreement. This is shown in Figs. 5.31*a* (peak to peak) and 5.31*b* (positive and negative peaks) where the largest discrepancy is with the calculated negative peak signal at bias currents between 12 and 14 mA. Calculations show that positive and negative peaks are each 470 µV at 13.6 mA, where the

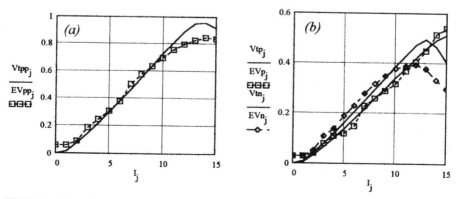

FIGURE 5.31 Comparison of calculated and experimental bias curves for a SAL-biased AMR head with conductive gap with MR = 120 Å and SAL = 58 Å: (*a*) peak-to-peak signal (mV) and (*b*) positive and negative signals (mV, 0-p).

experimental result is 400 µV at $I_{2S0} = 12$ mA. The experimental signal goes through a maximum of 850 µV (peak to peak) at 14 mA bias; some of the loss at higher current arises from device heating and some from magnetic saturation. Computations show a maximum signal of 958 µV at 14 mA (12.7% higher than experiment) with heating included; without heating, computations show 1250 µV at 15 mA, dropping to 1056 µV as a result of saturation at higher currents.

MR/SAL Moment Ratio and the Operating Current

The MR/SAL moment ratio and device operating current are basic design considerations. Bias curves for signal and asymmetry are given in Figs. 5.32a,b, respectively, for moment ratios of 1.0, 1.2, 1.5, and 2.0 (MR layer is constant at 120 Å and the SAL thickness is varied; $M_s = 800$ emu/cm³ for this hypothetical SAL). The general trend is clear: A thick SAL (low moment ratio) shunts lots of

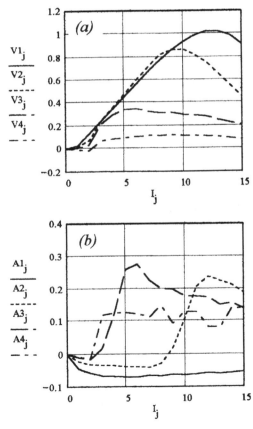

FIGURE 5.32 Calculated bias curves for various MR/SAL moment ratios: V1 = 120/60, V2 = 120/80, V3 = 120/100, V4 = 120/120: (*a*) signal (mV, p-p). (*b*) asymmetry.

current, saturates the MR layer at low bias current, and substantially reduces the output signal. At high moment ratios (thin SAL), little shunt current flows in the SAL, the MR saturates at higher bias currents, and the signal is not reduced much. In a range of operating currents (roughly 8–12 mA), signal asymmetry goes from small negative to large positive values for the selected parameters. [Recall that readback signal and asymmetry scale with the input signal field, which goes approximately as $4M_r \delta/(d+a)$. The physically correct relation for signal flux is given in (1.7).]

Figures 5.33a–d show additional details to assist in the design effort: Fig. 5.33a plots the MR layer saturation current (with positive signal flux) versus moment ratio, Fig. 5.33b shows how the asymmetry (at I values of 8, 10, and 12 mA) switches from positive to negative with increasing moment ratio, Fig. 5.33c plots the maximum peak-to-peak signal (which occurs at about 12 mA for this scenario) for any moment ratio from 1.0 to 2.0, and Fig. 5.33d shows how the bias angle (at $I = 8$ mA) depends on the moment ratio. A bias angle of about 45° provides reasonable negative asymmetry and dynamic range, whereas above 50° the device is at or above the saturation threshold and below 40° the asymmetry may be more negative than desired. From Figs. 5.33b,d it appears that overbiasing can be avoided by designing with a moment ratio of 1.7 or greater.

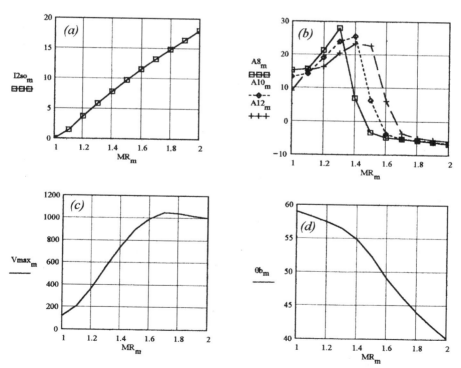

FIGURE 5.33 Calculations of MR head attributes versus moment ratio: (a) MR layer saturation current; (b) asymmetry (%) for $I = 8, 10, 12$ mA; (c) signal (mV, p-p); (d) bias angle at $I = 8$ mA.

Operating Bias Current and the Melting Current The operating bias current will be determined by various requirements for readback signal, asymmetry, sensor resistance, and allowable temperature rise (long-term reliability). As a first-pass estimate, one can use (2.28) for the thermal resistance R_τ and solve for current from the relation $\Delta T \simeq I^2 R R_\tau$ to obtain

$$I \simeq \sqrt{\frac{\Delta T}{RR_\tau}} \quad \text{A.} \tag{5.51}$$

With alumina gaps $g_1 \simeq g_2 = 0.08$ μm ($\kappa_0 = 1.0$ W/m-K) in a 3-Gbit/in.^{2} design ($W = 1.25$ μm, $H = 0.9$ μm), 50 K allowable rise, and a sensor resistance of 25 Ω, the nominal operating current becomes nearly 8 mA; output signal is 750 μV (or 600 μV/μm of track width) and asymmetry is close to −6%. A moment ratio greater than 1.5 (see Fig. 5.33b) keeps the asymmetry negative for operating currents greater than 8 mA. The melting current [see the discussion leading to (2.50)] is

$$I_m \simeq \left[\frac{T_m \kappa_0 (g_1 + g_2) t_f h^2}{g_1 g_2 \rho_0 (1 + \alpha T_m)}\right]^{1/2}, \tag{5.52}$$

which is $I_m = 19.1h$ (in milliamperes, with h in micrometers) with the parameters given above; for alumina gaps $I_m \simeq 17.2$ mA with $h = 0.9$ μm. Because of element heating, output signal goes through a maximum level with reductions in h. With the NiFeRh SAL design given above, signal versus h is plotted in Fig. 5.34a; at $h = 0.4$ μm the signal is maximum at 6.0 mA, which is just below the melting current of 7.6 mA. Largest output occurs for $h = 1.0$ μm at a current of 15 mA ($I_m = 19.1$ mA), and the temperature rise, shown in Fig. 5.34b, is about 200 K; for h in the range of 0.4 to 0.9 μm, the temperature fluctuations arise from computational artifacts related to uncertainty in the current at which the signal maximizes.

Increasing the sensor resistance (thinner SAL, higher SAL resistivity) requires improved thermal resistance; otherwise the temperature rise exceeds requirements and current must be reduced with no payback in improved signal or reliability. In the design example given above, the operating current can be increased with AlN gap material by the factor $[\kappa_0(\text{AlN})/\kappa_0(\text{Al}_2\text{O}_3)]^{0.5} = 3^{0.5}$ to 13.5 mA and maintain $\Delta T \simeq 50$ K; output nearly doubles (1450 μV or 1160 μV/μm of track width), and asymmetry can be held to about −6% if the MR/SAL moment ratio is ∼2. A design based on a SAL with high M_s (> 800 emu/cm^3) and high resistivity (∼100 μΩ-cm), along with high-thermal-conductivity gap material ($\kappa_0 = 3.0$ W/m-k), offers improved signal at higher operating currents when compared with other SAL materials (e.g., NiFeRh with $M_s = 725$ emu/cm^3, $\rho = 66$ μΩ-cm). Table 2.1 lists $Co_{90}Zr_9Cr_1$ with high magnetization (1075 emu/cm^3) and resistivity (95 μΩ-cm), low magnetostriction (2.6×10^{-7}), and an anisotropy field ($H_k = 14$ Oe) which is somewhat high. For a moment ratio of 2.0, sensor resistance increases to about 27 Ω and signal improves 8% to 1560 μV or 1250 μV/μm with larger asymmetry (∼ −8%) at $I = 13.5$ mA. Because less current is shunted by the SAL, the bias

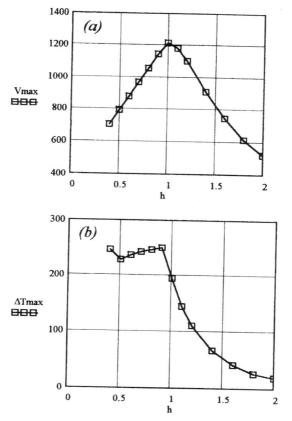

FIGURE 5.34 (*a*) Signal (μV, p-p) and (*b*) temperature rise (K) versus stripe height (μm).

point shifts in the underbiased direction, and asymmetry becomes more negative. The loss of shunt bias can be compensated by moving the MR layer closer to the center of the gap (moving from positive gap asymmetry toward zero or even to negative gap asymmetry). In other words, the design can be fine-tuned by shifting the moment ratio, shunt biasing, and shield biasing budget to an operating point that produces acceptable asymmetry with some room to prevent overbiasing the sensor. In the present design example, this amounts to adjusting gap thicknesses in the 25–50 Å range, which is at the edge of process capability for sputtered films.

Monte Carlo Analysis of Design and Process Control Requirements

This chapter closes with a Monte Carlo analysis of hypothetical SAL-biased AMR heads designed with a conductive spacer for use at 3.0 Gbits/in.² To better understand processing sensitivities, only the head is analyzed here, but it is possible to vary head, disk, and system parameters in a complete analysis. The most important nine head parameters are statistically varied to assess the signal, asymmetry, and temperature variations of the nominal design and to obtain their correlations with the head parameters under "reasonable" process control tolerances. Table 5.1 lists the nominal (mean) head parameters and their standard deviations. The SAL is $Co_{90}Zr_9Cr_1$, and with a Permalloy MR layer, the nominal moment ratio is 1.75. Gap insulation is AlN with $\kappa_0 = 3.0$ W/m-K, and the operating sense-bias current is 13.0 mA. Medium and system properties are $M_r = 350$ emu/cm^3, $\delta = 200$ Å, $d = 650$ Å, and $a = 400$ Å.

Of the thirty-six parameters introduced and defined thus far, variations in nine head parameters and four medium/spacing parameters exert noticeable influence on output signal variations of a SAL-biased AMR recording head. Monte Carlo techniques are nicely suited for analyzing complex systems involving a number of parametric variations over well-defined ranges and for assigning useful guidelines to limits on process control of critical variables. For those readers unacquainted with Monte Carlo analysis, a brief review of the technique is helpful:

1. Create a table of Gaussian random numbers (mean = 0, sigma = 1.0)
2. Assign mean (μ_i) and standard deviation (σ_i) values to $i = 1, 2, \ldots, n$ parameters.
3. Select at random a number N from the Gaussian table.
4. Multiply σ_1 by N and add the result to μ_1; the new parameter is $X_1 = \mu_1 + N\sigma_1$.
5. Repeat steps 3 and 4 for n variables.
6. Enter the set of new parameters X_i ($i = 1, 2, \ldots, n$) into a parameter table.
7. Calculate the responses and place them in a response table.
8. Repeat steps 2–6 for M trials.

TABLE 5.1 Head Parameters for Monte Carlo Analysis

Parameter	Mean	Standard Deviation
SAL thickness, t_1	50 Å	3 Å
Spacer thickness, g	75 Å	3 Å
MR thickness, t_2	120 Å	3 Å
Insulator gap, g_1	800 Å	10 Å
Insulator gap, g_2	800 Å	10 Å
Stripe height, h	1.0 µm	0.10 µm
Magnetic read width, W	1.25 µm	0.10 µm
SAL anisotropy field, H_{k1}	14.0 Oe	2.0 Oe
MR anisotropy field, H_{k2}	4.0 Oe	0.50 Oe

9. Find the mean and standard deviation of responses for M trials.
10. Create histograms and cumulative distributions for the responses.
11. Find the correlation coefficients between responses and each of the n parameters.

A Gaussian random-number table with 1000 entries will be sufficient for $M \sim 500$ trials; for a table of this size, the range in the random number z will be between -3.5 and $+3.5$, approximately. Statistical trends and correlations are well defined for about 350 or more trials. The standard deviation values assigned above are chosen for tutorial purposes, but the tolerances are not unreasonable for a large-volume manufacturing process. One can always argue that tolerances ought to be tighter, but the point of this exercise is to gain insight regarding variations in device behavior and to form judgments about the relative importance of each tolerance. A Monte Carlo of 500 trials is first performed with the variations of Table 5.1 assigned to the nine parameters; the mean responses and standard deviations for signal, asymmetry, and temperature are computed and a table of correlation coefficients is formed. Histograms and cumulative distributions for signal, asymmetry, and temperature rise are shown in Figs. 5.35a–c. To serve as cross-checks on the correlation coefficients, additional runs of 500 trials each are performed for extremely small ($\sigma \sim 0$) standard deviations of selected parameters.

Table 5.2 lists the results of 500 Monte Carlo trials with variations in all nine parameters according to the assignments in Table 5.1. Table 5.3 lists the correlation coefficients $R = \text{cov}(x, y)/[(\text{var } x)(\text{var } y)]^{1/2}$ for the analysis. Table 5.4 shows the results of additional Monte Carlo simulations in which the standard deviation of the nine parameters are either at the values given in Table 5.1 or set to zero; rows 1 through 9 are arranged from largest influence on signal (variations in all nine parameters) down to the smallest influence on the signal (variations in H_{k1} and H_{k2} only). Standard deviations for asymmetry and ΔT are also listed.

With variations in all parameters, the standard deviation (σ_V) for the signal is 194 μV (p-p) or 11.4% of the mean, σ_A (asymmetry) is 3.58% (56% of the mean), and $\sigma_{\Delta T}$ is 13.1 K (24.1% of the mean). Variations in magnetic track width (W) and stripe height (h) (see row 4) exert major influence on signal variations; rows 5 and 6 show that W is more influential than h regarding signal variations. This is followed by variations in the SAL/spacer/MR layer thicknesses (row 7). The influences of insulation thicknesses (row 8) and SAL/MR anisotropy fields (row 9) on signal amplitude are each almost nil in this design. High asymmetry in Fig. 5.35b is dominated by variations in SAL thickness ($t_1 > 55.6$ Å for 7 outlier devices with asymmetry $> +10\%$) and temperature rise in Fig. 5.35c is overwhelmed by stripe height variations ($h < 0.788$ μm for 11 devices with $\Delta T > 90$ K). In this particular simulated design, the strategy for process improvement is obvious: Place nearly equal emphasis on improved tolerances on magnetic track width and stripe height. There seems to be little penalty in using a SAL with high-anisotropy field, provided that high bias current (13 mA) is supported with good thermal conductivity ($\kappa_0 \simeq 3$ W/m-K) gap material.

156 AMR HEAD DESIGN AND ANALYSIS

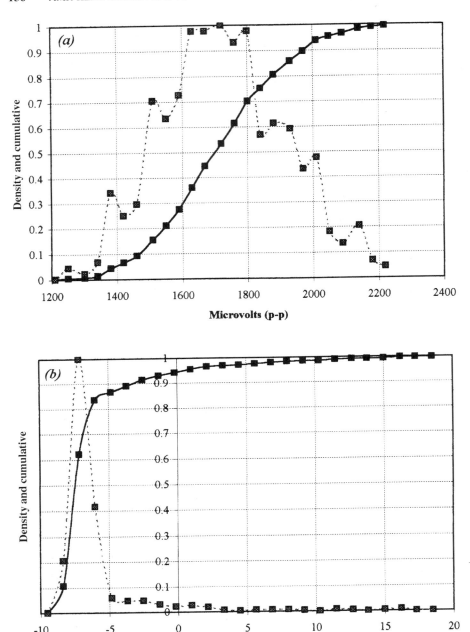

FIGURE 5.35 Monte Carlo simulation histograms and cumulative distributions for (*a*) signal, (*b*) asymmetry, and (*c*) temperature rise for a hypothetical SAL-biased AMR head.

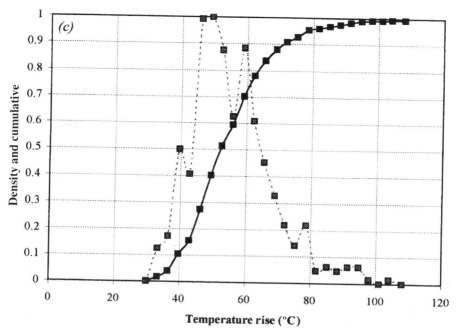

FIGURE 5.35 (*Continued*)

TABLE 5.2 Signal, Asymmetry, and Temperature Variations

Response	Mean	Standard Deviation	Minimum	Maximum
Signal, μV (p-p)	1710	194	1210	2260
Asymmetry, %	−6.39	3.58	−9.51	19.7
ΔT, K	54.4	13.1	29.9	111

TABLE 5.3 Correlation Coefficients for Monte Carlo Analysis

Response	t_1	g	t_2	g_1	g_2	h	W	H_{k1}	H_{k2}
Signal	0.082	0.044	−0.211	−0.058	0.005	−0.625	0.690	0.139	0.086
Asymmetry	0.588	0.063	−0.289	−0.105	0.153	0.280	0.008	0.064	0.025
ΔT	−0.055	0.043	−0.087	0.064	0.015	−0.959	0.025	0.029	0.072

TABLE 5.4 Influence of Tolerances on Response Variations

Row Number	t_1 (Å)	g (Å)	t_2 (Å)	g_1 (Å)	g_2 (Å)	h (μm)	W (μm)	H_{k1} (Oe)	H_{k2} (Oe)	σ_V (μV)	σ_A (%)	$\sigma_{\Delta T}$ (K)
1	3	3	3	10	10	0.1	0.1	2.0	0.5	194	3.58	13.1
2	3	3	3	10	10	0.1	0.1	0	0	189	3.94	14.2
3	3	3	3	0	0	0.1	0.1	0	0	184	3.23	13.3
4	0	0	0	0	0	0.1	0.1	0	0	168	0.34	13.6
5	0	0	0	0	0	0	0.1	0	0	132	0.032	~0
6	0	0	0	0	0	0.1	0	0	0	110	0.359	13.3
7	3	3	3	0	0	0	0	0	0	75.5	3.38	1.26
8	0	0	0	10	10	0	0	0	0	25.5	0.12	0.50
9	0	0	0	0	0	0	0	2.0	0.5	21.8	0.084	~0

REFERENCES

Anthony, T. C., Naberhuis, S. L., Brug, J. L., Bhattacharyya, M. K., Tran, L. T., Hesterman, V. W., and Lopatin, G. G., *IEEE Trans. Magn.*, **MAG-30**, 303 (1994).

Beaulieu, T. J. and Nepela, D. A., U.S. Patent 3,864,571, Feb. 4, 1975.

Bertram, H. N., *Theory of Magnetic Recording*, Cambridge University Press, Cambridge 1994.

Bhattacharyya, M. K. and Simmons, R. F., *IEEE Trans. Magn.*, **MAG-30**, 201 (1994).

Guo, Y., Hsu, Y., and Ju, K., *IEEE Trans. Magn*, **MAG-32**, 31 (1996).

Hsu, Y., Han, C. C., Chang, H., Ching, G., Hernandez, D., Chen, M. M., Ju, K., Che, C., and Fitzpatrick, J., *IEEE Trans. Magn.*, **MAG-31**, 2621 (1995).

O'Connor, D. J., Shelledy, F. B., and Heim, D. E., *IEEE Trans. Magn.*, MAG-21, 1560 (1985).

Potter, R. I., *IEEE Trans. Magn.*, **MAG-10**, 502 (1974).

Shelledy, F. B. and Brock, G. W., *IEEE Trans. Magn.*, **MAG-11**, 1206 (1975).

Smith, N., *IEEE Trans. Magn.*, **MAG-23**, 259 (1987).

Smith, N., Smith, D. R., and Shtrikman, S., *IEEE Trans. Magn.*, **MAG-28**, 2295 (1992).

Thompson, D. A., *AIP Conf. Proc.*, **24**, 528 (1974).

Voegeli, O., U.S. Patent 3,860,965, Jan. 14, 1975.

Zhu, J-G. and O'Connor, D. J., *IEEE Trans. Magn.*, **MAG-32**, 3401 (1996).

6
GMR HEAD DESIGN AND ANALYSIS

Three-layer GMR sensors with useful levels of $\Delta R/R$ ($\approx$ 4–9%) at room temperature and low applied fields (10–20 Oe) were invented by Grünberg (1990) and introduced as "spin valves" in a later publication by Dieny et al. (1991). The possibility of commercial exploitation stimulated vigorous research activity, and many publications followed giving details about the dependence of the GMR ratio $\Delta R/R$ on thicknesses of the ferromagnetic (FM) and nonmagnetic (NM) layers, the antiferromagnetic (AFM) material systems, processing considerations, and behavior of $\Delta R/R$ with temperature. A number of these papers are cited in the section on giant magnetoresistance effect in Chapter 2 and the reader is encouraged to study them. Notwithstanding the announcements in 1998 of AMR devices capable of 5 Gbits/in.[2] and greater, production volumes of GMR spin valve recording heads were introduced in IBM disk drives at areal densities of about 3 Gbits/in.[2], and today (mid-1999) virtually all new head designs are based on GMR technology. Rapid adoption of GMR sensors can be attributed to the combination of increased signal amplitude, improved linearity, and fewer complications in the application of these devices to hard disk drives. This chapter takes the analytical tools developed in Chapters 3 and 4, applies them to the basic "top" and "bottom" spin valve structures introduced in Chapter 2, and then progresses to analyze more complicated structures such as the synthetic, bias compensation layer, specular reflection spin valves, and dual synthetic spin valves (DSSVs). Specific design parameters and computations are initially centered on basic top and bottom spin valves appropriate for areal densities of 5–6 Gbits/in.[2]; using the scaling rules in Chapter 1, this translates to about 21 ktpi, 290 kbpi, GMR read width $W = 0.80/\text{tpi} \simeq 1.0\,\mu\text{m}$, stripe height $h = 0.7W = 0.7\,\mu\text{m}$, and a pulsewidth $PW50 \simeq 0.20\,\mu\text{m}$. The invention of more complicated and sensitive spin valve designs has led to incremental improvements in areal density; therefore, this chapter also introduces the design parameters and values for devices that have successfully operated at 10, 15, 20, and 25 Gbits/in.² or higher. In historical sequence, spin valves with NiFe layers (which have a substantial AMR effect) were first studied, Co films were found to increase the GMR effect when dusted onto thicker NiFe layers, and then much later, NiFe layers have been replaced with CoFe films. Laboratory demonstrations of 36 Gbits/in.² were performed in late 1999 with DSSVs in which all of the magnetic layers were CoFe; the output was 5.7 mV/μm of read track width and the asymmetry was very low at all (±) bias currents up to 6 mA. This chapter discusses the problems of bias current polarity and signal asymmetry that are associated with the undesired AMR effect in NiFe layers but are absent in CoFe layers. Monte Carlo simulation results for selected devices are given to emphasize the level of process control required for successful manufacturing at high volumes. Because device heating is fundamentally important, the thermal analysis given in Chapter 2 is applied to the GMR ratio and the dependence of exchange pinning fields on the blocking temperature of various AFM/FM pinned-layer materials. The empirical nature of the GMR ratio is made quite obvious by the use of parameter fits to experimental results, and the reader will appreciate this is typical of rapidly evolving technologies. In spin valves, the conducting film thicknesses are less than the mean free path for electron scattering, and thus the relation for ohmic resistance $R = \rho W/th$ (with ρ = resistivity, W = track width, t = film thickness, and h = stripe height) must be interpreted with caution; the resistivity of one layer within many layers depends on scattering processes at the interfaces, and the resistance variations of the GMR effect depend on the thickness and spin polarization of adjacent FM layers (as well as upon the nature of electron reflection at FM layer interfaces). In addition, most devices require thermal annealing at some point in the process, and thus the magnetic and electrical properties of very thin films may change substantially after diffusion processes have occurred.

BASIC SPIN VALVES: TOP AND BOTTOM STRUCTURES

The simple spin valve structure shown in Fig. 6.1 is composed of four layers on a substrate. The first and third layers are FM metals, the second is an NM noble metal, and the fourth is an AFM material (either high resistivity or insulating) that pins the magnetization of the adjacent FM layer along the direction of the exchange field H_{ex} of the AFM layer. The terminology given to this structure is *free layer/spacer/ pinned layer/AFM layer*, and it is a top spin valve because the AFM layer is on the top of the structure. In a bottom spin valve first the AFM layer is deposited, then the pinned layer is put down. The review paper by Coehoorn et al. (1998) discusses the microstructure and magnetic properties of films and their dependence on the properties of the underlying layer. Because of the different underlayers, top and bottom spin valves exhibit different exchange fields and blocking temperatures T_B (see Chapter 2). Coehoorn et al. also underscore the importance of "buffer" layers of Ta (~30 Å), which enhance a crystalline [111] columnar structure with improved

Simple Spin Valve

MnFe — exchange layer
NiFe — pinned layer
Cu — Cu spacer
NiFe — free layer

FIGURE 6.1 Geometry of a basic GMR spin valve structure.

magnetic properties for films of NiFe or AFM materials. "Capping" layers of Ta are also used to protect the top AFM or FM layer from oxidation and corrosion.

In their first publication on spin valve systems, Dieny et al. (1991) used a Si substrate followed by the top spin valve structure 50 Å Ta/60 Å $Ni_{80}Fe_{20}$/20 Å Cu/45 Å $Ni_{80}Fe_{20}$/70 Å $Fe_{50}Mn_{50}$/50 Å Ta, and they achieved a room temperature GMR ratio $\Delta R/R$ of 5% at 10 Oe applied field. The AFM layer served only to pin the magnetization of the adjacent NiFe layer with $H_{ex} \sim 300$ Oe, and they demonstrated that $\Delta R/R$ for large film samples depended upon the *relative* angle between the magnetization vectors in each film according to the relation

$$\frac{\Delta R}{R} = \left(\frac{\Delta R}{R}\right)_0 \left(\frac{1-\cos(\theta_F - \theta_P)}{2}\right). \quad (6.1)$$

Unlike the AMR effect, the GMR effect is *independent* of the angle between current density and the magnetization vector in either film; the change in resistance arises from scattering of spin-polarized electrons diffusing from an FM spin-up layer through the NM spacer into another FM layer whose spin is up or down. The high- and low-resistance states arise from ↑↓ and ↑↑ polarizations, respectively. Subsequent work by Dieny et al. (1992) revealed the dependence of the GMR ratio on layer thicknesses, and they were able to estimate the mean free path λ^+ for ↑↑ scattering in NiFe at about 70–80 Å whereas the shorter mean free path λ^- for ↑↓ scattering was roughly 4 Å. Neglecting the resistance of the Cu/NiFe/FeMn structure, the measured GMR ratio of the free layer is assigned to an *active* resistance R_0 (where spin scattering occurs) in parallel with an *inactive* resistance R_1 shunting the structure. That is,

$$\frac{\Delta R}{R} = \frac{R\uparrow\downarrow - R\uparrow\uparrow}{R\uparrow\uparrow}, \quad (6.2)$$

where the low-resistance state is

$$R\uparrow\uparrow = \frac{R_0 R_1}{R_0 + R_1} \quad (6.3)$$

and the high-resistance state is

$$R\uparrow\downarrow = \frac{(R_0 + \Delta R_0)R_1}{R_0 + R_1 + \Delta R_0}, \qquad (6.4)$$

which reduces to the expression

$$\left(\frac{\Delta R}{R}\right)_{\text{measured}} = \left(\frac{\Delta R_0}{R_0}\right)\frac{1}{1 + R_0/R_1}. \qquad (6.5)$$

They experimented with various thicknesses of free FM layers of Co, NiFe, and Ni and found the GMR ratio increases rapidly, goes through a broad maximum, and then falls off hyperbolically with increasing film thickness; their results at room temperature are shown in Fig. 6.2. This behavior is compatible with the interpretation that incoming electrons with a long mean free path λ^+ have a scattering probability proportional to $1 - \exp(-t_F/t_0)$, where t_F is the free-layer thickness and t_0 (which is roughly equal to λ^+) is the thickness of the active part of this layer. Electrons with a short mean free path $\lambda^- \approx \lambda^+/20$ have a high scattering probability (~ 1.0), so almost all of the electron flow between the free and pinned layers through the thin NM spacer arises from electrons with $\uparrow\uparrow$ polarization. The rest of the structure shunts the active part of the spin valve, and their experiments showed the sheet conductivity (G) of the total structure increases almost linearly with t_F. That is, $G \simeq G_{\text{rest}} + t_F/\rho_F$, and the GMR ratio can be written as

$$\frac{\Delta R(t_F)}{R} = A\frac{1 - \exp(-t_F/t_0)}{1 + t_F/(G_{\text{rest}}\rho_F)}, \qquad (6.6)$$

which fits the data of Fig. 6.2 with two adjustable parameters, A and t_0. Their experiments with Cu spacer thickness t_S showed that A varied approximately as $A_0 \exp(-t_S/\lambda_S)$ and that the spacer served only to decouple the free and pinned FM layers; the characteristic diffusion length for electrons through a Cu spacer is $\lambda_S \simeq 47$ Å.

Dieny et al. (1992) also studied the temperature dependence of their spin valve systems over the range $83\,\text{K} \leq T \leq 293\,\text{K}$; the effective shunting thickness $t_{\text{shunt}} = G_{\text{rest}}\rho_F$ increased almost linearly with T and the active thickness t_0 for Co and NiFe each reduced about 20%. The adjustable parameter A depends on temperature, and the data (at $t_S = 22$ Å) shown in Fig. 6.3 fit the relation $A(T) = A_0 \exp(-T/T_0)$, where A_0 and T_0 depend on the free-layer material. Instead of using the form $A_0 \exp(-T/T_0)$, Dieny et al. fitted second-order polynomials to their temperature data and showed that $\Delta R/R$ for NiFe and Co free layers extrapolated to zero at a temperature $T_{0\text{SV}} \simeq 530$ K. Based on the published experiments, the behavior of the GMR ratio as a function of film thickness and

BASIC SPIN VALVES: TOP AND BOTTOM STRUCTURES   163

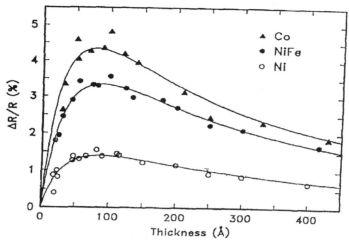

FIGURE 6.2 Variation of the magnetoresistance versus the thickness of the free magnetic layer for the materials Co, NiFe, and Ni at room temperature. The lines are two-parameter fits according Eq. (6.6). Reprinted with permission from B. Dieny, P. Humbert, V. S. Speriosu, S. Metin, B. A. Gurney, P. Baumgart, and H. Lefakis, *Phys. Rev. B*, **45**, 806. Copyright © 1998 by the American Physical Society.

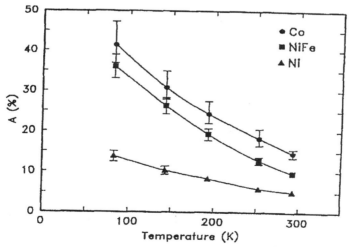

FIGURE 6.3 Thermal variation of the characteristic MR parameter *A* for Co, NiFe, and Ni. Reprinted with permission from B. Dieny, P. Humbert, V. S. Speriosu, S. Metin, B. A. Gurney, P. Baumgart, and H. Lefakis, *Phys. Rev. B*, **45**, 806. Copyright © 1998 by the American Physical Society.

temperature can be approximately expressed as the product of simple exponentials and a hyperbola:

$$\frac{\Delta R(t_F, t_S, T)}{R} \simeq A_0 \exp\left(-\frac{t_S}{\lambda_S}\right) \exp\left(-\frac{T}{T_0}\right) \left[\frac{1 - \exp(-t_F/t_0)}{1 + t_F/t_{\text{shunt}}}\right]. \quad (6.7)$$

The quantities A_0, T_0, t_0, and $t_{\text{shunt}} = G_{\text{rest}} \rho_F$ are given in Table 6.1 for each of the FM free layers of their study. Figure 6.4a is a plot of (6.7) using the parameters for NiFe from Table 6.1; the computed curve fits the data points, which have been extracted from Fig. 6.2 for $t_S = 22$ Å and $T = 293$ K. Parkin (1993) demonstrated the change of $\Delta R/R$ with Co "dusting" of NiFe layers at the Cu interfaces; the GMR effect more than doubles with about 10 Å of Co dusting. Figure 6.4b shows Parkin's empirical fit to the room temperature data $\Delta R/R(\%) = 2.9 + 3.5[1 - \exp(-t_{\text{Co}}/2.3$ Å$)]$ for the structure Si/NiFe(53 Å $- t_{\text{Co}}$)/ Co(t_{Co})/Cu 32 Å/ Co(t_{Co})/NiFe(22 $- t_{\text{Co}}$)/FeMn 90 Å/Cu 10 Å. For 53 Å of NiFe, (6.7) gives $\Delta R/R \simeq 2.8\%$, which agrees closely with Parkin's result of 2.9% at $t_{\text{Co}} = 0$.

Many spin valve recording heads produced today exploit the improved signal of Co dusting or use a free/spacer/pinned-layer structure similar to $Ni_{81}Fe_{19}$/ $Co_{90}Fe_{10}$/Cu/$Co_{90}Fe_{10}$/AFM where the Co alloy is about 20 Å thick; these structures give a GMR ratio of about 6%. The thickness of the pinned layer t_P is determined by a trade-off between the exchange pinning field strength $H_{\text{ex}} = E_{\text{ex}}/M_S t_P$ and the magnitude of the GMR ratio, which follows the monotonically increasing relation shown above [see (2.14); $M_S \simeq 1540$ emu/cm^3 for CoFe and $E_{\text{ex}} \simeq 0.1$ erg/cm^2 with FeMn at room temperature]. Since a large pinning field is desired, the pinned layer will be thinner than the free layer but thick enough to avoid pin holes.

SHIELDED GMR RESPONSE (TRANSFER CURVE)

The fundamental relation for the output signal of a GMR spin valve is derived from (1.5) and (6.1), where the voltage is the product of the sense current I and the change in resistance $\Delta R(H)$. Experimentally, a spin valve structure is characterized by the GMR ratio $\Delta R/R$ and the sheet resistance R_S (Ω/square) measured in the low-resistance ↑↑ state; thus the sensor output is written [see (2.20) and Heim et al. (1994)]

$$\text{Signal} = I \, \Delta R = I \frac{\Delta R}{R} R_S \frac{W}{h} \left(\frac{1 - \langle \cos(\theta_F - \theta_P) \rangle}{2}\right), \quad (6.8)$$

where R_S is the sheet resistance, W and h are the track width and stripe height, respectively, and the brackets $\langle \cdots \rangle$ indicate an average value over h. The factor of 2 in (6.8) normalizes the output for the dynamic range given by the factor $-1 \leq \langle \cos(\theta_F - \theta_P) \rangle \leq +1$. The AMR effect (which is associated with the NiFe layer in GMR sandwiches) is ignored here, but its influence is significant and is

TABLE 6.1 GMR Parameters for Free Layers

FM Material	A_0 (%)	T_0 (K)	t_0 (Å)	t_{shunt} (Å)
Co	98	215	72	65
$Ni_{80}Fe_{20}$	98	160	72	85
Ni	33	215	85	65

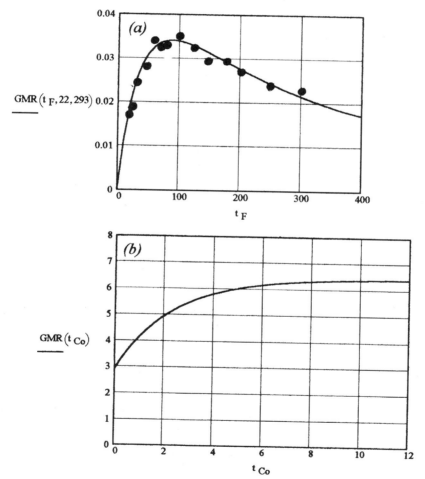

FIGURE 6.4 (*a*) A plot of Eq. (6.7) using the parameters for NiFe from Table 6.1. The data points are extracted from the work of Dieny et al. (1992) shown in Figure 6.2. (*b*) Plot of the empirical Co-dusted GMR effect $\Delta R/R$ from Parkin (1993) as a function of Co thickness.

166 GMR HEAD DESIGN AND ANALYSIS

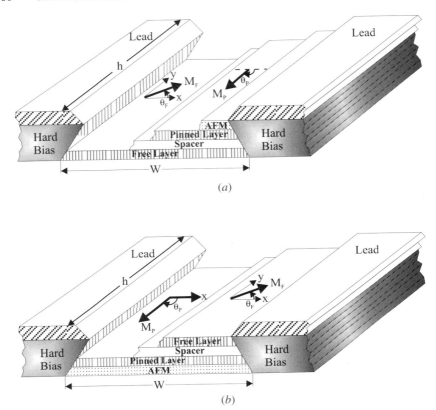

FIGURE 6.5 Exploded views of (*a*) top and (*b*) bottom spin valves with hard-bias stabilization and overlay leads.

included later in this chapter to explain the experimental results of spin valve devices. Exploded views of top and bottom spin valves with hard-bias stabilization and overlay leads are shown in Figs. 6.5*a,b*; the reference direction for angular displacement is along the *x*-direction and the pinned-layer magnetization $\mathbf{M}_P$ is aligned with the *y*-axis ($\theta_p = \pm \frac{1}{2}\pi$). Important differences between top and bottom spin valves arise from the junction angle between the hard-bias layer and the spin valve sandwich; the free layer of the top structure is wider and less well defined than the bottom structure.

The magnetization vector $\mathbf{M}_F$ is free to rotate in response to the total magnetic field from all sources; if one thinks of the total field as arising from *independent* sources, then $H_T = H_{\text{sig}} + H_I + H_{\text{cpl}} + H_{Dp} + H_K + H_{\text{HB}}$, where the subscripts refer to fields from the medium transition signal, sense current *I*, interlayer coupling between free and pinned layers, effective demagnetizing (or fringing) field from the pinned layer, effective anisotropy field (from pair-ordering, shape, and stress anisotropies) of the free layer, and the hard-bias stabilizing field. In this view of independent sources, the *transverse* fields H_I, H_{cpl}, and H_{Dp} influence the bias point, whereas the *longitudinal* fields H_K, and H_{HB} influence the slope (sensitivity) of the transfer curve. The uniform rotation (Stoner–Wohlfarth model) analysis introduced in Chapter 2 gives useful but limited insight regarding the transfer curve of an *unshielded* spin valve. Normalized solutions to (2.6) for the hard-axis magnetization $m_y(\theta, H_x, H_y)$ are plotted in Fig. 6.6 as a function of H_y for several values of easy-axis field H_x applied to an "ideal" film. The solutions for $H_x > 0$ do not have an analytic form, but they are well approximated by the function $\tanh[H_y/(H_K + H_x)]$, which is also given along with the numerical solutions in Fig. 6.6. For vanishingly small H_x the $m_y(H_y)$ magnetization curve is essentially linear with a normalized slope $1/H_K$ up to a field $H_y \simeq H_K$; applying a longitudinal field H_x, such as hard bias, reduces the slope of $m_y(H_y)$ approximately as $1/(H_K + H_x)$ near $H_y \sim 0$.

With coupled films placed between magnetic shields, the uniform rotation model must be replaced by a more appropriate model where the magnetization is a function of position within the layer. In general terms, the pinned and free layers would be described by the relations

$$M_P(y, I, T) = M_{SP} \frac{H_{IP}(I) + H_{ex}(T)}{H_{KP}} f(y), \qquad (6.9a)$$

$$M_F(y, I, T) = M_{SF} \frac{H_{IF}(I) + H_{cpl}(T)}{H_{KF}} f(y), \qquad (6.9b)$$

where $M_{SP,F}$ are the saturation magnetizations of the pinned and free layers, $H_{IP,F}(I)$ are the fields that depend on current flow, $H_{ex}(T)$ is the temperature-dependent

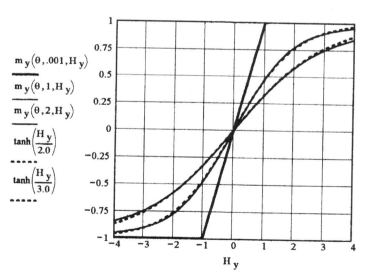

FIGURE 6.6 Normalized Stoner–Wohlfarth solutions for the hard-bias magnetization m_y as a function of uniform external field for several values of easy-axis field H_x. These solutions are well approximated by the hyperbolic tangent functions also shown in the figure.

exchange pinning field, $H_{cpl}(T)$ is a temperature-dependent interlayer coupling field, $H_{KP,F}$ are the effective anisotropy fields of the pinned and free layers, and $f(y)$ is a function describing the nonuniform magnetization within a layer. These terms and factors are discussed later in this chapter.

Sense Current Distribution and Bias Point Shift

The field H_{IF} experienced by the free layer (which arises from the sense current flowing in all of the layers) shifts the operating point of the sensor transfer curve. In the absence of shields, this field is determined by estimating the current density $J_x(z)$ in each of the conducting layers, integrating along the z-direction through the layer sandwich, and finding the average field at the free layer. Bulk values of film resistivity cannot be used to form an accurate parallel-resistor model in very thin films because electron scattering at the film interfaces increases the resistivity. A rigorous analysis of current density in a spin valve sandwich would include spin-dependent scattering of electrons as they diffuse through the Cu spacer, but such an analysis is complex and goes beyond the scope of this book. The theory of Fuchs (1938) and Sondheimer (1952), which is briefly reviewed in Chapter 2, is useful for estimating the dependence of resistivity on film thickness and for extracting "bulk" resistivity and mean free path parameters from experimental results. Camley and Barnas (1989), Johnson and Camley (1991), and Camblong (1995) extended the Fuchs–Sondheimer Boltzmann transport equation analysis to include spin-dependent scattering and multiple layers, and these references may be consulted by the interested reader. The following analysis treats the spin valve structure as parallel resistors in which the resistivity of each layer is a function of the layer thickness, and the small change in the distribution of current density when switching between high- and low-resistance states is ignored. With this simplification, the estimate of the mean field at the free layer will be in error by several percent, but the essential idea of bias point shifting with sense current emerges in a straightforward manner.

The experimental results for the resistivity of thin films of Cu and NiFe (given in Chapter 2, Tables 2.3 and 2.4) are convenient for building a parallel-resistance (or conductance) model from which the current density is estimated for each layer in a sandwich. A simple *top* spin valve structure with capping layers Ta 50 Å/NiFe 60 Å/Cu 25 Å/NiFe 45 Å/FeMn 80 Å/Ta 50 Å will serve as an example. The resistivities of the Ta and FeMn layers are about 200 and 140 μΩ-cm, respectively (see Tables 2.2 and 2.5), and the resistivities for the NiFe 60 Å/Cu 25 Å/NiFe 45 Å sandwich, estimated with the ρ_0 and λ parameters in Table 2.4, become 34.1, 12.0, and 37.7 μΩ-cm, respectively. Because the current density $J = \sigma E$, a more direct approach is to find the sheet conductances for each layer (G_i, $i = 1, 2, \ldots, n$) and add the results. That is,

$$G_{\text{total}} = \sum_1^n G_i = \sum_1^n \frac{t_i}{\rho_i(t_i)} \, (\Omega^{-1}), \tag{6.10}$$

which, for the present case, is $G_{\text{total}} = 0.0025 + 0.0176 + 0.0208 + 0.0119 + 0.00615 + 0.0025 = 0.0615 \, \Omega^{-1}$. The sheet resistance $R_S = 1/G_{\text{total}} = 16.3$

Ω/square, and the fraction of current flowing in each layer is G_i/G_{total} = 0.041/0.286/0.338/0.194/0.10/0.041. The magnetic field associated with a slab of thickness d and uniform current density J_x follows the relation

$$H_y = \int_0^d J_x \, dz \tag{6.11}$$

and the fields of individual layers add algebraically. The normalized current density and field of the top spin valve structure are shown in Figs. 6.7a,b. Positive current is along the $+x$-axis and positive field is along the $+y$-axis. For this particular *unshielded* example, the average normalized field over the thickness of the free layer

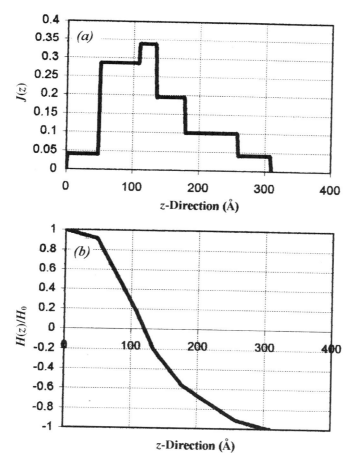

FIGURE 6.7 Normalized (*a*) current density and (*b*) magnetic field through the thickness of a top spin valve structure with six layers.

is $Q_F = H(z)/H_0 = +0.522$, where the normalizing constant $H_0 = 2\pi I/h$ (in oersteds, with I in milliamperes and h in micrometers); thus the bias point is shifted by the amount $H_I = Q_F \cdot 2\pi I/h$. At the interface between the AFM and pinned layer, the normalized field $Q_P = -0.59$, so a positive sense current field *opposes* a positive pinning field and *aids* a negative pinning field. Shifting of the transfer curve for negative and positive sense currents is shown in Fig. 6.8; in this simple uniform rotation model, the slope of the unshielded free-layer transfer curve is determined by the sum of the anisotropy and free-layer demagnetizing fields ($H_K + H_{Df} = 4.0 + 23$ Oe in this example, where $h = 0.7$ µm and $W = 1.0$ µm). (*Note:* The influence of the pinned-layer demagnetizing field is ignored here; it is treated later in this chapter.)

The situation is somewhat different for a *bottom* spin valve system defined by the layers Ta 50 Å/IrMn 50 Å/CoFe 20 Å/Cu 25Å/CoFe 20 ÅNiFe 30 Å/Ta 50 Å. The CoFe layers provide the enhanced GMR effect discussed by Parkin (1993), and the NiFe layer reduces the effective anisotropy field of the free layers. The resistivity of IrMn is about 325 µΩ-cm (see Table 2.2) and the film is thin, so the current density distribution is significantly different from the top spin valve system defined earlier. The sheet conductance (in millimhos) of the layers is $G_{\text{total}} = 2.5 + 1.67 + 4.31 + 24.2 + 4.31 + 6.66 + 2.5 = 46.2$ mΩ^{-1} for which the sheet resistance is $R_S = 21.7$ Ω/square, and the fraction of current flowing in each layer is approximately 0.095/0.064/0.066/0.46/0.066/0.15/0.095. The normalized field as a function of distance z through the layers is plotted in Fig. 6.9; inspection of this figure shows that $Q_F \simeq -0.60$ and $Q_P = +0.68$, so positive sense current *aids* a positive pinning field with this structure. Top and bottom spin valves each suffer from a basic trade-off between H_I polarity that either aids the pinning field and adds

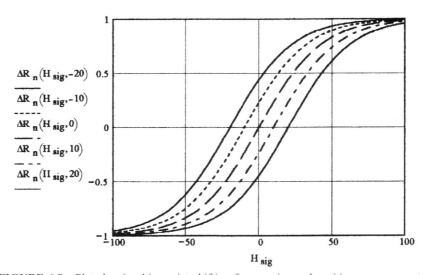

FIGURE 6.8 Plot showing bias point shifting for negative and positive sense currents.

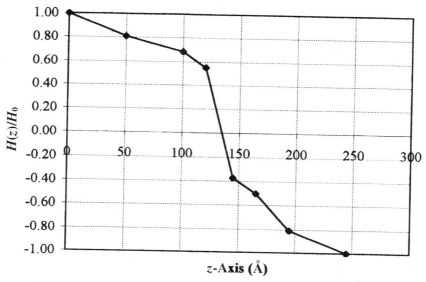

FIGURE 6.9 Normalized field arising from the current density distribution through a bottom spin valve system with seven layers.

to the effect of H_{Dp} on the free layer or opposes the pinning field and subtracts from H_{Dp} at the free layer. If the pinning field is insufficient at normal operating temperatures, H_I would be chosen to aid pinning and the loss of dynamic operating range field would be accepted.

Effective Anisotropy Field of the Free Layer

The effective anisotropy field H_K of an *isolated* layer arises from pair-ordering, shape, and stress factors; this subject is discussed in Chapter 2, where it is shown that the anisotropy energy density $K_T = K_u + (N_y - N_x)M_S^2/2 + 3\lambda_S \Delta\sigma/2$ and thus $H_K = 2K_T/M_S = H_{Ku} + (H_{Dy} - H_{Dx}) + H_\sigma$ for an unshielded and isolated sense layer. In a spin valve, the free and pinned layers are coupled; thus the demagnetizing factors N_x, N_y have no real meaning, and the equations of coupled films must be used to define the highly nonuniform demagnetizing field and its influence on the transfer curve. In a shielded device the equations describing flow of magnetic flux must be redefined to include the pathways through shields as well as the coupled free and pinned layers. The effective anisotropy field of the coupled free layer is therefore expressed by the relation $H_{KF} = H_{Ku} + 3\lambda_S \Delta\sigma/M_S$ (in oersteds), and the role of demagnetizing energy is included in the characteristic length λ for flux decay in coupled films. The anisotropy field arising from pair ordering H_{Ku} will normally be 3–4 Oe for thin Permalloy films with low magnetostriction constants ($\lambda_S \sim 5 \times 10^{-7}$

or less). Stress anisotropy $\Delta\sigma$ can be compressive or tensile depending on processing conditions and substrate material; however, it is quite often found to be tensile and of the order to $10^9 \, \text{dyn/cm}^2$, and thus the effective field arising from stress $H_\sigma = 3\lambda_S \, \Delta\sigma / M_S$ could raise or lower the effective anisotropy field by a significant amount. For example, with $\lambda_S = \pm 1 \times 10^{-6}$ and $\Delta\sigma = 1 \times 10^9 \, \text{dyn/cm}^2$, $H_\sigma = \pm 3.75 \, \text{Oe}$ for $M_S = 800 \, \text{emu/cm}^3$. The CoFe alloys have high positive magnetostriction ($\lambda_S \sim 1 \times 10^{-5}$ to $\lambda_S \sim 6 \times 10^{-5}$), so anisotropic stress could dominate the behavior of spin valve free layers made with this material. The measured H_K of Cu/Co$_{90}$Fe$_{10}$ 20 Å/Cu is about 8 Oe, which presumably includes a contribution from stress; with Ta/Co$_{90}$Fe$_{10}$ 20 Å/Ta the measured H_K increases above 20 Oe. Uncontrolled stresses could lead to instabilities in the output signal arising from changes in direction of the easy axis and to multiple domains in the free layer. (See Fig. 7.20 for an example of domain instability.) Moreover, it is conceivable that stress anisotropy could relax over time and cause gradual changes in device sensitivity.

Flux Coupling between the Pinned and Free Layers

The simple uniform rotation model with independent field sources ignores the *coupled-film* nature of the interaction between the pinned and free layers, and it must be abandoned in favor of an approach that leads naturally to a more general understanding of the behavior of spin valve sensors whether shielded or unshielded. In this regard, the pinned and free layers form a coupled-film system where the highly nonuniform fringing (demagnetizing) field is confined to the nonmagnetic region between the magnetic layers. This problem is treated in Chapter 3, and the pinned and free GMR layers behave in a manner similar to SAL-biased AMR devices treated in depth in Chapter 5; the physics of the GMR and AMR *effects* are entirely different, but the magnetostatic interactions of the various layers are nearly the same. The pinned layer is a permeable film whose magnetization $\mathbf{M}_P$ responds to the sum of the AFM layer exchange pinning field H_{ex} (which is of very short range and acts only on the pinned layer) and the field H_I from current flow, and thus the pinned layer magnetizes with $H_{\text{ex}} + H_I$ and the demagnetizing flux couples through the free layer. With sufficient field ($\pm$), $\mathbf{M}_P$ is pinned upward or downward ($\theta_2 \simeq \pm \frac{1}{2}\pi$) and saturated over much of the stripe. The equations describing coupled films with partial saturation are found in (3.43)–(3.49) and in Chapter 5 in the section on SAL biasing with a conducting spacer. For convenience to the reader, these equations are rewritten here with small changes in terminology; the pinned layer is functionally equivalent to a SAL and the free layer functions like the AMR layer of Chapter 5. The bottom spin valve structure shown in Fig. 6.5b will be used for this design example; the field acting at the AFM–pinned layer interface is $H_1(I, T)$, which includes the sense current field and the temperature-dependent pinning field. The free-layer centerline is acted upon by $H_2(I)$, which includes the sense current field and the interlayer coupling field H_{cpl} (which will be discussed later in this chapter). Using the analysis leading to (3.49a), the normalized y-component of the pinned-layer magnetization m_P is written

$$m_P(y, I, T) = \frac{M_P}{M_{SP}} = \frac{H_1(I, T)}{H_{KP}}\left[1 - \frac{\cosh[(|y| - y_S)/\lambda]}{\cosh[(h/2 - y_S)/\lambda]}\right], \quad (6.12)$$

where the saturated location $|y_S|$ above and below $y = 0$ is given by (3.47a) and rewritten here as

$$|y_S| = \frac{h}{2} - \lambda \cosh^{-1}\left(\frac{|H_1(I, T)|}{|H_1(I, T)| - H_{KP}}\right). \quad (6.13)$$

The applied field is calculated for shielded conductors using (4.14a,b,c) for current sheets and (2.14) for exchange coupling; this field is given by the relation

$$H_1(I, T) \simeq \frac{2\pi}{hG_{ss}}[I_1(g_1 + t_A - t_S - t_F - g_2) + 2I_2(t_A + g_1)]$$

$$+ \frac{2\pi}{hG_{ss}}[2I_3(t_A + g_1) + 2I_4(0.5t_A - t_P - t_S - t_F - g_2)] + H_{ex}(T), \quad (6.14)$$

where $G_{ss} = g_1 + t_A + t_p + t_S + t_F + g_2$ is the shield-to-shield distance and $I_1, I_2, I_3,$ and I_4 flow in the pinned, free, spacer, and AFM layers, respectively. (The Ta buffer and capping layers are ignored here to keep the terminology uncluttered. Their influence would be included for more accurate estimates of H_I.) The current in each layer is calculated from the sheet conductances [see (6.10)]:

$$I_i = \frac{G_i}{G_{total}} I_{sense}, \quad (6.15)$$

where I_{sense} is the total sense current of the device. The term in I_1 of (6.14) shows the contribution of shield images and disappears for symmetric placement of the pinned layer between the shields. The pinning field is given by the relation

$$H_{ex}(T) = \frac{E_{ex}(T)}{M_{SP}t_P}, \quad (6.16)$$

where the exchange surface energy $E_{ex}(T)$ temperature dependence is experimentally determined. Heating, and its impact on exchange pinning, is discussed in the sections on thermal analysis of GMR heads and AFM layer and distribution of blocking temperatures, which follow shortly.

The flux decay characteristic length λ is derived in Chapter 3 for nonidentical films; reading from (3.42), the decay length for coupled pinned (P) and free (F) layers is given by the relation

$$\lambda^2 = \frac{\mu_P t_P \mu_F t_F t_S}{\mu_P t_P + \mu_F t_F} + \frac{A_P}{M_{SP} H_{KP}} \left(1 - \frac{1}{\cosh(t_P/2\lambda_{Pe})}\right)^2 + \frac{A_P}{M_{SP} H_{KP}}$$

$$+ \frac{A_F}{M_{SF} H_{KF}} \left(1 - \frac{1}{\cosh(t_F/2\lambda_{Fe})}\right)^2 + \frac{A_F}{M_{SF} H_{KF}}. \quad (6.17)$$

As discussed in Chapter 3 in the section on the influence of exchange energy in coupled films, the five terms in (6.17) are related to the demagnetizing energy, the exchange energy for tilting M_P within the pinned layer, the exchange energy for rotating M_P, and the tilting and rotation exchange energies for M_F, respectively. At this point, it is useful to assign values to each of the above parameters and see how the energy budget is divided. Assume a bottom spin valve system defined as IrMn 100 Å/CoFe 20 Å/Cu 25 Å/NiFe 100 Å with $M_{SP} = 1540 \text{ emu/cm}^3$, $M_{SF} = 800 \text{ emu/cm}^3$, $A_P = 1.5 \times 10^{-6} \text{ erg/cm}$, $A_F = 1.0 \times 10^{-6} \text{ erg/cm}$, $H_{KP} = 8 \text{ Oe}$, $H_{KF} = 4 \text{ Oe}$, $\lambda_{Pe} = 35 \text{ Å}$, and $\lambda_{Fe} = 50 \text{ Å}$ [see (3.36) for the last two parameters]. The calculation for λ^2 works out to the result (in square micrometers)

$$\lambda^2 = \lambda_{\text{demag}}^2 + \lambda_{P,\text{ex}}^2 + \lambda_{F,\text{ex}}^2 = (0.102)^2 + (0.111)^2 + (0.187)^2 = (0.240)^2, \quad (6.18)$$

or $\lambda = 0.240 \, \mu\text{m}$, and it is seen that exchange energy dominates the system; this is not surprising because the layers are very thin. A close inspection shows that the exchange energy of the pinned layer arises almost completely from rotation within the film and is split about 88%/12% between rotation and tilting in the thicker free layer. Figure 6.10 shows magnetization profiles of a pinned layer for $I_{\text{sense}} = \pm 6 \text{ mA}$ with a room temperature pinning field $H_{\text{ex}}(T = 25°C) = -217 \text{ Oe}$ from an IrMn AFM layer. [*Note:* The exchange field of AFM and FM materials depends significantly on processing details. Nogués and Schuller (1999), for example, show the interface exchange energy ranges from 0.01 to 0.19 erg/cm² for IrMn and from 0.02 to 0.32 erg/cm² for PtMn.] The curves reveal the influence of positive or negative sense currents, which create a field H_I that *aids* or *opposes* the negative pinning field from the AFM layer.

The normalized magnetization of the free layer m_F is determined by the demagnetizing flux of the pinned layer and the influence of the fields H_2 and H_{cpl}. With a very thin spacer layer t_S relative to g_1 and g_2, one can assume essentially all of the pinned-layer demagnetizing flux biases the free layer; under this assumption [see (5.30) and (5.31)], the normalized magnetization of the free layer becomes

$$m_F(y, I, T) = \frac{-M_{SP} t_P}{M_{SF} t_F} m_P(y, I, T) + \frac{H_2(I)}{H_{KF}} \left[1 - \frac{\cosh(y/\lambda_{\text{sh}})}{\cosh(h/2\lambda_{\text{sh}})}\right], \quad (6.19)$$

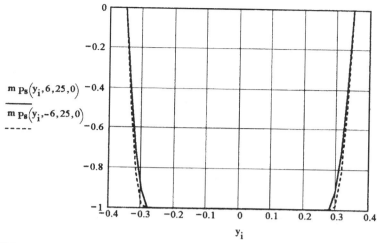

FIGURE 6.10 Normalized magnetization profiles of a pinned layer for $I_{\text{sense}} = \pm 6$ mA with a pinning field of -217 Oe from IrMn AFM layer.

and $H_2(I)$, which includes the interlayer coupling field, is derived from (4.14a,b,c) and written as

$$H_2(I) = \frac{2\pi}{hG_{ss}}[-2I_1(0.5t_F + g_2) + I_2(g_1 + t_A + t_P + t_S - g_2)]$$
$$+ \frac{2\pi}{hG_{ss}}[-2I_3(0.5t_F + g_2) - 2I_4(0.5t_F + g_2)] + H_{\text{cpl}}. \quad (6.20)$$

The term in I_2 shows the influence of shield biasing, which disappears with symmetrical placement of the free layer between shields. The term H_{cpl} is assumed to be uniform in (6.20), but it is actually proportional to M_P, which is uniform only in the saturated region of the stripe; the errors arising from this assumption will likely be small because the boundary conditions imposed on the free layer are $m_F(y = \pm \frac{1}{2}h) = 0$ in the absence of signal excitation flux. The free layer carries the pinned-layer flux plus the additional flux that flows through the shields and closes back through the free layer. The flux decay length for coupling between the free layer and shields is derived from (4.19) and is written

$$\lambda_{\text{sh}} = \sqrt{\frac{\mu_F t_F(g_1 + t_A + t_P + t_S)g_2}{G_{ss}}}. \quad (6.21)$$

This relation assumes the pinned layer is saturated and that flux conducted through the shields passes directly through the pinned layer and closes through the free layer. This assumption greatly simplifies the analysis and leads to results that are in reasonable agreement with experiment. An estimate for the free-layer thickness t_F that produces an optimal bias point can be obtained by averaging $m_F(y, I, T)$ over the stripe height h, setting the result to zero and solving the average $\langle m_F \rangle$ for t_F in terms of the other design parameters. Since λ_{sh} is also a function of t_F, (6.21) and the average of (6.19) can be solved simultaneously for the free-layer thickness. With the parameters of the present design example, one finds that a moment ratio $M_{SF}t_F/(M_{SP}t_p)$ of about 2.5–3.5 is required and fairly thick values of t_F (~110–160 Å) can be justified, but this leads to loss of output signal. The value of 100 Å represents a compromise in favor of improved signal level with a somewhat suboptimal bias point. On the other hand, a very thin free layer (~50 Å) leads to higher resistance, increased signal for positive I with greater risk of thermal unpinning, and saturation of the free layer with large negative asymmetry at negative I. The effective magnetization of a composite free layer such as CoFe/NiFe is found using the relation $[M_{SF}]_{\text{eff}} = (M_1 t_1 + M_2 t_2)/(t_1 + t_2)$. With CoFe 20 Å/NiFe 50 Å the effective moment/volume is $M_{SF} = 1011 \, \text{emu/cm}^3$ and the thickness is 70 Å.

Sense Current Polarity for Top and Bottom Spin Valves

Magnetization profiles of the free layer are calculated with (6.19) for $I_{\text{sense}} = -6, 0, +6$ mA and the results are plotted in Fig. 6.11; for positive current, the field H_I opposes the bias shift from H_{Dp} and helps establish a good bias profile, whereas negative current *aids* H_{Dp} and shifts the profile to a large positive level. For zero current, the bias level arises only from the pinned-layer demagnetizing flux and the interlayer coupling field H_{cpl}; for these curves, $H_{cpl} = -10$ Oe, which will be discussed and justified in the next section. The ideal bias point is achieved when the transfer curve of the free layer has no net offset when averaged along the

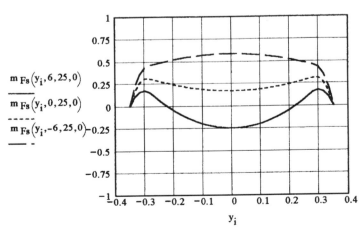

FIGURE 6.11 Normalized magnetization profiles of the free layer of a bottom spin valve for $I_{\text{sense}} = -6, 0, +6$ mA.

transverse (y) direction; this condition is obtained when $H_I + H_{cpl} + H_{Dp} = 0$, and the maximum linear dynamic range is then available for excitation by the signal field H_{sig}. When the sense current polarity and resultant H_I are chosen to balance the bias offset from the pinned layer, the field H_I *opposes* the pinning field H_{ex}. As will be shown in a following section on thermal analysis of GMR heads, H_{ex} reduces in magnitude with increasing device temperature, so unpinning can be reduced if H_I *aids* the pinning field. This is one of the many engineering trade-offs encountered with device design and analysis: H_I opposing H_{ex} gives good linear dynamic range to the signal but loses signal amplitude with the degree of unpinning, whereas H_I aiding H_{ex} gives improved pinning but degrades the linear dynamic range. Figures 6.12a,b compare the field directions for top and bottom spin valves; positive current flows into the page and M_P points along the negative y-axis in both cases, thus the selected pinning direction and sense current polarity will be different for top and bottom spin valve designs. With the top spin valve case, H_I and H_{Dp} are in the same direction and only H_{cpl} helps restore the ideal bias point, whereas H_I and H_{cpl} both oppose H_{Dp} in the bottom spin valve design, and thus the ideal bias point is more nearly achieved.

Interlayer Coupling Field

The free-layer response is influenced by short-range magnetic coupling with the pinned layer through the intervening nonmagnetic spacer; the interaction can be positive, which favors parallel alignment of M_P and M_F, or negative, which favors antiparallel alignment. Kools et al. (1995) and Leal and Kryder (1996) describe mechanisms for interlayer coupling: direct ferromagnetic exchange through pin holes in the spacer, Néel-type topological coupling [correlated waviness or "orange-peel" coupling; see Néel (1962) and Zhang and White (1996)], and indirect oscillatory exchange coupling (similar to a Ruderman–Kittel–Kasuya–Yodsida, or RKKY, exchange interaction). Leal and Kryder (1996) showed that oscillatory interaction dominates for very smooth surfaces and topological coupling dominates for rough surfaces. The problem of direct exchange coupling through pin holes (which spoils the spin valve effect) is ignored in this chapter, and attention is confined to oscillatory exchange (AFM and FM) and positive (FM) topological interactions through a nonmagnetic spacer. The correlated waviness interaction energy, given by (2.11) and rewritten here for convenience (with a small change in terminology), is

$$E_{topo} = \frac{\pi^2}{\sqrt{2}} \frac{r^2}{\lambda} M_P M_F \exp\left(\frac{-2\pi\sqrt{2} t_S}{\lambda}\right) \qquad \text{ergs/cm}^2, \qquad (6.22)$$

where r is the waviness amplitude of each film, λ is the wavelength of the surface variations, and t_S is the spacer thickness. With the pinned and free layers uniformly saturated, $M_P = 1540$, $M_F = 800$ emu/cm^3 (CoFe/Cu/NiFe), $r = 8$ Å, $\lambda = 150$ Å,

178 GMR HEAD DESIGN AND ANALYSIS

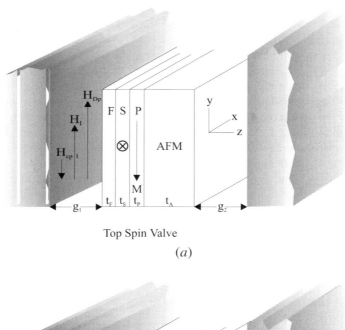

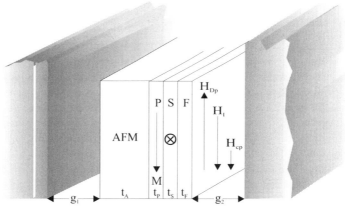

FIGURE 6.12 Sketches showing geometry and field polarities of (*a*) top spin valve and (*b*) bottom spin valve designs.

and $t_S = 25$ Å, the coupling energy is about 8.3×10^{-3} erg/cm^2, and the corresponding field

$$H_{\text{topo}} = \frac{E_{\text{topo}}}{M_{SF} t_F} \qquad \text{(Oe)} \qquad (6.23)$$

would be about 10.4 Oe for a free-layer thickness $t_F = 100$ Å.

The oscillatory interaction for the free layer is of the form

$$E_{ex,F} = \frac{E_0}{(k_0 t_S)^2} \sin\left(\frac{2\pi t_S}{\Lambda} + \phi\right), \tag{6.24}$$

where the surface energy E_0, wavenumber k_0, wavelength Λ, and phase shift ϕ parameters are extracted from experiment [see (2.13) and the discussion in Chapter 2]. The measured interlayer coupling field H_{cpl} for spin valve sandwiches with a structure $Co_{90}Fe_{10}$ 20 Å/Cu t_S Å/$Co_{90}Fe_{10}$ 20 Å is shown in Fig. 6.13a [unpublished data from Tong and Funada (1999)]; the solid line is a curve fit based on the sum of two interactions

$$H_{cpl}(t_S) = \frac{E_{ex,F}(t_S) + E_{topo}(t_S)}{M_{SF} t_F}, \tag{6.25}$$

where the fitting parameters are $M_{SF} = 1540$ emu/cm^3, $t_F = 20$ Å (the free-layer thickness), $E_0 = 1.35$ ergs/cm^2 (which gives a coupling strength $E_{ex} = 0.0135$ erg/cm^2 at 10 Å of Cu thickness), $k_0 = 1.0$ Å^{-1}, $\Lambda = 10.7$ Å, $\phi = 1.6\pi$ rad, $r = 4.2$ Å, and $\lambda = 103$ Å. Plots of the total coupling energy per square centimeter $E_{cpl} = E_{ex,F} + E_{topo}$ and the individual terms are shown in Fig. 6.13b. With smooth layers, H_{cpl} influences the bias point less than H_{Dp}, but it is large enough for consideration in the design process and in the analysis of root causes in spin valve device behavior. In Fig. 6.13a at $t_S = 25$ Å, the experimental coupling field is about 15 Oe, of which 10 Oe is from topological coupling and 5 Oe is from oscillatory exchange.

Output Signal and Asymmetry Versus Sense/Bias Current

The output signal of a spin valve includes GMR and AMR effects; Fig. 2.11 shows greatly simplified transfer curves for GMR, AMR, and a combination of the two effects; the AMR contribution shifts the operating point upward from the ideal midpoint, and the slope for negative excitation is therefore greater than that for positive excitation and the signal asymmetry is shifted toward negative values. The GMR portion is defined relative to the total sense current flowing through the device while the AMR portion depends only on the free-layer magnetization $\mathbf{M}_F$ and current I_2. Because the GMR effect depends on the difference in alignment between $\mathbf{M}_F$ and $\mathbf{M}_P$ and the AMR effect arises from the difference in alignment between $\mathbf{M}_F$ and the current density, the relation for signal voltage (6.8) can be rewritten in a convenient form that includes both contributions to the signal:

$$\text{Signal} = I_{sense} R_S \frac{W}{h} \left[\frac{\Delta R}{R} \left(\frac{-\langle \cos \Delta\theta \rangle}{2} \right) + \frac{\Delta \rho_{AMR}}{\rho_F} (1 - \langle \sin^2 \theta_F \rangle) \right], \tag{6.26}$$

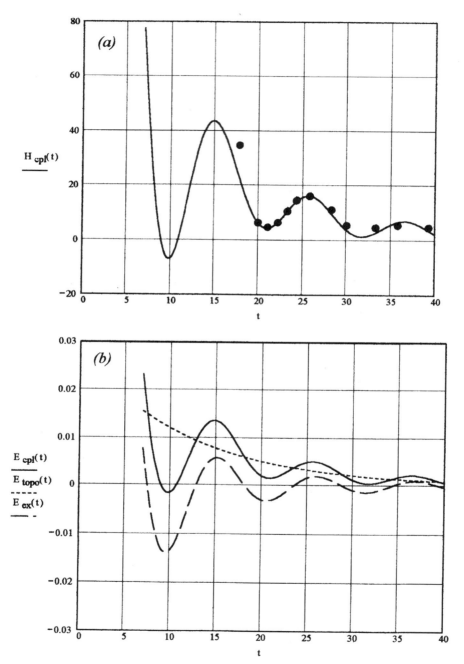

FIGURE 6.13 (a) Experimental interlayer coupling field for CoFe/Cu/CoFe spin valve structures as a function of Cu thickness. (b) Calculated fitting curves based on RKKY-like oscillatory interaction and a ferromagnetic topological interaction. The solid curves in (a) and (b) are the sum of the two separate interactions.

where $\langle\cos\Delta\theta\rangle$ is the average cosine of $\Delta\theta(y) = \theta_F(y) - \theta_P(y)$ over the stripe height and the AMR response curve is given by $\langle\cos^2\theta_F\rangle = 1 - \langle\sin^2\theta_F\rangle = 1 - \langle m_F^2 \rangle$. The signal flux from a recording medium was discussed in Chapter 5 and the equations of that analysis [(5.4)–(5.6)] are applicable to the shielded GMR sensor. The normalized magnetizations of the pinned and free layers given in (6.12) and (6.19) are modified to include the excitation from a magnetic storage disk. One cannot assume that the pinned layer is insensitive to signal flux at the bottom (input) region of the stripe, and thus the equations describing the excitation flux are written to allow for sharing of this flux on a proportional basis. With this understanding, the signal flux is defined with (5.23a,b), which were used for the dual-stripe MR (DSMR) analysis and are repeated here for convenience to the reader. With small changes in terminology, the excitation for each of the layers is

$$\Delta M_P^{\text{signal}}(y) \simeq \left(\frac{\phi(y_e)t_P}{4\pi W(t_P + t_F)^2}\right)\frac{\sinh[(h/2-y)/\lambda_{\text{sh}}]}{\sinh(h/\lambda_{\text{sh}})}, \quad (6.27a)$$

$$\Delta M_F^{\text{signal}}(y) \simeq \left(\frac{\phi(y_e)t_F}{4\pi W(t_P + t_F)^2}\right)\frac{\sinh[(h/2-y)/\lambda_{\text{sh}}]}{\sinh(h/\lambda_{\text{sh}})}, \quad (6.27b)$$

which is based on the understanding that the sensor magnetization is shifted in direct proportion to the injected signal flux, which gradually leaks to the shields along the stripe and returns through the head–medium interface back to the medium. Assuming an arctangent transition in a recording medium, the flux flowing into a shielded sensor is given by the relation derived by Potter (1974) and later modified by Bertram (1995) (see Chapter 1 discussion); reading from (5.5), this flux is

$$\phi(y_e) = \frac{2}{\pi}4\pi M_r\,\delta W\frac{y_e}{g}\left[f\!\left(\frac{g+t/2}{y_e}\right) - f\!\left(\frac{t/2}{y_e}\right) + f\!\left(\frac{-g-t/2}{y_e}\right) - f\!\left(\frac{-t/2}{y_e}\right)\right]$$

$$= \frac{16 M_r\,\delta W\,y_e}{g}\left[f\!\left(\frac{g+t/2}{y_e}\right) - f\!\left(\frac{t/2}{y_e}\right)\right], \quad (6.28)$$

where $f(z) = z\tan^{-1}(z) - 0.5\ln(1+z^2)$, g is the distance from the sense layer to a shield ($g \simeq g_2 \simeq g_1$), and the effective readback spacing $y_e = d + a + \frac{1}{2}\delta$ (see Chapter 1 for the definition of these variables). It is more convenient to work with the normalized excitation, so (6.27a,b) become

$$m_P^{\text{sig}}(y, M_r) = \frac{\Delta M_P^{\text{signal}}(y)}{M_{SP}}, \quad (6.29a)$$

$$m_F^{\text{sig}}(y, M_r) = \frac{\Delta M_F^{\text{signal}}(y)}{M_{SF}}, \quad (6.29b)$$

and the total normalized magnetization of each layer becomes

$$m_{tP}(y, I, T, M_r) = m_P(y, I, T) + m_P^{\text{sig}}(M_r), \quad (6.30a)$$

$$m_{tF}(y, I, T, M_r) = m_F(y, I, T) + m_F^{\text{sig}}(M_r), \quad (6.30b)$$

neither of which can ever exceed the saturation level (± 1.0). Plots of (6.30b) are given in Fig. 6.14 for selected values of each parameter; the unexcited curves

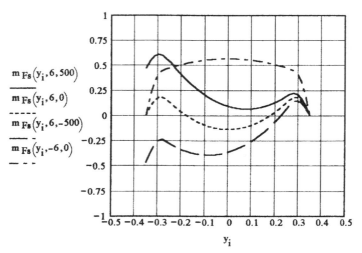

FIGURE 6.14 Normalized bias curves for a bottom spin valve free layer showing positive, zero, and negative signal excitations for $I_{\text{sense}} = +6\,\text{mA}$ and for $I_{\text{sense}} = -6\,\text{mA}$ without signal excitation.

($M_r = 0$) show the bias shift with current polarity, and the curves at $+6\,\text{mA}$ with $M_r = \pm 500\,\text{emu/cm}^3$ show the modulation of m_F with excitation (with a medium thickness $\delta = 200\,\text{Å}$, the normalized medium excitation $M_r\delta = 1.0 \times 10^{-3}\,\text{emu/cm}^2 = 1.0\,\text{memu/cm}^2$).

Since the normalized magnetization $m_i = \sin\theta_i$, the angle at any position y in the stripe height is found from the relation $\theta_i = \sin^{-1} m_i$, and the required angular difference is

$$\Delta\theta(y, I, T, M_r) = \sin^{-1}[m_F(y, I, T, M_r)] - \sin^{-1}[m_P(y, I, T, M_r)]. \qquad (6.31)$$

The averages required for (6.26) become

$$\langle \cos[\Delta\theta(I, T, M_r)] \rangle = \frac{1}{h}\int_{-h/2}^{+h/2} \cos[\Delta\theta(y, I, T, M_r)]\,dy, \qquad (6.32a)$$

$$\langle \sin^2\theta_F(I, T, M_r) \rangle = \frac{1}{h}\int_{-h/2}^{+h/2} [m_F(y, I, T, M_r)]^2\,dy, \qquad (6.32b)$$

and finally, the device voltage transfer curve (GMR + AMR) is expressed as

$$\text{Signal} = I_{\text{sense}}[\Delta R_{\text{GMR}}(-\tfrac{1}{2}\langle\cos\Delta\theta\rangle) + \Delta R_{\text{AMR}}(1 - \langle\sin^2\theta_F\rangle)], \qquad (6.33)$$

where $\Delta R_{\text{GMR}} = (\Delta R/R)R_S W/h$ and $\Delta R_{\text{AMR}} = (\Delta\rho_{\text{AMR}}/\rho_F)R_S W/h$. Coehoorn et al. (1998) indicate that typical MR ratios for NiFe 80 Å/Cu 25 Å/NiFe 60 Å sandwiches are 4.5% GMR and 1.1% AMR, so the AMR effect is a perturbation of about 24% relative to the GMR effect. Uehara et al. (1996) showed an AMR effect of about 16% of the total measured ΔR for spin valves with a sandwich structure given by Ta 100 Å/NiFe 60 Å/CoFe 25 Å/Cu 36 Å/CoFe 55 Å/FeMn 100 Å/Ta 100 Å. In the form given by (6.33) the signal goes from a low to a high state with the DC component removed. Figure 6.15a gives plots of normalized transfer curves for positive and negative sense currents with the excitation given in units of memu per square centimeter; the inflection on the +6-mA curve at excitation of 2 memu/cm²

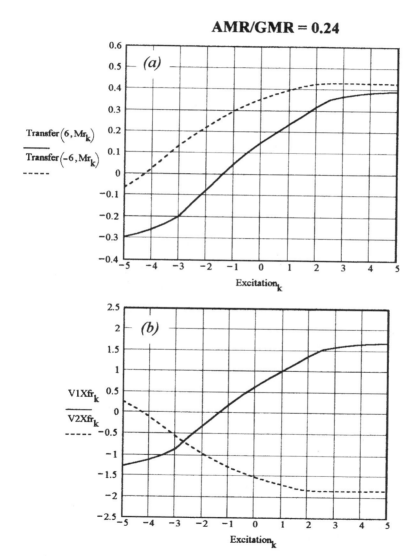

FIGURE 6.15 (a) Normalized transfer curves at $I_{\text{sense}} = \pm 6$ mA with the relative effects of 1.0 GMR and 0.25 AMR. (b) Voltage transfer curves (mV) computed with 3.5% GMR and 0.84% AMR effects (a 25% relative effect). (c) Voltage transfer curves (mV) with and without an AMR effect; GMR effect = 3.5%.

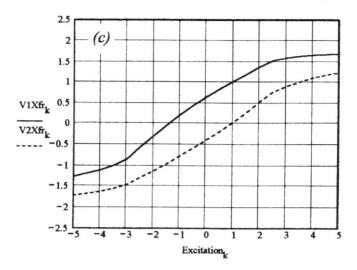

FIGURE 6.15 (*Continued*)

arises from the unpinning behavior of the pinned layer in its reponse to the signal field. A higher exchange field eliminates this problem. The structure is a bottom spin valve IrMn 100 Å/CoFe 20 Å/Cu 25 Å/NiFe 100 Å with $g_1 = g_2 = 650$ Å and the film properties are unchanged from the values used to estimate λ in (6.18). With positive sense current, the free layer operates *above* the optimal bias point because the AMR term adds an offset of 24% in these calculations. With no AMR contribution, the operating point is *below* the optimal. At negative sense currents, the transfer curve slope and dynamic range are each reduced. Figure 6.15b is a plot of the corresponding voltage transfer curves (6.33) computed with a GMR effect of 3.5% [taken from (6.6)], an AMR effect of 0.84% (0.24 × 3.5%), and a sheet resistance R_S of 16.4 Ω/square obtained from (6.9). The film resistivities are estimated with (2.23a) using the parameters of Table 2.4; this gives a maximum available $\Delta R_{GMR} = 0.78$ Ω and $\Delta R_{AMR} = 0.19$ Ω for a track width $W = 1.0$ μm and stripe height $h = 0.70$ μm. Figure 6.15c shows transfer curves at $I = +6$ mA with and without the AMR effect; at zero excitation $I \Delta R_{AMR}$ gives 1.1 mV of positive shift in the operating point and the curvature (convex) produces *negative* asymmetry, which is in stark comparison with the *positive* asymmetry arising from concave behavior where no AMR effect is present. The transfer curves for ±6 mA reveal the bias point shift and dynamic range trade-off mentioned earlier. At $I = -6$ mA, the free layer enters saturation at about 1.5 memu/cm² excitation because H_I aids H_{Dp}; with $I = +6$ mA (and H_I opposing H_{Dp}) saturation occurs at excitations near -3 and $+2.0$ memu/cm². Details of the free-layer saturation are revealed by looking at the magnetization angle $\theta_F(y) = \sin^{-1}[m_F(y)]$ at several positions along the stripe; in Fig. 6.16 saturation occurs in the bottom half of the stripe where the excitation level is strongest.

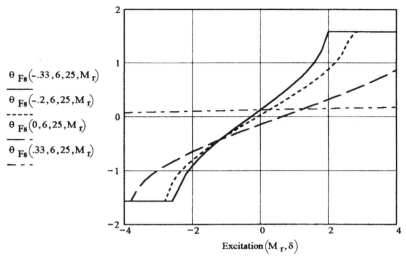

FIGURE 6.16 Plots of the free-layer magnetization angle versus excitation (memu/cm²) for several positions along the stripe $-0.33\,\mu m \leq h \leq +0.33\,\mu m$.

Positive, negative, and peak-peak output signals are defined using the voltage transfer curves in Fig. 6.15b:

$$V_p(I, T, M_r) = |V_{Xfr}(I, T, M_r) - V_{Xfr}(I, T, 0)|, \tag{6.34a}$$

$$V_n(I, T, M_r) = |V_{Xfr}(I, T, 0) - V_{Xfr}(I, T, -M_r)|, \tag{6.34b}$$

$$V_{pp}(I, T, M_r) = V_p(I, T, M_r) + V_n(I, T, M_r). \tag{6.34c}$$

Signal asymmetry is defined by the relation

$$\text{Asymmetry} = \frac{V_p(I, T, M_r) - V_n(I, T, M_r)}{V_{pp}(I, T, M_r)}. \tag{6.35}$$

With this convention, the positive and negative signals are defined by their respective *amplitudes*, which are always positive. Peak-to-peak amplitude as a function of sense current is plotted in Fig. 6.17a for a simple free-layer NiFe 100 Å with and without the AMR effect, and Fig. 6.17b shows the asymmetry in amplitudes. At +6 mA the peak-to-peak signal is 806 µV/µm with a signal asymmetry of −7.6%. With no AMR effect, the peak-to-peak amplitude is greatly improved at negative currents and the asymmetry at all currents is small and positive. The fluctuations in asymmetry arise from computational artifacts traced to numerical averaging over the stripe height; increasing the number of increments and/or increasing the pinning field smooths the variations in $m_P(y, I)$ and $m_F(y, I)$. The improved output arising from CoFe at each Cu interface is shown in the plots of Figs. 6.17c,d. The free-layer structure CoFe 20 Å/NiFe 60 Å replaces the simple NiFe 100 Å layer used in Figs.

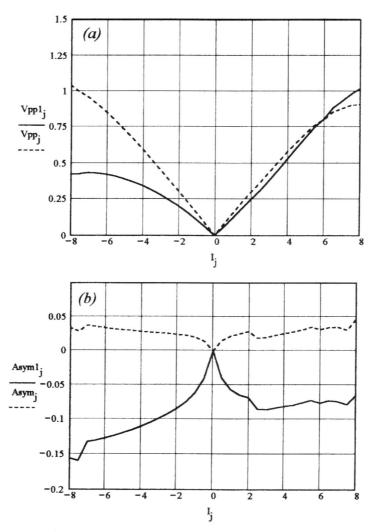

FIGURE 6.17 Calculated (*a*) signal (mV, p-p) and (*b*) asymmetry bias curves for a bottom spin valve with 100 Å NiFe free layer with 3.5% GMR effect and an AMR effect of 0.84% or zero at an ambient temperature of 25°C. Comparison of calculations with experiment (*c*) and (*d*) for a top spin valve with a compound free layer CoFe 20 Å/NiFe 60 Å with GMR = 6.4% and AMR of 1.1% and 1.5%. The signal (*c*) agrees best with 1.1% AMR while asymmetry (*d*) agrees better with AMR = 1.5%. Ambient temperature is 25°C.

6.17*a,b*. The effective properties of the free layer are $M_{SF} = 985 \, \text{emu/cm}^3$, $H_{KF} = 6 \, \text{Oe}$, and $t_F = 80 \, \text{Å}$ with a GMR effect of 6.4% (see Fig. 6.4*b*) and relative AMR effects of 17 and 23% (i.e., $\Delta \rho_{AMR}/\rho_F$ equals 0.011 and 0.015.) The sheet conductance of the composite free layer is $G_F = 0.0176 + 0.0043 = 0.0219 \, \Omega^{-1}$, so the average resistivity of the free layer is $\rho_F = t_F/G_F = 36.5 \, \mu\Omega\text{-cm}$ and $\Delta \rho_{AMR}$ of

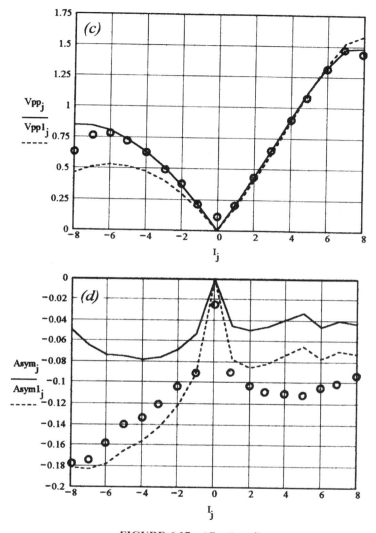

FIGURE 6.17 (*Continued*)

0.4 and 0.6 $\mu\Omega$-cm; with these properties, the composite free layer gives $\Delta R_{\text{GMR}} = 1.6\,\Omega$ and ΔR_{AMR} of 0.27 and 0.41 Ω at 25 °C. For a sense current of $+6\,\text{mA}$, the output is 1500 $\mu\text{V}/\mu\text{m}$ with asymmetries of -4.4 and -7.8% for the two cases of AMR effect. The experimental points (open circles) for a head with $W = 0.9\,\mu\text{m}$ show the signal agrees better with the 1.1% AMR case, whereas the experimental asymmetry agrees better with the 1.5% AMR case. The main point of the calculations is that asymmetry is very sensitive to the exact nature of the transfer curves, and the AMR effect has considerable influence in this regard. With the assumed AFM properties, unpinning becomes apparent at $|I| \sim 7\,\text{mA}$. Thermal unpinning is analyzed in the next two sections.

The published experiments of Tsang et al. (1994) for a top spin valve with $W = 2.0\,\mu m$ and $h = 1.0\,\mu m$ gave 750–$1000\,\mu V/\mu m$ at $+6\,mA$ and $M_r\delta = 1.25\,memu/cm^2$ excitation. Their film thicknesses were NiFe 100 Å/Cu 25 Å/Co 22 Å/FeMn 110 Å and the shield-to-shield gap was 0.25 μm. With the parameters of Tsang et al., the model developed above gives a peak-to-peak output of $815\,\mu V/\mu m$ at $T = 25°C$, which agrees with their experiment. The asymmetry results differ somewhat: At I values of -6 and $+6\,mA$ the experiment gave -25 to -35 and -12% asymmetry, respectively, where the present model gives -18 and -11%. The differences can be explained by an AMR effect that is about 30% of the GMR effect instead of the assumed 24% in the model calculations.

Thermal Analysis of GMR Heads

The temperature rise of an energized spin valve is found using the thermal analysis of MR heads given in Chapter 2; with appropriate averages over the spin valve sandwich, this analysis is directly applicable to the present situation. Reading from (2.35), (2.37), and (2.44), the temperature rise profile of a spin valve sensor is written

$$T_{\text{rise}}(x) = \frac{\lambda_f^2 \rho_f J^2}{\kappa_f} \left[1 - \frac{\cosh(x/\lambda_f)}{\cosh(W/2\lambda_f) + (\kappa_f t_f h \lambda_L / \kappa_L t_L h_L \lambda_f) \sinh(W/2\lambda_f)} \right], \quad (6.36)$$

where the characteristic lengths for heat decay in the sensor film (f) and lead (L) regions are

$$\lambda_f = \sqrt{\frac{\kappa_f g t_f}{2\kappa_0 - \alpha_f \rho_f J^2 g t_f}}, \quad (6.37)$$

$$\lambda_L = \sqrt{\frac{\kappa_L g t_L}{2\kappa_0 - \alpha_L \rho_L J^2 g t_L}}, \quad (6.38)$$

respectively. In the form given by (6.36), heating in the leads is ignored, but the full expression reduces to (2.45) for long leads with insignificant heating. In this analysis the bottom spin valve is treated as a single thin-film resistor and heat is conducted through the leads and insulating gaps. In this simplified treatment the effective resistivity $\rho_f = t_f/G_t$ (where $t_f = \sum t_i$ is the total thickness of all conducting films), the average thermal conductivity $\kappa_{\text{av}} = t_f / \sum (t_i/\kappa_i)$, and the average temperature coefficient of resistivity is $(1/n) \sum \alpha_i$ (the value that would be measured for the complete device). Thermal conductivity is estimated using the Lorentz number $L = \kappa/\sigma T = 2.45 \times 10^{-8}\,(V/K)^2$, which is fairly constant in the region of 300 K (see Chapter 2 and the references therein). The electrical conductivity $\sigma = 1/\rho$. (*Note:* The high resistivity of thin films, especially for Cu, is discussed at length in Chapter 2. See Table 2.4 for resistivity and mean free path parameters of films.) With the IrMn 100 Å/CoFe 20 Å/Cu 25 Å/NiFe 100 Å structure, the resistivities (in microhm-centimeters) are approximately 325/43/11.3/30. The estimated thermal conductivities (in watts per meter-Kelvin) are 2.3/17/65/24.5, respectively;

the average thermal conductivity is thus $\kappa_{av} \simeq (2.45 \times 10^{-8} \text{ m})/(4.90 \times 10^{-9} \text{ m}^2 \times \text{K/W}) = 5.0 \text{ W/m-K}$, and the effective resistivity $\rho_f = 2.45 \times 10^{-6} \text{ cm}/0.061 \, \Omega^{-1}) = 40.2 \, \mu\Omega\text{-cm}$. Experimentally, the average temperature coefficient of resistivity is about $2 \times 10^{-3} \text{ K}^{-1}$ including the leads.

The temperature profile across the sensor width is plotted in Fig. 6.18a with Al_2O_3 ($\kappa_0 = 1.0 \text{ W/m-K}$) insulating gaps $g_1 = g_2 = g = 650 \text{ Å}$ and leads with

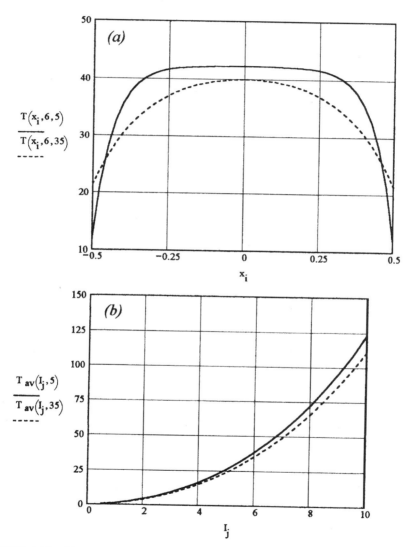

FIGURE 6.18 (a) Calculated temperature profiles across a sensor width $W = 1.0 \, \mu\text{m}$ for a stripe $h = 0.7 \, \mu\text{m}$ with thermal conductivity $\kappa_{mr} = 5$ and 35 W/m-K. (b) Comparison of average temperature of a sensor as a function of bias current for $\kappa_{mr} = 5$ and 35 W/m-K.

$t_L = 500$ Å and $\kappa_L = 15$ W/m-K. Figure 6.18b is a plot of the average temperature (2.47) as a function of sense current; the average temperature rise exceeds 50 K at a sense current greater than 7 mA. Two values of κ_f (5 and 35 W/m-K) are used to demonstrate the fairly insensitive nature of these temperature estimates to the exact value of average thermal conductivity of the GMR sandwich. The AFM layer has high resistivity (for a conductor) and its low thermal conductivity (2.3 W/m-K) combined with the layer thickness (100 Å) substantially reduces the average thermal conductivity of the sandwich. The internal ambient temperature of operating disk drives can reach 60–70°C, so an energized GMR element could be at a temperature $T = T_{\text{ambient}} + T_{\text{rise}} \simeq 120°$C or more, depending on the operating current. Some reliability lifetime tests (to be discussed in Chapter 8) stress GMR elements at temperatures in the neighborhood of 180–190°C, so thermal unpinning (i.e., degradation of the exchange field with temperature) is a major concern when designing spin valves and choosing AFM materials.

AFM Layer and Distribution of Blocking Temperatures

The relation for the exchange energy $E_{\text{ex}}(T)$ between AFM/FM layers is given in (2.15) and is useful for estimating the exchange pinning field at device operating temperatures. Using the relation in (2.14), $H_{\text{ex}}(T)$ can be written as

$$H_{\text{ex}}(T) \simeq \frac{E_{\text{ex0}}/(M_{SP}t_P)}{\exp[(T - T_0)/\Delta T] + 1}, \qquad (6.39)$$

and the distribution parameters of the AFM/pinned-layer behavior can be extracted from experimental curves. Figure 6.19 shows exchange field data from Lederman (1999) for NiFe 110 Å/NiMn, NiFe 110 Å/PdPtMn, and NiFe 170 Å/IrMn and from Fuke et al. (1997) for IrMn 80 Å/CoFe 20 Å along with fits to (6.39); the exchange surface energies, T_0, ΔT, and T_B parameters for each AFM material are given in Table 6.2. (See the discussion in Chapter 2 on exchange coupling with AFM films.) According to Lederman (1999), the minimum AFM layer thicknesses for saturation of H_{ex} are about 150, 250, and 50–100 Å for NiMn, PdPtMn, and IrMn, respectively. (The range for IrMn shows the process sensitivity of AFM materials: 50 and 100 Å are for DC magnetron and radio frequency (RF) diode sputtering, respectively.) Fuke et al. (1997) annealed their magnetron-sputtered samples for 5 min at 280°C in a magnetic field and vacuum. It is instructive to explore the loss of exchange pinning with a material having somewhat degraded properties. Of these four samples, the lowest exchange energy and blocking temperature are found with the IrMn 145 Å sample; the magnitude of the exchange field for a thin CoFe layer ($M_{SP} = 1540$ emu/cm^3, $t_P = 20$ Å) is plotted in Fig. 6.20a as $|H_{\text{ex}}|$ versus T_{av} (the average *rise* above ambient) and Fig. 6.20b as $|H_{\text{ex}}|$ versus I.

Unpinning and Repinning the Pinned Layer The impact of thermal unpinning is shown in curves of output signal and asymmetry versus sense current; Figs. 6.21a,b are calculated for the IrMn 80 Å and IrMn 145 Å materials of Table 6.2 with a

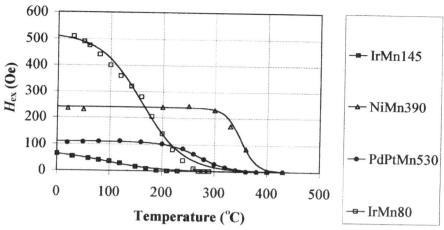

FIGURE 6.19 Exchange field data versus temperature for NiFe 110 Å/NiMn 390 Å, NiFe 110 Å/PdPtMn 530 Å, NiFe 170 Å/IrMn 145 Å. [Data points from Lederman (1999); IrMn 80 Å/CoFe 20 Å data points from Fuke et al. (1997.)] Curves are empirical fits to Eq. (6.39) with the parameters listed in Table 6.2.

simulated disk drive ambient of 70°C to which the average thermal rise of the sensor is added. At ±8 mA the sensor temperature rise is 81°C above ambient, so the sensor is at 151°C (424 K) and the device with IrMn 145 Å unpins at +8 mA because H_{ex} has degraded to −63 Oe and H_I opposes H_{ex}. With the IrMn 80 Å parameters, $H_{ex} = -287$ Oe at 8 mA. The fluctuations in the asymmetry curves in Fig. 6.21b arise from computational artifacts (increasing the number of elements for numerical averaging along the stripe reduces the fluctuations).

Figure 6.22 plots the total field $H_1(I)$ at the IrMn 145 Å pinned layer; the total temperature versus current $T(I)$ and $H_{ex}[T(I)]$ are included implicitly in $H_1(I)$. At currents greater than +8 mA (but less than the melting current), the field applied to the pinned layer would be positive, so repinning in the opposite sense becomes possible. Pulses of sense current of short duration (microseconds or less) can be exploited to repin an unpinned layer; diffusion processes require long times (hours, days, weeks) to accumulate irreversible changes to the interlayer metallurgy. This subject is discussed in greater depth in Chapter 8.

The exchange field curve $H_{ex}(T)$ is reversible with temperature provided the maximum temperature does not exceed the original pinning procedure where a large

TABLE 6.2 Distribution of Blocking Temperatures

Material	E_{ex0} (ergs/cm^2)	T_0 (°C)	ΔT (°C)	$T_B \simeq T_0 + 2\Delta T$ (°C)
IrMn 80 Å	0.16	160	45	250
IrMn 145 Å	0.085	90	50	190
PdPtMn 530 Å	0.097	270	30	330
NiMn 390 Å	0.211	350	15	380

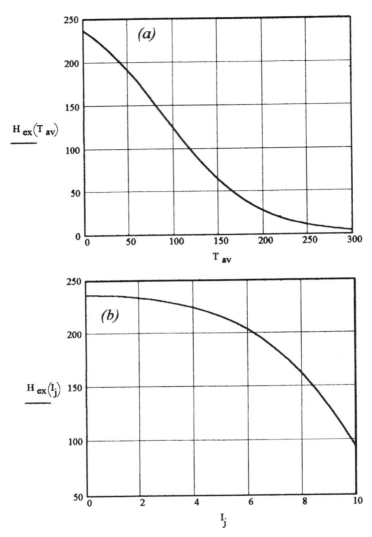

FIGURE 6.20 Calculated exchange field of the IrMn 145 Å sample in Table 6.2 with a CoFe 20 Å pinned layer plotted versus (a) average temperature rise (°C) and (b) bias current (mA).

field (~10 kOe is sufficient to overcome the large demagnetizing fields of the thick shield layers) is applied and the device is annealed at a temperature $T \sim T_B$ for a number of hours. The temperature cannot be much greater than about 270°C, otherwise interface diffusion occurs between the various layers and device function is seriously degraded, if not outright destroyed. If H_I opposes the original direction of H_{ex}, unpinning and repinning in the opposite direction can occur if the

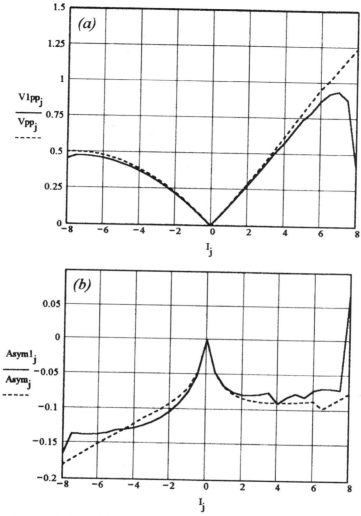

FIGURE 6.21 Calculated (*a*) signal and (*b*) asymmetry bias curves comparing the IrMn 145 Å and IrMn 80 Å AFM materials pinning a CoFe 20 Å with a NiFe 100 Å free layer. The simulated disk drive ambient is 70°C and; at $I = 8$ mA the sensor is at $T = 151$°C. The IrMn 145 Å AFM begins to unpin ($H_{ex} = -63$ Oe) while the IrMn 80 Å layer remains pinned with $H_{ex} = -287$ Oe.

temperature is sufficiently high, but not so high as to degrade the device. The temperature profile is not uniform, so unpinning will normally occur in the vicinity of the track center and spread toward the track edges with increasing temperature. The subject of microtrack profiles and nonuniform readback response from partial unpinning is discussed in Chapter 7, where one finds a more thorough treatment of

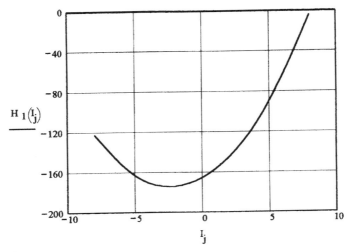

FIGURE 6.22 Total magnetic field at the interface between a pinned layer and an IrMn AFM layer plotted as a function of bias current with drive at ambient $+70°C$.

this important subject. The average temperature rise of the stripe is plotted versus sense current in Fig. 6.23 for stripe heights h of 0.4, 0.6, 0.8, and 1.0 µm; the shortest stripe exceeds 300°C at 8 mA so damage from interdiffusion should be expected if the device is energized for long periods. The melting current, which was introduced

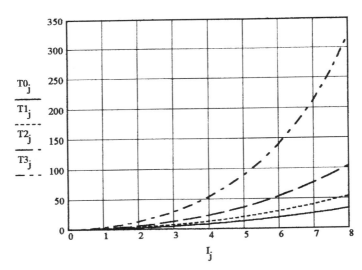

FIGURE 6.23 Family of curves showing the average temperatures $T0$, $T1$, $T2$, and $T3$ of sensors as a function of bias current for stripe heights h of 0.4, 0.6, 0.8, and 1.0 µm, respectively.

and defined in Chapter 2 in (2.50), is given here for convenience:

$$I_{\text{melt}} = \left[\frac{T_{\text{melt}} \kappa_0 (g_1 + g_2) t_f h^2}{g_1 g_2 \rho_f (1 + \alpha_f T_{\text{melt}})} \right]^{1/2}, \qquad (6.40)$$

and with the present design, $I_{\text{melt}} = 26.4h$ (in milliamperes, with h in micrometers). In other words, if the melting temperature is 1450°C (which is a rough estimate for the sandwich), the melting current is between 10.6 and 26.4 mA over the range of stripes between 0.4 and 1.0 μm high.

Signal and Sensor Temperature Dependence on Stripe Height Stripe height control is a major processing challenge in large-volume manufacturing, and thus it is important to have some idea of the readback signal dependence on stripe height variations. Under the condition of constant sense/bias current (which is found in a variety of disk drive applications for spin valves), devices with short stripes will have higher resistance and greater heating than devices with long stripes. High temperature serves to unpin the pinned layer and to reduce the magnitude of the GMR and AMR effects. As discussed in Chapter 2, both of these effects monotonically decrease with increasing temperature, and this behavior can be conveniently subsumed under the approximate relation

$$\Delta R_{\text{GMR}}(T) \simeq \Delta R_{\text{GMR0}} \exp(\alpha_{\text{GMR}} T), \qquad \Delta R_{\text{AMR}}(T) \simeq \Delta R_{\text{AMR0}} \exp(\alpha_{\text{AMR}} T), \qquad (6.41)$$

where experimentally it is found that $\alpha_{\text{GMR}} \simeq \alpha_{\text{AMR}} \simeq -2 \times 10^{-3} \text{ K}^{-1}$. The experimental evidence for the exponential drop in the AMR effect is given in Fig. 2.15c. The data for the GMR effect are not clearly exponential in nature, but the curves are not linear either, so the exponential approximation is used as a computational convenience. Output signal versus stripe height is given in Fig. 6.24a and the temperature (25°C ambient + average temperature rise) at 6 mA is given in Fig. 6.24b. Unpinning occurs at 4.5, 5.5, and 6.0 mA for stripes of 0.4, 0.5, and 0.6 μm, so the signals for these stripes are calculated at the unpinning current, and the results for longer stripes are calculated at 6.0 mA.

Monte Carlo Analysis of Basic Spin Valves

The Monte Carlo technique introduced in Chapter 5 is a useful tool for learning about processing sensitivities of devices having numerous critical variables. The structure of spin valves is similar to SAL-biased AMR heads and the number of important geometrical variables is nearly the same. This discussion of basic spin valves closes with a comparison of two designs: Both are bottom spin valves with the structure AFM/CoFe 20 Å/Cu 25 Å/CoFe 20 Å/NiFe 60 Å. The first case studied has a thin 70-Å IrMn AFM layer with the properties $\rho_A = 325\ \mu\Omega\text{-cm}$, $E_{\text{ex0}} = 0.16\ \text{erg/cm}^2$, $T_0 = 160°C$, $\Delta T = 45°C$, and $T_B = 260°C$ (see Table 6.2 for

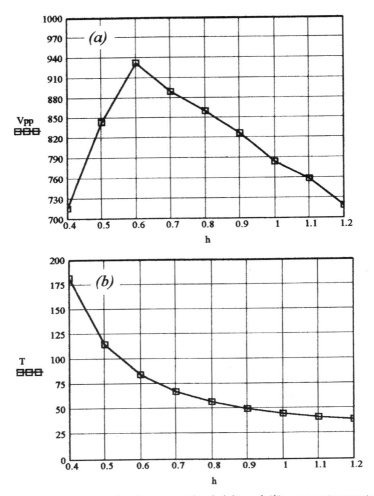

FIGURE 6.24 (a) Output signal versus stripe height and (b) average temperature versus stripe height for a basic spin valve.

the "IrMn 80 Å" entry), which gives (downward) pinning fields H_{ex} of -519 and -480 Oe at 25 and 70°C, respectively. The second design uses a much thicker 400-Å PdPtMn AFM layer whose properties are $\rho_A = 185\,\mu\Omega$-cm, $E_{ex0} = 0.10$ erg/cm^2, $T_0 = 270$°C, $\Delta T = 30$°C, and $T_B = 330$°C (see Table 6.2 entry for "PdPtMn 530 Å".) The pinning fields for this material are -315 and -314 Oe at 25 and 70°C, respectively. That is, the hypothetical IrMn layer has a higher exchange field than the PdPtMn layer at room temperature, but IrMn operates somewhat hotter because its sheet resistance is 19.7 Ω/square, compared with 14.2 Ω/square for PdPtMn. In addition, PdPtMn has a larger blocking temperature, so it is not clear which AFM material would perform better at a simulated ambient of 70°C, nor is it clear what the

TABLE 6.3 Spin Valve Parameters for Monte Carlo Analysis

Parameter	Mean	Standard Deviation
Free-layer NiFe/CoFe	80 Å	1.0 Å
Cu spacer	25 Å	1.0 Å
Pinned-layer CoFe	20 Å	1.0 Å
AFM layer IrMn	70 Å	0
AFM layer PdPtMn	400 Å	0
Insulation g_1 with IrMn	595 Å	10 Å
Insulation g_2 with IrMn	710 Å	10 Å
Insulation g_1 with PdPtMn	430 Å	10 Å
Insulation g_2 with PdPtMn	545 Å	10 Å
Stripe height	0.70 μm	0.10 μm
Read track width	1.00 μm	0.10 μm
Free-layer H_K	6.0 Oe	0.8 Oe
Pinned-layer H_K	8.0 Oe	1.0 Oe
Interlayer coupling H_{cpl}	−10 Oe	4 Oe

upper limit of sense current should be. Monte Carlo simulations are very useful for answering these kinds of questions, and in the exercises that follow, the designs are nearly identical except g_1 and g_2 are adjusted to maintain the free layer centered between the shields (zero-gap asymmetry) and the nominal shield-to-shield gap G_{ss} is 1500 Å in both cases. Each spin valve design is subjected to 1000 Monte Carlo trials at different values of sense/bias current, and the signal, asymmetry, and temperature statistics are noted. A threshold in sense current (and temperature) is reached where unpinning behavior is observed in at least one device out of the 1000 random variations in parameters, and at higher currents the number of devices that unpin dramatically increases. For this series of computer experiments, a device has "unpinned" if the amplitude is low, the temperature is at the blocking point or greater, and the asymmetry is either positive or less than −15%. Devices with long stripe heights can have positive asymmetry and a useful signal and are not unpinned. Table 6.3 lists the nominal head parameters and the standard deviations chosen for this analysis.

The results of 1000 trials at 6.0 mA for the IrMn spin valve are given in Table 6.4. The excitation is $M_r\delta = 1.0$ memu/cm^2 at an effective spacing $y_e = 0.095$ μm and the ambient temperature is 70°C. The ranges for signal and asymmetry are large, and

TABLE 6.4 IrMn SV Signal, Asymmetry, and Temperature

Response	Mean	Standard Deviation	Minimum	Maximum
Signal (μV, p-p)	1760	295	79	3400
Asymmetry (%)	0.52	4.4	−30.6	+92.3
Temperature (°C)	127	23.2	91.4	377

TABLE 6.5 Correlation Coefficients for IrMn Monte Carlo Analysis

Response	t_F	t_S	t_P	g_1	g_2	h	W	H_{KF}	H_{KP}	H_{cpl}
Signal	−0.077	−0.116	−0.064	−0.017	0.062	−0.600	0.610	0.088	0.056	−0.145
Asymmetry	−0.051	−0.073	0.029	−0.052	0.006	−0.338	−0.004	0.025	0.025	−0.034
Temperature	−0.015	−0.073	0.020	0.006	−0.006	−0.875	0.008	0.019	0.029	0.050

thus at least one device has unpinned at this operating current (the probability of unpinning at a given level of sense current is explored below). The mean temperature at 6.0 mA sense current is 127°C and the exchange field has reduced to approximately −360 Oe. The correlation coefficients (see Chapter 5) are given in Table 6.5. These correlations show that signal variations are dominated by track width and stripe height variations, where asymmetry and sensor temperature are dominated by stripe height variations, and thus the tolerances on film thicknesses are acceptable and improvements on stripe height and width control are justified. The interlayer coupling field variation ($\sigma = 4$ Oe with a mean of −10 Oe), which is quite large in this analysis, demonstrates that its influence might be measurable in extreme cases. In this design H_{cpl} assists in restoring the bias point to the optimal location, and thus the mean signal increases with larger (negative) coupling fields.

The responses for 1000 trials with PdPtMn SV at 6.0 mA and 70°C ambient are given in Tables 6.6 and 6.7. Excitation and effective spacing are unchanged from the IrMn case. Ranges for signal and asymmetry are not excessive, and thus unpinning has not occurred at 6 mA operating current. Mean signal and temperature each scale with device sheet resistance, which is about 30% lower than the IrMn design. At 99°C the exchange field is −324 Oe. The PdPtMn spin valve behavior is similar to the IrMn spin valve: signal is dominated by stripe height and track width variations, and asymmetry and temperature are dominated by stripe height variations. The

TABLE 6.6 PdPtMn SV Signal, Asymmetry, and Temperature

Response	Mean	Standard Deviation	Minimum	Maximum
Signal (μV, p-p)	1320	212	774	2340
Asymmetry (%)	0.45	1.4	−3.1	5.6
Temperature (°C)	99	10.6	82	188

TABLE 6.7 PdPtMn SV Signal, Asymmetry, and Temperature

Response	t_F	t_S	t_P	g_1	g_2	h	W	H_{KF}	H_{KP}	H_{cpl}
Signal	−0.087	−0.101	−0.094	0.052	0.019	−0.745	0.566	0.005	−0.035	−0.160
Asymmetry	0.018	−0.050	−0.168	−0.011	−0.026	−0.358	0.020	0.081	−0.096	0.020
Temperature	−0.015	−0.045	0.011	0.072	0.014	−0.914	−0.015	−0.067	−0.006	0.006

improved pinning of this (hypothetical) AFM material maintains useful signal with short, hot stripes; the correlation coefficient of signal with h is somewhat greater in this case because the devices with short stripes do not unpin as readily as in the IrMn study. Scatterplots of signal versus stripe height are shown for both designs in Figs. 6.25a,b at an ambient temperature of 25°C and in Figs. 6.25c,d for 70°C.

Probability of Device Unpinning Versus Operating Current With an additional 1000 Monte Carlo trials for each of the two designs, three IrMn devices unpinned and no PdPtMn devices unpinned at 6mA sense current and 70°C ambient. (The seed number for the random-number table is different from that used for the results posted in Tables 6.4 and 6.6, and thus the results are somewhat different even though the operating conditions are unchanged.) This observation naturally opens up the question regarding the probability of unpinning at any reasonable operating current. Using a Monte Carlo approach, the hypothetical unpinning behavior can be studied and quantified, and "failed" devices can be analyzed for root causes. Scatterplots of ordered pairs of signal and asymmetry reveal unpinning in a particularly clear manner; Figs. 6.26a,b,c show signal versus asymmetry of the IrMn design at 4.0, 6.0, and 8.0 mA sense current, which is just below the unpinning threshold up to a current where unpinning is likely. At 4.0 mA, the signals below 800 µV arise from long (0.85–1.1-µm) stripes that are relatively cool and pinned, and all data are confined to a range of asymmetries between −5 and +5% with signals in the range of 680–1990 µV. At 6.0 mA the range in asymmetry expands from −88 to +16% and the signal range is 50–3020 µV; the 3 devices with amplitudes below 1000 µV are unpinned, the stripes are very short (between 0.38 and 0.42 µm), the widths are close to nominal (0.99–1.13 µm), and the signal polarities have reversed in sign. The 14 devices with signals above 2500 µV are hot but unpinned (temperatures are between 166 and 231°C), stripes are between 0.44 and 0.53 µm, and the widths are between 0.89 and 1.25 µm. At 8.0 mA, asymmetry ranges between −93 and +97% with signal covering the range of 70–3550 µV. More devices fail from unpinning (77 parts with short, hot stripes) all of which suffer from reversal of the pinned-layer magnetization and signal polarity; mean temperature is 326°C with stripes between 0.35 and 0.57 µm. Qualitatively, the behavior of PdPtMn is much the same, but the threshold of unpinning occurs at about 9.0 mA. A comparison of the number of unpinned devices per thousand trials is given in Fig. 6.27a, and the mean temperatures are plotted in Fig. 6.27b, each as a function of the sense current. Without exception, the unpinned devices had very short stripe heights in the range of 0.35–0.57 µm, and there was no apparent correlation to the other randomized parameters of the study. With a mean stripe height at 0.70 µm and the assigned standard deviation of 0.10 µm, the failed devices fall in the -1.3σ to -3.5σ end of the distribution. (The Gaussian random-number table of 1000 digits was limited to −3.5 and +3.5 standard deviations about the mean.) At 8.0 mA sense current 21 IrMn devices out of 1000 had asymmetry greater than +20%, of which 4 devices had a signal greater than 1000 µV; these 4 parts were at temperatures between 236 and 238°C, which is close to the blocking temperature (250°C). At 11.0 mA sense current, 40 PdPtMn devices unpinned and 12 others had positive asymmetry with

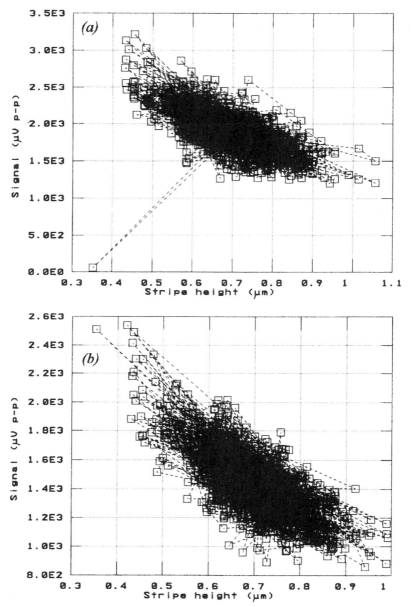

FIGURE 6.25 Scatterplots of computed signal versus stripe height for two different AFM materials: (a) IrMn at $T_{amb} = 25°C$; (b) PdPtMn at $T_{amb} = 25°C$; (c) IrMn at $T_{amb} = 70°C$; (d) PdPtMn at $T_{amb} = 70°C$.

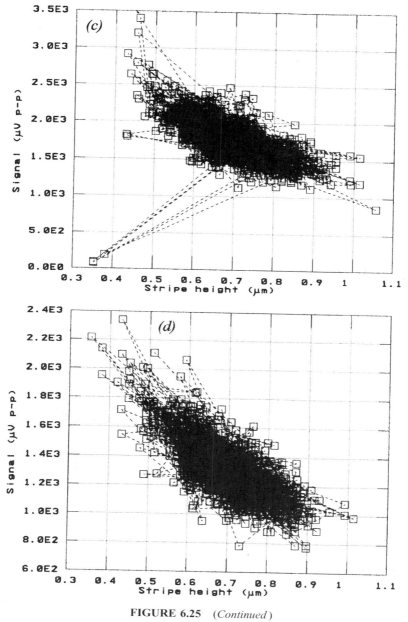

FIGURE 6.25 (*Continued*)

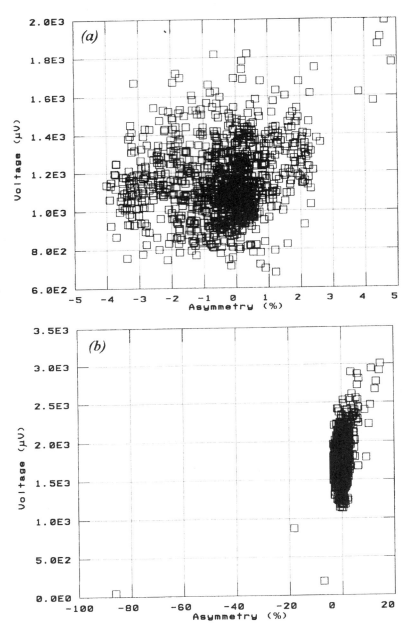

FIGURE 6.26 Scatterplots of computed signal versus asymmetry for an IrMn AFM design at (a) 4, (b) 6, and (c) 8 mA; $T_{amb} = 70°C$ in each case.

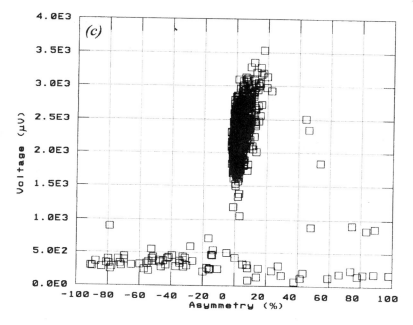

FIGURE 6.26 (*Continued*)

signals between 1300 and 1700 µV; these devices had long stripes (0.93–1.06 µm) and the temperatures were low (114 and 124°C).

The exchange field versus temperature for these hypothetical AFM layers is plotted in Fig. 6.28; the cross-over temperature is about 140°C. The benefit of well-ordered AFM materials, such as the characteristics shown here for PdPtMn, is realized in the reduced likelihood of unpinning at elevated ambient temperatures and in the ability to use an increased operating current. The unpinning thresholds are about 5.5 and 9.0 mA for these IrMn and PdPtMn design cases. At 6.0 mA and 70°C ambient, the IrMn design has greater signal (1760 µV) than the PdPtMn design (1320 µV), but the probability of unpinning with IrMn is about 0.8%, which represents an unacceptable level of risk in a disk drive.

The evolution of unpinning behavior is revealed in the theoretical voltage transfer curves shown in Fig. 6.29a; the four plots are for a "nominal" IrMn device ($h = 0.70$ µm, $W = 1.0$ µm) at temperatures of 122, 250, 280, and 300°C at a constant 6.0 mA sense current (for which the temperature rise is 26°C above ambient temperatures of 96, 224, 254, and 274°C, respectively). At $T = T_B = 250$°C with excitation between -2.5 and -1.5 memu/cm^2, the transfer curve develops interesting reversals in slope arising from the independent variations in θ_P and θ_F with excitation; at 280°C the transfer curve kink has moved toward the positive-excitation regime at higher temperatures, and at 300°C the kink has disappeared, and the slope is small but positive in the vicinity of 0 ± 1.0 memu/cm^2 excitation. Figure 6.29b

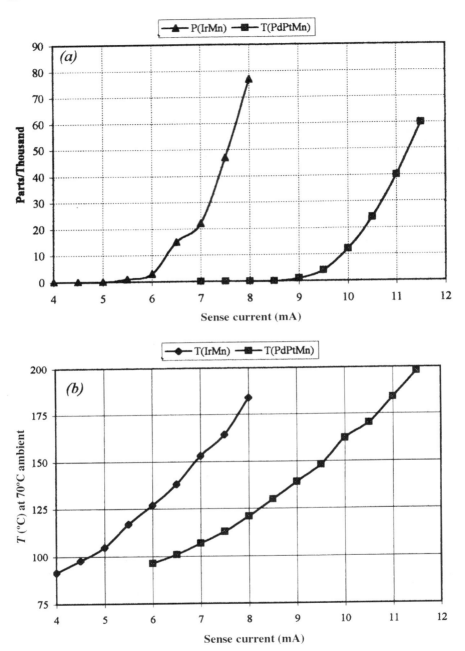

FIGURE 6.27 (*a*) Projected number of unpinned devices and (*b*) average temperature versus sense current for IrMn and PdPtMn AFM layers based on Monte Carlo computations.

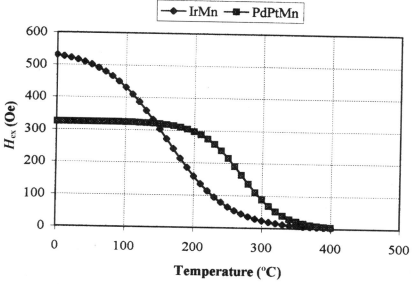

FIGURE 6.28 Exchange field versus temperature for the hypothetical IrMn and PdPtMn AFM layers of the Monte Carlo analysis of basic spin valve systems.

compares the nominal device transfer curve at $T = 122°C$ with that of a device at $T = 232°C$ (arising from $h = 0.48$ and $W = 1.1$ µm); the short stripe is just below the blocking temperature and the signal would be about 3300 µV (p-p) for an excitation of ±1.0 memu/cm² as compared to about 1900 µV (p-p) for the nominal device.

CRITIQUE OF BASIC TOP AND BOTTOM SPIN VALVE DESIGNS

The basic spin valve devices share a number of common design and processing issues. Because interlayer diffusion must be nil, processing temperatures exceeding 300°C are avoided and AFM/FM pinning is limited to blocking temperatures below this value. This leaves a basic device where unpinning could be a concern, and thus the operation mode might have H_I aiding the exchange field of the pinned layer and adding with the effective demagnetizing flux to reduce the useful dynamic range. The thickness of the pinned layer cannot be much less than 20 Å without inheriting yield problems from pin holes, and thus the effective demagnetizing field H_{Dp} (which scales with $M_{SP}t_P$) forces the free-layer moment $M_{SF}t_F$ to large values ($\simeq 3M_{SP}t_P$) to provide a useful dynamic range about the bias point. With a thick (~100-Å) NiFe free layer, the peak-to-peak signal will be limited to about 1000 µV/µm of track width, the AMR effect could be 25% (or more) of the GMR effect (which leads to significant negative asymmetry), and areal densities greater than about 5 Gbits/in.² cannot be readily achieved. A thin free layer of CoFe

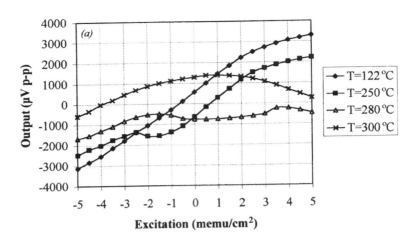

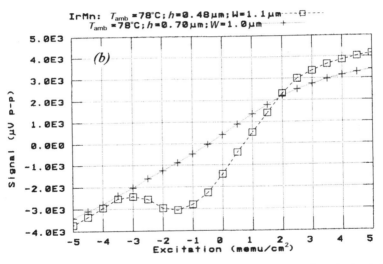

FIGURE 6.29 (*a*) Voltage transfer curves showing the evolution of unpinning behavior in an IrMn spin valve at 6 mA sense current at device temperatures of 122, 250, 280, and 300°C. (*b*) Comparison of the voltage transfer curve of a *nominal* device at $T = 122°C$ with a device at $T = 232°C$ arising from $h = 0.48\,\mu m$ and $W = 1.1\,\mu m$. The hot device is just below the blocking temperature, and thus the operating slope increases substantially.

or Co "dusting" of NiFe layers at the Cu interface increases the GMR effect to about 6%, and signals of about $1500\,\mu V/\mu m$ can be obtained; however, a thin free layer requires some means of reducing the net demagnetizing flux from the pinned layer. The angle of the hard bias and lead junction with the sense layers produces a top spin valve with a wider effective free-layer track width than a bottom spin valve; with very narrow track widths (<1 μm) the bottom spin valve design might be somewhat easier to control in large-volume production. In any case, to reach areal densities of 10 Gbits/in.^{2} and greater, it is necessary to find ways of improving the output to levels of 2000 μV/μm and more. Recent developments such as the synthetic antiferromagnetic sandwich for exchange biasing and the bias compensation layer (BCL) have led to new devices with greatly improved output characteristics; these devices are called synthetic and BCL spin valves.

SYNTHETIC SPIN VALVES

In 1993 Heim and Parkin (1995) applied for a U.S. patent covering spin valve sensors based on the concept of a laminated pinned layer; the claims cited various ferromagnetic materials and alloys coupled antiferromagnetically through a very thin (<10-Å) layer of Ru, Cr, Rh, Ir, and their alloys. This invention directly attacked the problem of unwanted biasing of the free layer and opened up design latitude for thinner films with improved output and linearity. Later, Speriosu et al. (1996) and Berg et al. (1996) discussed the concept of "synthetic" or "artificial" antiferromagnetically coupled systems formed with structures of Co/NM/Co, where NM is a layer of Cu or Ru 6–10 Å thick. Leal and Kryder (1998) showed virtually no magnetostatic coupling with spin valve free layers deposited on synthetic pinned Co layers of equal thickness; their structures were Cr 21 Å/Co 30 Å/Ru 7 Å/Co 30 Å/Cu 25 Å/Co 30 Å/NiFe 28 Å/Ru 21 Å on Corning glass substrates. The coupling in synthetic trilayers FM/Ru/FM arises from an oscillatory exchange interaction (RKKY-like) and must be saturated with a large field [~6000 Oe; see Parkin et al. (1990)]. With Co 20 Å/Ru multilayers, the exchange coupling energy was about −5 ergs/cm^2, dropping rapidly with Ru thickness to about −2 ergs/cm^2 at 8 Å; at 11 Å the interaction was ferromagnetic and the period of FM/AFM oscillation was about 12 Å. The unpublished data of Anderson and Huai (1999), which are plotted in Fig. 6.30, show the necessity for tight control of Ru thickness for CoFe 20 Å/t_{Ru} Å/CoFe 23 Å synthetic structures; Hs is the saturation field.

With the correct NM thickness, the antiparallel magnetization vectors of the two layers yield a net moment $M_1 t_1 - M_2 t_2$ that can be exploited as a design variable for free-layer biasing. The antiparallel alignment of M_1 and M_2 has no intrinsic preferred direction within the plane of the films, so it becomes necessary to create a stable direction by exchange pinning M_1 to an AFM layer as described earlier. Therefore, the synthetic spin valve structure shown in Fig. 6.31 is similar to the basic spin valve discussed earlier, but it has the advantage of adjustable biasing of the free layer and increased effective pinning strength.

Bias Point and Transfer Curves

The tools necessary for designing and analyzing synthetic spin valves are readily developed from the preceding results for basic spin valves. Referring to Fig. 6.31, the fields arising from exchange and current flow must be redefined for the AFM/P1/Ru/P2 interfaces. The field at the AFM/P1 interface is already defined

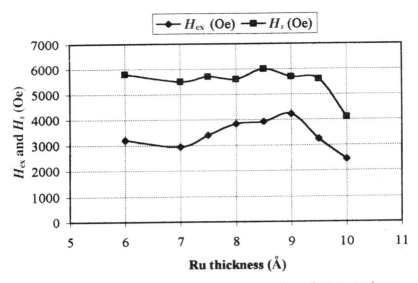

FIGURE 6.30 Experimental pinning field for CoFe 20 Å/t_{Ru} Å/CoFe 23 Å films versus thickness of the Ru interlayer (from Anderson and Huai, 1999).

in (6.14) and the field at the Ru/P2 interface is of the same form except a distinction must be made such that the exchange field arises from the RKKY-like interaction through the thin Ru layer. In this regard, (6.14) is rewritten as

$$H_{11}(I, T) = \frac{2\pi}{hG_{ss}}[I_1(g_1 + t_A - t_S - t_F - g_2) + 2I_2(t_A + g_1)]$$

$$+ \frac{2\pi}{hG_{ss}}[2I_3(t_A + g_1) + 2I_4(0.5t_A - t_P - t_S - t_F - g_2)] + H_{ex1}(T)$$

(6.42)

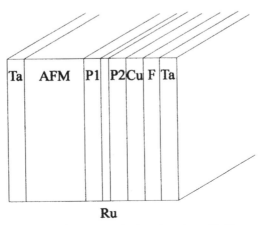

FIGURE 6.31 Geometry of a bottom synthetic spin valve with Ta capping layers.

for the AFM/P1 interface, where $H_{ex1}(T)$ is defined by (6.16) with appropriate values assigned for $E_{ex1}(T)$, M_{SP1}, and t_{P1}. The currents I_1, I_2, I_3, and I_4 refer to the P1/Ru/P2, free, spacer, and AFM layers, respectively. The Ru layer (8 Å) conductivity is less than $10^{-3}\,\Omega^{-1}$ (see Table 2.4) so its influence on current flow is ignored here. At the Ru/P2 interface, the field is written

$$H_{12}(I, T) = \frac{2\pi}{hG_{ss}}[I_1(g_1 + t_A - t_S - t_F - g_2) + 2I_2(t_A + t_S + g_1)]$$

$$+ \frac{2\pi}{hG_{ss}}[2I_3(t_A + g_1) + 2I_4(0.5t_A - t_S - t_F - g_2)] + H_{ex2}(T), \quad (6.43)$$

with a corresponding definition for $H_{ex2}(T)$. The temperature dependence of the exchange interaction between P1 and P2 through Ru is not accurately known, but preliminary experimental results show a loss of about 1.7×10^{-3} parts/°C or roughly 10 Oe/°C; in this regard, the exchange field completely dominates at the Ru/P2 interface, so the influence of H_I contributions are relatively unimportant. In (6.42) the pinned layer thickness t_P can be taken as the combined thickness $t_{P1} + t_{Ru} + t_{P2}$ with insignificant damage to the accuracy of the estimated total field since the exchange fields will normally be much greater than the current field. The magnetizations of P1 and P2 are defined by (6.12) and (6.19), respectively, with appropriate changes in terminology. The free-layer magnetization is similarly defined with an alteration in the net moment of the P1–P2 coupled layers. That is,

$$m_F(y, I, T) = \frac{M_{SP1}t_{P1} - M_{SP2}t_{P2}}{M_{SF}t_F} m_{P2}(y, I, T) + \frac{H_2(I)}{H_{KF}}\left[1 - \frac{\cosh(y/\lambda_{sh})}{\cosh(h/2\lambda_{sh})}\right]. \quad (6.44)$$

It is immediately clear from (6.44) that the bias point shift arising from P1 and P2 can be adjusted to any reasonable level by choosing the parameters of each layer; this is shown in Fig. 6.32a,b for a synthetic spin valve (SSV) NiO 200 Å/CoFe 20 Å/Ru 8 Å/CoFe t_{P2} Å/Cu 25 Å/CoFe 10 Å/NiFe 40 Å, where t_{P2} is 35, 30, and 20 Å. Magnetization profiles are plotted in Fig. 6.32a and voltage transfer curves at $I = +4$ mA (labeled V1Xfr, V2Xfr, and V3Xfr, all in millivolts) are shown in Fig. 6.32b; the top curves have the greatest bias shift ($t_{P2} = 35$ Å) and in the bottom curves (where $H_{Dp} = 0$) biasing arises only from H_2, which includes shield imaging, and $H_{cpl} = -10$ Oe. A free-layer AMR effect of 0.38 µΩ-cm (with $\rho_F = 50.0$ µΩ-cm) is included with the 6.4% GMR effect; that is, $\Delta R_{AMR} = 0.20\,\Omega$ and $\Delta R_{GMR} = 1.68\,\Omega$ at room temperature, so the relative AMR effect is 11.9%. The sheet resistance is a high value of 21.9 Ω/square because the AFM layer is an insulator and Ta buffer and capping layers have been omitted. The exchange fields are $H_{ex1} \simeq +1000$ Oe (upward pinning on P1) and $H_{ex2} \simeq -5000$ Oe for P2 at a device temperature of 50°C. For the SSV structure one can assume the P1/Ru/P2 sandwich is rigidly coupled and not capable of conducting flux from a written transition, and thus all of the excitation flux enters the free layer and leaks off to the

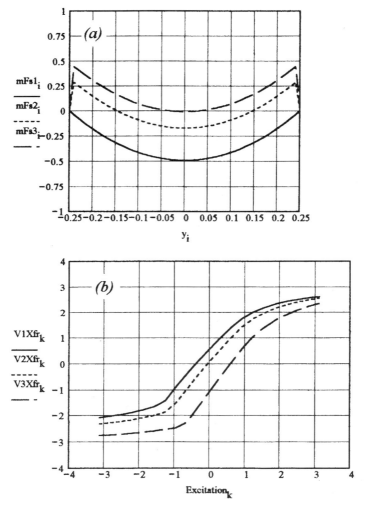

FIGURE 6.32 (a) Computed magnetization bias profiles of a synthethic bottom spin valve for various thicknesses of the second pinned layer, t_{P2}: 35, 30, and 20 Å. (b) Voltage transfer curves V1Xfr, V2Xfr, and V3Xfr for pinned layer, t_{P2}: 35, 30, and 20 Å, respectively. Sense current $I = +4$ mA and voltages are in mV (p-p).

shields. The modulation of the free-layer magnetization is then given by a modified form of (6.27b),

$$\Delta M_F^{\text{signal}}(y) = \left(\frac{\phi(y_e)}{4\pi W t_F}\right) \frac{\sinh[(0.5h - y)/\lambda_{\text{sh}}]}{\sinh(h/\lambda_{\text{sh}})}. \tag{6.45}$$

The decay length λ_{sh} is found with (6.21) as before. The voltage transfer curve is calculated with (6.33) and the positive, negative, and peak-to-peak signals and asymmetry are found with (6.34a,b,c) and (6.35).

Output Signal and Sense/Bias Curves

Tsang et al. (1999) were able to demonstrate areal densities of 12 Gbits/in.^{2} and outputs of 2100 μV/μm with synthetic bottom spin valve structures of the form NiO/M_1/Ru/M_2/Cu/M_3/Ta. The layers M_i were composed of Co and NiFe with enhancement of the GMR effect ($\Delta R/R \simeq 6.5\%$ and $R_S \simeq 20$ Ω/square) through Co "dusting" (see Parkin, 1993) at each interface with the Cu spacer. A stable reference direction was provided by pinning M_1 to an AFM layer NiO 400 Å thick, which gave $H_{ex} \sim 1000$ Oe with a blocking temperature of about 200°C (~470 K). Track and linear densities were about 34 ktpi and 350 kbpi, repectively; the read track width $W \simeq 0.6$ μm, stripe height $h \simeq 0.5$ μm, and the shield-to-shield gap length was 0.14 μm. Assuming the structure and properties of the Tsang et al. (1999) device are given by NiO 400 Å/CoFe 20 Å/Ru 8 Å/CoFe 30 Å/Cu 25 Å/CoFe 10 Å/NiFe 40 Å with $E_{ex1} = 0.35$ erg/cm^2, $E_{ex2} = 2.5$ ergs/cm^2, $\rho_P = 43$ μΩ-cm, $\rho_F = 50$ μΩ-cm, $\Delta\rho_{AMR} \simeq 0$, $\rho_S = 11.3$ μΩ-cm, $g_1 = 200$ Å, $g_2 = 700$ Å, and the magnetic properties of the pinned and free layers unchanged from above, the normalized transfer curves for positive and negative sense currents (given in Fig. 6.33) show the +4-mA curve is fairly linear, has a little negative bias offset ($H_{cpl} = -5$ Oe and no AMR effect), and has useful dynamic range over the excitation interval -1 to $+1$ memu/cm^2. The simulated sheet resistance $R_S = 21.9$ Ω/square and the GMR effect $\Delta R_{GMR} = 1.61$ Ω at $T = 49$°C (4 mA sense current, 25°C ambient). The output signals and asymmetry as a function of sense current are plotted in Fig. 6.34a,b along with experimental points from the paper published by Tsang et al. (1999). With an excitation of 0.35 memu/cm^2 at $d + a = 480$ Å, the signal calculations are 18% greater than experiment (about 2480 μV/μm calculated and 2100 μV/μm experimental) whereas the calculated asymmetry is smaller than experiment (about +1% calculated and 1–5% experimental). The reasons for this

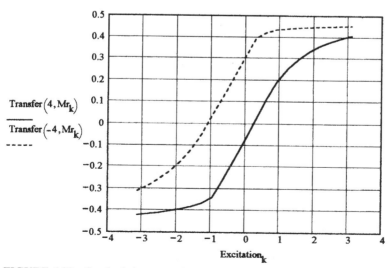

FIGURE 6.33 Synthetic bottom spin valve voltage transfer curves for positive and negative sense currents ($I = \pm 4$ mA).

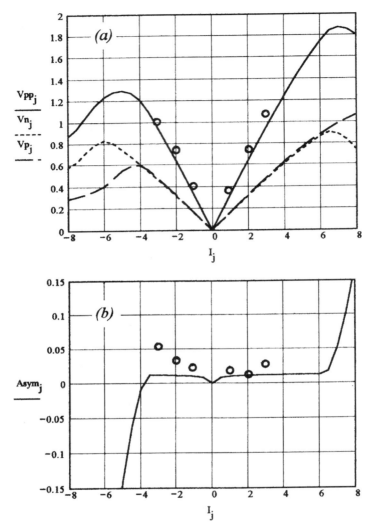

FIGURE 6.34 (*a*) Output signal and (*b*) asymmetry for a bottom synthetic spin valve. Plotted lines are computed results and open circles are experimental points from Tsang et al. (1999).

disagreement are not clear. The signal field at the bottom of the free layer is about 300 Oe; the signal flux decay length $\lambda_{sh} = 0.57\,\mu$m so the stripe height of 0.5 μm is about right for this device.

With positive sense current (H_I aiding the pinned direction) saturation of the free layer occurs for negative excitation. The saturation current is estimated from (6.44) by setting $m_F(0, I, T)$ to -1.0 (saturation in the negative direction at the stripe center) and solving the equation for $H_2(I)$. Using (6.15) and some algebra, the free-layer saturation current $I_{sense} = I_{SF}$ becomes

$$I_{SF} \simeq \frac{hG_{ss}}{2\pi}$$

$$\times \left[H_{KF} \cdot \frac{\left[\frac{\cosh(h/2\lambda_{sh})}{\cosh(h/2\lambda_{sh}) - 1}\right] \left(\frac{M_{SP1}t_{P1} - M_{SP2}t_{P2}}{M_{SF}t_F} - 1 + \frac{\Delta m}{2}\right)}{\frac{G_2}{G_t}(g_1 + t_A + t_P + t_S - g_2) - 2\left(\frac{G_1 + G_3 + G_4}{G_t}\right)(0.5t_F + g_2)} - H_{cpl} \right].$$

(6.46)

The term $\frac{1}{2}\Delta m$ accounts, in an approximate way, for the average shift in free-layer magnetization arising from the medium excitation; it is essentially $1/M_{SF}$ times the average of (6.45) over the stripe height. This approximation is valid for $h/\lambda_{sh} \leq 1.0$ and becomes less accurate with long stripes. Figures 6.34a,b show saturation at about 7 mA and (6.46) estimates saturation at about 7.1 mA. With negative sense/bias currents, the free layer saturates in the positive (upward) direction, so $m_F(0, I, T) = +1.0$ is substituted into (6.44) and (6.46) is modified accordingly; this substitution predicts saturation at $I_{SF} = -3.8$ mA and Figs. 6.34a,b show reasonable agreement with this value. The influence of interlayer coupling H_{cpl} is relatively weak in comparison with the term whose prefactor is H_{KF}; in the present example, changing H_{cpl} from -10 Oe to zero shifts the saturation current by 0.1 mA.

Monte Carlo Analysis of a SSV Design

The SSV design discussed above has sufficient signal to support areal densities in the range of 10–15 Gbits/in.² It is useful at this point to briefly examine this design

TABLE 6.8 SSV Parameters for Monte Carlo Analysis

Parameter	Mean	Standard Deviation
Free-layer NiFe/CoFe	50 Å	1.0 Å
Cu spacer	25 Å	1.0 Å
P1 layer CoFe	20 Å	1.0 Å
Ru layer	8 Å	0
P2 layer CoFe	30 Å	1.0 Å
AFM layer NiO	400 Å	0
Insulation g_1 Al$_2$O$_3$	200 Å	10 Å
Insulation g_2 Al$_2$O$_3$	700 Å	10 Å
Stripe height	0.50 μm	0.07 μm
Read track width	0.50 μm	0.07 μm
Free-layer H_K	6 Oe	1.0 Oe
Pinned-layer H_K	8 Oe	1.0 Oe
Interlayer coupling H_{cpl}	-10 Oe	4.0 Oe

using a Monte Carlo analysis with realistic values assigned to most of the critical variables. Table 6.8 lists the nominal design parameters and the respective standard deviations for this analysis. The interlayer coupling field H_{cpl} is included to demonstrate its weak influence on signal and asymmetry variations with this design. The Ru thickness is held at 8 Å to maintain antiferromagnetic coupling between P1 and P2; at 10 Å the exchange field is about 2500 Oe and drops to roughly half this value at 11 Å. The unpinning behavior of basic spin valves has received ample discussion earlier in this chapter and will not be repeated here.

The results of 1000 Monte Carlo trials at an ambient temperature of 25°C are given in Table 6.9. The signal distribution is fairly broad (stripe height and track width variations) whereas asymmetry is a tight distribution (good pinning) except for two outliers that have saturated (long stripes). Table 6.10 shows the correlation coefficients for each response and the selected parameters of the study. As with basic spin valves with similar process controls, the behavior of SSVs is dominated by variations in stripe height and track width. The interlayer coupling field shows essentially no influence on signal or asymmetry variations; this is to be contrasted with the basic spin valve, which has greater bias shift from the pinned-layer demagnetizing field such that H_{cpl} is more influential in restoring the bias operating point toward an optimal location.

Critique of SSV Designs

The synthetic pinned-layer concept solves the major problem of unwanted bias point shifting from the effective demagnetizing field of a single pinned layer and reduces the risk of unpinning because the net moment of the synthetic structure is small. The P2 moment $M_{SP2}t_{P2}$ is adjusted to produce an optimal bias point, and the reduced effective demagnetizing field H_{Dp} opens up the possibility of much thinner free layers, higher sheet resistance, and increased output signals with acceptable asymmetry. A thin free layer with 6.4% GMR effect (CoFe 25 Å for example) can

TABLE 6.9 SSV Signal, Asymmetry, and Temperature

Response	Mean	Standard Deviation	Minimum	Maximum
Signal (µV, p-p)	1260	261	567	2370
Asymmetry (%)	1.21	0.21	1.09	7.4
Temperature (°C)	49.6	8.2	34.8	88.9

TABLE 6.10 Correlation Coefficients for SSV Monte Carlo Analysis

Response	t_F	t_S	t_{P1}	g_1	g_2	h	W	H_{KP}	H_{KF}	H_{cpl}
Signal	−0.117	−0.075	−0.030	−0.001	0.027	−0.669	0.668	−0.033	0.027	−0.003
Asymmetry	−0.141	−0.022	0.033	0.031	0.040	0.201	0.030	−0.006	0.149	−0.009
Temperature	−0.043	−0.049	−0.022	0.093	0.016	−0.936	−0.035	−0.047	−0.071	−0.037

lead to $R_S \simeq 23\,\Omega$/square and signals of $3.5\,\text{mV}/\mu\text{m}$ at $5\,\text{mA}$ sense current. This level of sensitivity can be exploited with read track widths of about $0.35\,\mu\text{m}$, track densities approaching $50\,\text{ktpi}$, and areal densities of $25\,\text{Gbits/in.}^2$

BIAS COMPENSATION LAYER SPIN VALVE

The bias compensation layer BCL spin valve (see, e.g., Kanai et al., 1998) addresses the problem of free-layer bias shifting in a different manner from the synthetic spin valve. Figure 6.35 is a bottom spin valve sketch showing a BCL added on top of the free layer, and the dotted lines depict the flow of demagnetizing flux from the pinned layer and BCL through the free layer. With positive pinning direction, positive sense current produces a field H_I that aids the exchange pinning field H_{ex} at the AFM/P interface, and the BCL is magnetized in the negative direction by the combined influences of H_I and the effective demagnetizing field H_{Dp} of the pinned layer. The free-layer bias profile is described by a modified form of (6.44), namely

$$m_F(y, I, T) = \frac{(M_{SB}t_B - M_{SP}t_P)}{M_{SF}t_F} m_P(y, I, T) + \frac{H_2(I)}{H_{KF}}\left[1 - \frac{\cosh(y/\lambda_{\text{sh}})}{\cosh(h/2\lambda_{\text{sh}})}\right]. \quad (6.47)$$

The BCL moment $M_{SB}t_B$ is selected to provide an optimal bias point; if $M_{SB}t_B$ matches the pinned-layer moment $M_{SP}t_P$, the net demagnetizing flux through the free layer vanishes. The bottom pinned-layer magnetization profile $m_P(y, I, T)$ is described by (6.12), and appropriate modifications arising from the presence of the BCL and spacer S2 are applied to $H_1(I, T)$ and $H_2(I)$. The equation for the characteristic decay length (6.17) is also modified to account for coupling to the BCL, but the details are omitted here. The layer S2 is selected to produce diffuse scattering of spin-polarized electrons (e.g., 30-Å Ta) flowing from the free layer to the BCL such that an additional (and poorly controlled) GMR effect is quenched. If

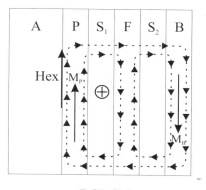

BCL SV

FIGURE 6.35 Geometry of a bottom spin valve with a bias compensation layer "B".

S2 is made from Cu or another noble metal, an additional ΔR is introduced with the BCL whose moment is not well pinned, and thus signal amplitude and asymmetry could degrade. In one sense, the BCL spin valve is an evolutionary step toward spin valves with specular reflection layers or toward the dual spin valve (DSV), which is a device having a free layer with two symetrically placed pinned layers.

ENHANCEMENT OF GMR RATIO WITH SPECULAR REFLECTION LAYERS

Based on theoretical concepts of electron scattering at interfaces, Hood and Falicov (1992) speculated that magnetoresistance experiments ought to show a dependence of the GMR effect on the reflection of electrons at the outer surfaces of basic trilayered spin valves. Somewhat later, Anthony et al. (1994) and Egelhoff et al. (1995) found large GMR ratios in spin valve structures containing interfaces of Co/NiO, and they offered the interpretation that NiO could provide some specular scattering of electrons back into the FM layers, thereby *extending the lifetime of spin polarized electrons*. Swagten et al. (1996) studied trilayered systems of Co/Cu/Co and NiFe/Cu/NiFe where the bottom FM layer was pinned by an NiO AFM layer and the top FM layer was covered by a sandwich of Cu/NiO. In their studies, the basic structure is a bottom spin valve similar to the sketch in Fig. 6.35 where the layer B (for BCL material) is replaced with an insulating NiO layer, so the structure A/P/S1/F/S2/B is defined by a sandwich, such as NiO 500 Å/Co 20 Å/Cu 20 Å/Co t Å/Cu 12 Å/NiO 100 Å. The thin S2 layer (Cu 12 Å) was used to suppress exchange coupling between the free layer (Co t Å) and the top insulating AFM layer (NiO 100 Å). At free layer thicknesses $t = 15$ Å, the room temperature GMR ratio was ~13%, increasing to ~15% at $t = 30$ Å, and dropping to ~14% at $t = 50$ Å. [These GMR ratios are more than twice that reported by Parkin (1993) for Co dusted layers.] They interpreted their results as a spin valve trilayer sandwiched between insulating layers, each of which specularly reflected all conduction electrons such that the trilayer mimicked an infinite multilayered spin valve system. The dependence of the GMR ratio on free layer thickness is qualitatively similar to that shown in Fig. 6.4a where the data are extracted from the work of Dieny et al. (1992) with all-metal spin valves. With specular reflection the maximum GMR ratio occurs in the thickness range of 20–40 Å, whereas with diffuse scattering the maximum occurs in the 50–100 Å range. Swagten et al. (1996) calculated the GMR ratio using the Fuchs–Sondheimer treatment of the Boltzmann equation (see Chapter 2) and obtained fair agreement with their experimental results.

The demand for higher output signal/unit track width continues to stimulate research in material systems with enhanced GMR ratios. Shimizu et al. (2000) presented results for bottom synthetic spin valves with the structure Ta/NiFe/ PdPtMn/CoFeB/Ru/CoFeB/Cu/CoFeB/oxide where the oxide layer was sputtered from targets of αFe_2O_3, NiO, or Al_2O_3. There was no layer to destroy exchange coupling between the free CoFeB layer and the oxide reflection layer, thus with an AFM layer of NiO the exchange field did not allow free rotation of the magnetization in the adjacent CoFeB layer, however with the nonmagnetic oxides (αFe_2O_3 and Al_2O_3) the GMR effect is enhanced through increased specular reflection at the top interface. Sakakima et al. (2000) inserted a specular oxide layer (the material was not disclosed) within the pinned layer of a bottom spin valve Ta/PtMn/CoFe/ oxide/CoFe/Cu/CoFe/Ta and the GMR ratio increased from 8.1% up to 11.2% with the insertion of ~6–14 Å of oxide. In additional experiments an oxide layer was also inserted in the free layer and the GMR ratio increased to ~14.5% with 14 Å of oxide. As in other work the increase in GMR ratio was attributed to specular scattering at the oxide/CoFe and CoFe/oxide interfaces. Tsuchiya et al. (2000) discusses "highly reflective magnetic oxide layers (MOL)" of undisclosed material. Their structure is a top spin valve NiFe 20 Å/MOL1/NiFe 10 Å/CoFe 20 Å/Cu 25 Å/CoFe 20 Å/MOL2/CoFe 10 Å/RuRhMn 100 Å with buffer and capping layers of a Ta 50 Å. The spin valve GMR ratio went from 8.2% (without MOL) up to 11.8% (with MOL), however the thermal stability of the RuRhMn AFM layer was poor. With MOL the sheet resistance was $R_S = 15.2\, \Omega$/square and the GMR resistance change was $\Delta R = 1.78\, \Omega$/square. Other devices, such as the dual spin valve (DSV) achieve improved GMR ratios by utilizing one free layer with symmetrically placed pinned layers. This chapter concludes with a discussion of dual synthetic spin valve (DSSV) design and analysis, a critique of designs based on this concept, and finally, projections are given for the output signal and areal densities which might be achieved with DSSV devices.

DUAL SYNTHETIC SPIN VALVE

The DSSV is a device with one free layer and two symetrically displaced synthetic pinned layers. Spin-polarized electrons diffuse through each Cu spacer and scatter with a long or short mean path (λ^+ for ↑↑ or λ^- for ↑↓ polarizations) in the free and pinned layers. The GMR effect is essentially doubled over that of a single pinned layer; $\Delta R/R$ values greater than 15% have been reported in 1999, and designs based on this approach have demonstrated signals greater than $5\,mV/\mu m$ and areal densities greater than 25 Gbits/in.² The structures are necessarily complicated because of the numerous thin layers. The structure AFM/CoFe 20 Å/Ru 8 Å/CoFe 20 Å/Cu 25 Å/CoFe 20 Å/Cu 25 Å/CoFe 20 Å/Ru 8 Å/CoFe 20 Å/AFM with thin Ta buffer and capping layers is perhaps the minimal DSSV architecture (see Fig. 6.36).

Various modifications to the free layer can be imagined, but Co and its alloys are found next to the Cu interfaces because of the enhanced GMR effect discussed by Parkin (1993). With AFM layers of IrMn 70 Å (resistivity about $325\,\mu\Omega$-cm) and Ta 30 Å buffer and capping layers (resistivities about $200\,\mu\Omega$-cm), the total sandwich thickness becomes 366 Å and the sheet resistance will be $\sim 12.6\, \Omega$/square (sheet conductance $G = 0.079\, \Omega^{-1}$). A 15% GMR effect thus produces a useful resistance change $\Delta R \simeq 1.9\, \Omega$ at room temperature. The symmetric structure places the free layer at the center of the current density distribution, and if the sandwich is centered between magnetic shields, the average field over the free layer will be zero. The

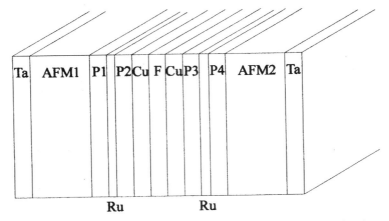

FIGURE 6.36 Geometry of a dual synthetic spin valve with Ta capping layers.

normalized field throughout a sandwich is shown in Fig. 6.37. The pinned layers P1 and P4 must be pinned in the same direction at the time of field annealing in the wafer process cycle; this places one of the pinned interfaces in the "H_I aiding" mode and the other in the "H_I opposing" mode. With matched pinned layers (equal moment–thickness products) the net demagnetizing flux is zero and the free layer is naturally at the optimal bias state ($\theta_F = 0$).

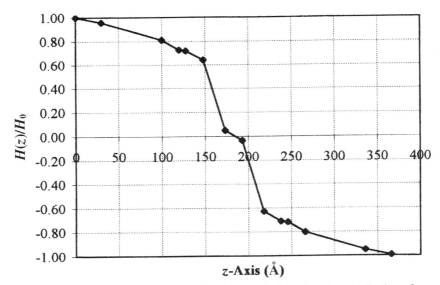

FIGURE 6.37 Plot showing the normalized magnetic field from the distribution of current densities within the layers of a DSSV with Ta capping layers.

The output signal is derived from a modification of (6.26), which includes separate contributions of spin scattering from the P2 and P3 pinned layers. Because the GMR effect is empirically determined with a completed sandwich, the relation given below assumes half of the total change in resistance comes from the angular difference $\Delta\theta_2(y) = \theta_F(y) - \theta_{P2}(y)$ and half comes from $\Delta\theta_3(y) = \theta_F(y) - \theta_{P3}(y)$. That is,

$$\text{Signal} = I_{\text{sense}} R_S \frac{W}{h} \left[\frac{\Delta R}{R} \left(\frac{-\langle\cos\Delta\theta_2\rangle - \langle\cos\Delta\theta_3\rangle}{4} \right) + \frac{\Delta\rho_{\text{AMR}}}{\rho_F} (1 - \langle\sin^2\theta_F\rangle) \right], \tag{6.48}$$

and as before, the brackets $\langle\cdots\rangle$ indicate averages over the stripe height. The original work by Dieny et al. (1991) showed the Cu spacer serves to decouple the GMR interaction between free and pinned layers; recall from (6.7) that this interaction attenuates as $A_0 \exp(-t_S/\lambda_S)$, where λ_S is the characteristic diffusion length (~45 Å) for electrons in Cu. It is reasonable that wafer-processing variations would shift the 50/50 split in ΔR contributions assumed here, but with 3σ variations of 3 Å in Cu thickness about a mean of 25 Å, the worst-case split would be 53.5/46.5, so little additional insight is gained by pursuing this idea further.

The free-layer magnetization is given by a modified form of (6.44) that includes the free-layer coupling with the P1/Ru/P2 and P3/Ru/P4 synthetic films but ignores coupling between the two synthetic structures:

$$m_F(y, I, T) = \frac{M_{SP1}t_{P1} - M_{SP2}t_{P2}}{M_{SF}t_F} m_{P2}(y, I, T) + \frac{M_{SP4}t_{P4} - M_{SP3}t_{P3}}{M_{SF}t_F} m_{P3}(y, I, T)$$
$$+ \frac{H_2(I)}{H_{KF}} \left[1 - \frac{\cosh(y/\lambda_{\text{sh}})}{\cosh(h/2\lambda_{\text{sh}})} \right]. \tag{6.49}$$

Because of the extra conducting layers, the magnetic field $H_2(I)$ at the centerline of the free layer is more complicated than (6.20), which is valid for a basic spin valve or a SSV. The total current may be split into seven parallel sheets with $I_{1,2,3,4}$ flowing in the P1/Ru/P2, free, S1, and A1 layers, respectively, and $I_{5,6,7}$ flowing in the S2, P3/Ru/P4 and A2 layers, respectively. With this terminology, the field at the free layer becomes

$$H_2(I) = \frac{2\pi}{hG_{ss}} [-2(I_1 + I_3 + I_4)Z_{\text{sh2}} + I_2(Z_{\text{sh1}} - Z_{\text{sh2}}) + 2(I_5 + I_6 + I_7)Z_{\text{sh1}}]$$
$$+ H_{\text{cpl2}} + H_{\text{cpl3}}, \tag{6.50}$$

where Z_{sh1} and Z_{sh2} are the distances from the free layer to shield 1 (left side) and shield 2 (right side), respectively. The currents are found with (6.15) as before. With a perfectly matched structure, the current-related fields cancel and what remains is the interlayer coupling fields from the pinned layers P2 and P3. Assuming that the dual synthetic layers P1/Ru/P2 and P3/Ru/P4 are well pinned, then

$m_{P2}(y, I, T) = m_{P3}(y, I, T) = -1.0$ and the free-layer magnetization is given by the relation

$$m_F(y, I, T) = \frac{M_{SP2}t_{P2} - M_{SP1}t_{P1} + M_{SP3}t_{P3} - M_{SP4}t_{P4}}{M_{SF}t_F} + \frac{H_2(I)}{H_{KF}}\left[1 - \frac{\cosh(y/\lambda_{sh})}{\cosh(h/2\lambda_{sh})}\right]. \tag{6.51}$$

The dual synthetic Mt moments are selected to compensate for the bias point shift that would typically arise from interlayer coupling.

A DSSV design appropriate for 20–25 Gbits/in.^{2} would have the following approximate nominal parameters: $W = 0.4\,\mu m$, $h = 0.4\,\mu m$, $G_{ss} = 0.12\,\mu m$, $g_1 = g_2 \simeq 400\,\text{Å}$, the DSSV structure AFM/CoFe 20 Å/Ru 8 Å/CoFe 22 Å/Cu 25 Å/CoFe 25 Å/Cu 25 Å/CoFe 22 Å/Ru 8 Å/CoFe 20 Å/AFM with Ta buffer, and capping layers about 30 Å thick. The sandwich thickness is 375 Å, which just fits in the required G_{ss}; recall from (1.10a) that PW50 $\simeq [(G_{ss}^2 + t_F^2)/2 + 4(d+a)(d+a+\delta)]^{1/2}$, and thus with $d = 250\,\text{Å}$, $a = 230\,\text{Å}$, and $\delta = 125\,\text{Å}$, the estimated PW50 is about 1360 Å or 136 nm. Assuming a channel density $U = 2.3 = $ PW50 $\cdot$ linear density [from (1.12)], the linear density would be about 429 kbpi. The track density would be about $0.8/W$ or 50.8 ktpi, thus this nominal design could support an areal density of nearly 22 Gbits/in.^{2} The sheet conductivity $G_t = 0.0825\,\Omega^{-1}$, sheet resistance $R_S = 12.1\,\Omega$/square, and the experimentally determined GMR effect is $\Delta R/R \simeq 15\%$, which produces a maximum $\Delta R = 1.82\,\Omega$ at room temperature. The voltage transfer curve for this device is shown in Fig. 6.38; the interlayer coupling fields from P2 and P3 are a total of $-10\,\text{Oe}$, and no AMR effect is included because the free layer is CoFe.

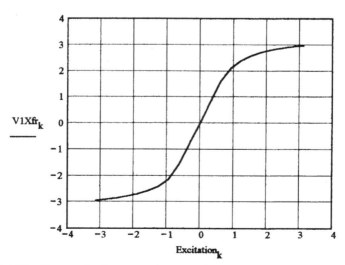

FIGURE 6.38 Computed voltage transfer curve (in mV, p-p) of a DSSV. The excitation is given in units of memu/cm^2.

Note that the bias point is well compensated by the choice of pinned-layer thicknesses. The correct moments for the pinned layers are approximately found with (6.51) by setting $m_F(0, I, T) = 0$ and solving for the appropriate differences in pinned-layer thicknesses. In the present case, all of the magnetic layers are CoFe with $M_S = 1540\,\text{emu/cm}^3$, so (6.51) reduces to the relation

$$\Delta t_P = t_F \left(\frac{H_{\text{cpl2}} + H_{\text{cpl3}}}{H_{KF}} \right) \left[1 - \frac{1}{\cosh(h/2\lambda_{\text{sh}})} \right], \qquad (6.52)$$

where Δt_P is positive (thicker) or negative (thinner) if the interlayer coupling direction is negative or positive, respectively. The dependence on stripe height is shown in Fig. 6.39 for a total of 10 Oe interlayer coupling in the negative direction; the ordinate is in angstroms and the abscissa is in micrometers. The output signals and asymmetry for an excitation of $M_r\delta = 0.35\,\text{memu/cm}^2$ (280 emu/cm^3 and a medium thickness of 125 Å) are shown as a function of sense current in Figs. 6.40a,b. At 5.0 mA sense current the signal is 2.15 mV (p-p) or 5.4 mV/μm sensitivity, and the asymmetry is +1.4%. With gaps of Al_2O_3 (for which the thermal conductivity is 1.0 W/m-K) the sensor temperature is 74°C (ambient = 25°C). At higher currents device heating causes a significant loss of signal. Output signal and stripe temperature are shown in Figs. 6.41a,b as a function of stripe height; below

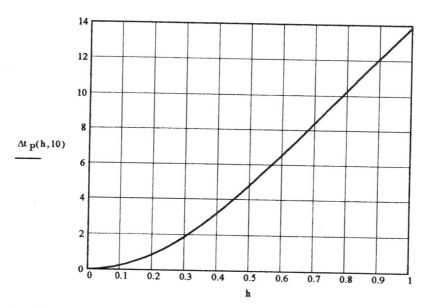

FIGURE 6.39 Plot of the required thickness difference for the pinned layers of a DSSV relative to the free-layer thickness such that an ideal bias point is achieved. The plot is given as a function of stripe height h. The total interlayer coupling field is 10 Oe and $H_{KF} = 8$ Oe.

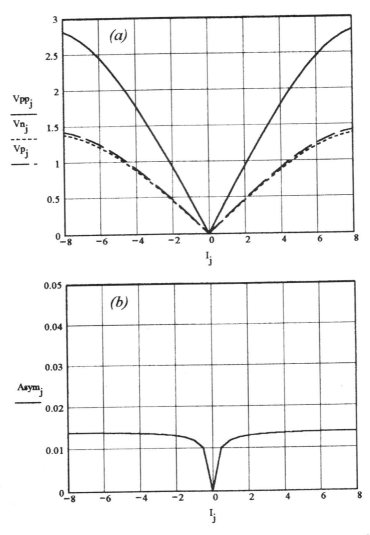

FIGURE 6.40 (a) Computed output signals (in mV) and (b) asymmetry as a function of sense current for a DSSV design capable of supporting 25 Gbits/in.^{2}

0.3 μm the temperature would be high enough to cause unpinning failures at the P4/AFM2 interface. Experimental signal and asymmetry from a DSSV are shown in Fig. 6.42; the device supported 26.5 Gbits/in.^{2} (504 kbpi and 52.6 ktpi) with design parameters close to those given above. The measured PW50 was 119 nm, which is 88% of the PW50 calculated above; the discrepancy can be attributed to uncertainties in actual values of G_{ss} and the transition parameter.

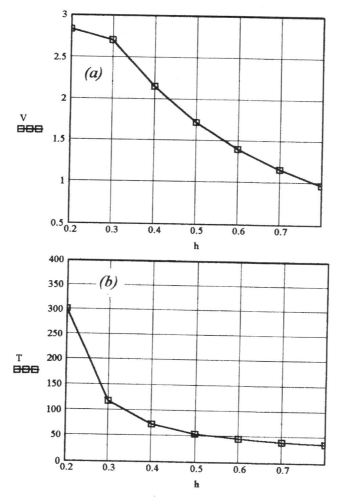

FIGURE 6.41 (a) Computed DSSV output signal (mV, p-p) and (b) average stripe temperature as a function of strip height (μm).

With good process control, the DSSV sandwich could be made thinner with a subsequent increase in sheet resistance to perhaps 20 Ω/square. With CoFe magnetic layers about 10–12 Å thick, Cu spacers of perhaps 20 Å, and AFM layers of 50 Å thickness, the output sensitivity could be increased to about 8 or 9 mV/μm. If this goal is realized in the near future, the DSSV structure would provide a useful signal of 1000 μV (p-p) at a track width $W \sim 0.125$ μm, or a track density in the neighborhood of 150 ktpi. At 600 kbpi (a bit cell ratio of 4; see Chapter 1), the areal density could be pushed up to 90 Gbits/in.^{2} or so.

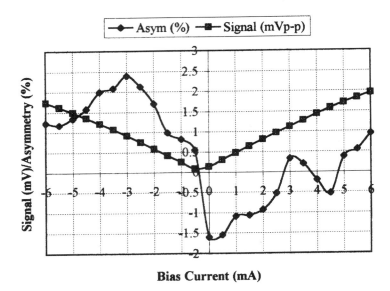

FIGURE 6.42 Experimental signal and asymmetry from a DSSV device supporting 26.5 Gbits/in.^{2} (504 kbpi and 52.6 ktpi).

REFERENCES

Anderson, G. and Huai, Y., Read-Rite, Internal data, Feb. 1999.
Anthony, T. C., Brug, J. A., and Zhang, S., *IEEE Trans. Magn.*, **MAG-30**, 3819 (1994).
Berg, H., Clemens, W., Gieres, G., Rupp, G., Schelter, W., and Vieth, M., *IEEE Trans. Magn.*, **MAG-32**, 4624 (1996).
Bertram, H. N., *Theory of Magnetic Recording*, Cambridge Univ. Press, Cambridge (1995).
Camblong, H. E., *Phys Rev. B*, **51**, 1855 (1995).
Camley, R. E. and Barnas, J., *Phys. Rev. Lett.*, **63**, 664 (1989).
Coehoorn, R., Kools, J. C. S., Rijks, Th. G. S. M., and Lenssen, K.-M. H., *Philips J. Res.*, **51**, 93 (1998).
Dieny, B., Speriosu, V. S., Parkin, S. S. P., Gurney, B. A., Wilhoit, D. R., and Mauri, D., *Phys. Rev. B*, **43**, 1297 (1991).
Dieny, B., Humbert, P., Speriosu, V. S., Metin, S., Gurney, B. A., Baumgart, P., and Lefakis, H., *Phys. Rev. B*, **45**, 806 (1992).
Egelhoff, W. F., Ha, T., Misra, R. D. K., Kadmon, Y., Nir, J., Powell, C. J., Stiles, M. D., McMichael, R. D., Lin, C.-L., Silvertsen, J. M., Judy, J. H., Takano, K., Berkowitz, A. E., Anthony, T. C., and Brug, J. A., *J. Appl. Phys.*, **78**, 4004 (1997).
Fuchs, K., *Proc. Camb. Phil. Soc.*, **34**, 100 (1938).
Fuke, H., Saito, K., Kamiguchi, Y., Iwasaki, H., and Sahashi, M., *J. Appl. Phys.*, **81**, 4004 (1997).
Grünberg, P., U.S. Patent 4,949,039, Aug. 14, 1990.

Heim, D. E. and Parkin, S. S. P., U.S. Patent 5,464,185, Nov. 7, 1995.

Heim, D. E., Fontana, R. E., Tsang, C., Speriosu, V. S., Gurney, B. A., and Williams, M. L., *IEEE Trans. Magn.*, **30**, 316 (1994).

Hood, R. Q. and Falicov, L. M., *Phys. Rev. B.*, **46**, 8287 (1992).

Johnson, B. L. and Camley, R. E., *Phys. Rev. B*, **44**, 9997 (1991).

Kanai, H., Yamada, K., Kanamine, M., and Toda, J., *IEEE Trans. Magn.*, **34**, 1498 (1998).

Kools, J. C. S., Rijks, Th. G. S. M., De Veirman, A. E. M., and Coehoorn, R., *IEEE Trans. Magn.*, **31**, 3918 (1995).

Leal, J. L. and Kryder, M. H., *IEEE Trans. Magn.*, **32**, 4642 (1996).

Leal, J. L. and Kryder, M. H., *J. Appl. Phys.*, **83**, 3720 (1998).

Lederman, M., *IEEE Trans. Magn.*, **MAG-35**, 794 (1999).

Néel, L., *Compte Rendus*, **255**, 1676 (1962).

Nogués, J. and Schuller, I. K., *J. Magn. Magn. Mater.*, **192**, 203 (1999).

Parkin, S. S. P., *Phys. Rev. Lett.*, **71**, 1641 (1993).

Parkin, S., More, N., and Roche, K., *Phys. Rev. Lett.*, **64**, 2304 (1990).

Potter, R. I., *IEEE Trans. Magn.*, **MAG-10**, 502 (1974).

Sakakima, H., Satomi, M., Sugita, Y., Kawawake, Y., and Adachi, H., *Intermag Conf. Proc.*, Toronto, Paper FA-08 (2000).

Shimizu, Y., Varga, L., Eguchi, S., and Tanaka, A., *Intermag Conf. Proc.*, Toronto, Paper FA-07 (2000).

Sondheimer, E. H., *Adv. in Phys.*, **1**, 1 (1952).

Speriosu, V. S., Gurney, B. A., Wilhoit, D. R., and Brown, L. B., Paper presented at INTERMAG'96 Conference, Seattle, 1996.

Swatgen, H. J. M., Strijkers, G. J., Bloeman, P. J. H., Willekens, M. M. H., and de Jonge, W. J. M., *Phys. Rev. B*, **53**, 9108 (1996).

Tsang, C., Fontana, R. E., Lin, T., Heim, D. E., Speriosu, V. S., Gurney, B. A., and Williams, M. L., *IEEE Trans. Magn.*, **MAG-30**, 3801 (1994).

Tsang, C., Pinarbasi, M., Santini, H., Marinero, E., Arnett, P., Olson, R., Hsiao, R., Williams, M., Payne, R., Wang, R., Moore, J., Gurney, B., Lin, T., and Fontana, R., *IEEE Trans. Magn.*, **MAG-35**, 689 (1999).

Tong, H. C. and Funada, S., Unpublished internal data, Read-Rite, Milpitas, CA, 1999.

Tsuchiya, Y., Sano, M., Araki, S., Morita, H., and Matsuzaki, M., *Intermag Conf. Proc.*, Toronto, Paper FA-09 (2000).

Uehara, Y., Yamada, K., and Kanai, H., *IEEE Trans. Magn.*, **32**, 3431 (1996).

Zhang, J. and White, R. M., *IEEE Trans. Magn.*, **32**, 4630 (1996).

7

MICROTRACK PROFILES

The remarkable improvements in storage capacity of hard disk drives arise from appropriate reductions in the dimensions of the magnetic bit cell, for which the scaling rules in magnetic recording systems were introduced in Chapter 1 and the concepts of "bit cell ratio" $R_{bc} = $ bpi/tpi and areal density $= $ bpi · tpi $= R_{bc} \cdot$ tpi^2 were defined. For example, an areal density of 6 Gbits/in.^{2} requires nearly 21 ktpi and 290 kbpi at $R_{bc} = 14$, and the scaling rules for track width [see (1.3) and (1.4)] would require a magnetic read width $W \simeq 0.8$ track pitch, or about 1.0 µm. At higher areal densities, the present trend (1999) is to emphasize $R_{bc} \sim 10$ (higher tpi and lower bpi) and take advantage of reduced bandwidth requirements; for 25 Gbits/in.^{2} (50 ktpi and 500 kbpi with $R_{bc} = 10$) the appropriate read width is about 0.40 µm. At narrow track widths, one must account for the three-dimensional nature of magnetic sensors because the side-reading behavior increases in its importance relative to the on-track signal. This chapter examines the sensitivity of AMR and GMR heads to flux excitation from very narrow written tracks and discusses the diagnostic information that may be extracted from careful study of track profiles. In normal applications, the write width is somewhat greater than the read width, such that the read sensor position can vary $\sim \pm 15\%$ from the nominal position and the signal is not greatly corrupted by unwanted flux from adjacent data tracks or noise between tracks. One of the standard evaluation tests of MR heads is to write a test track whose width is much narrower than the read width, scan the read sensor across this "microtrack," and measure the signal versus off-track position; the result of this test is a "microtrack profile" that reveals important details about the magnetic state of the MR sensor. This chapter introduces concepts regarding ideal sensors and three-dimensional sensitivity functions, then examines the anisotropic propagation of signal flux through a magnetic layer and shows how this phenomenon leads to asymmetric side reading in a sensor whose magnetization is biased at an angle relative to the magnetic easy axis. A number of experimental microtrack profiles are shown, and discussions about root causes are presented. Some of the important references to the literature on this subject are found in Heim (1994), Wallash et al. (1991), Yuan et al. (1994), and Yuan and Bertram (1994). Additional useful information is also found in Heim (1989), Kanai et al. (1989), Feng (1990), and Lee et al. (1991).

IDEAL TRACK PROFILES

The ideal written track is a step function at the track edges; that is, the magnetization at any position across the track is constant up to an edge and abruptly drops to zero at all positions beyond either edge. The ideal sensitivity function for a readback sensor would be a step function also. A written track can be prepared with reasonably sharp edges by writing data along the track, then stepping the head away from the track centerline, passing DC current through the write head coil, and then repeating the operation along the opposite track edge; if such a track is narrower than the reader width, it is called a "microtrack." The signal profile derived by scanning a read sensor across a written track is the convolution product of the write and read functions. That is, if $W(x)$ and $R(x)$ are the write and read profiles, then the read signal $S(x)$ across the track (x-direction) is given by the relation

$$S(x) = W(x) * R(x) = \int_{-\infty}^{\infty} W(\tau) R(x - \tau) \, d\tau. \tag{7.1}$$

The sensor and microtrack geometry are shown in Fig. 7.1, and hypothetical track profiles $S(x, R, W, \eta)$ are shown in Fig. 7.2: ideal (rectangular) written tracks, whose widths W are 0.2, 0.5, 1.0, and 1.5 µm, are convolved with an ideal (rectangular) read sensitivity function of fixed width $R = 1.0$ µm whose side-reading parameter η is vanishingly small. (Side reading is discussed in the next section.) The results are trapezoidal profiles where the maximum signal is determined by the available magnetic flux from the written track up to the limit where $W > R$. Since R is rectangular, the absense of side reading limits the maximum amplitude to 1.0 even when $W \gg R$. The full width at half maximum (FWHM) is R for $R \leq W$ and W for $W > R$. With microtracks ($W < R$), the maximum amplitude drops in proportion to W/R and the read profile trapezoidal base is *broadened by W/2 at each side* of the read track such that the full indicated read width is $R + W$ at the zero-amplitude level; the excess width $\frac{1}{2}W$, or "*skirt*" at each edge, is an important artifact of the microtrack test. In other words, microtrack analysis requires very narrow ($W/R < 0.1$) microtracks to resolve details about a nonideal read track sensitivity function, but small W produces very low signal amplitude, which normally requires signal averaging to attenuate white noise. Typically, $W \sim 0.05$ µm is the smallest practical microtrack with the available instrumentation (late 1999).

MICROTRACK PROFILES

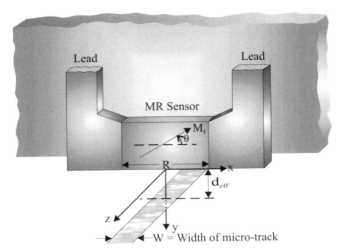

FIGURE 7.1 Geometry of a sensor located over a microtrack. The active width of the sensor is R, located a distance d_{eff} over a microtrack whose width is W.

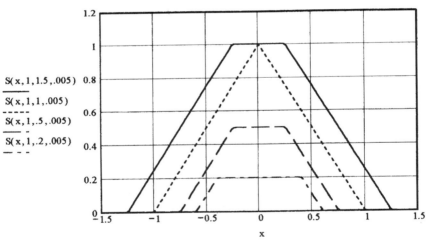

FIGURE 7.2 Hypothetical microtrack profiles for a read sensor with $R = 1.0\,\mu\text{m}$ and written microtracks with $W = 0.2, 0.5, 1.0, 1.5\,\mu\text{m}$ Side reading is negligible ($\eta = 0.005\,\mu\text{m}$) for these computations, so the profiles approach the ideal shape.

NONIDEAL TRACK PROFILES

The read sensitivity function for shielded MR heads is not ideal; at each edge of the sensor there exists side reading whose extent and character depend on details of the device, the effective spacing between the medium and the bottom of the sensor, its stripe height, how it is biased, and its gap length (if G is large). The side-reading parameter η introduced above is a useful concept in understanding the behavior of devices, and it can be defined for a variety of read sensitivity functions. While not based on correct physics, perhaps one of the simplest side-reading functions is constructed with hyperbolic tangent functions of the form $f(x) = \tanh[(x - x_0)/\eta]$, where η serves to define the slope at $x = x_0$; as $\eta \to 0$, the edge approaches the sharpness of a step. The read sensitivity function thus becomes

$$R(x) = \frac{1}{2}\left[\tanh\left(\frac{x + R/2}{\eta}\right) - \tanh\left(\frac{x - R/2}{\eta}\right)\right], \quad (7.2)$$

which can be normalized by $\tanh(R/2\eta)$. This function is convolved with a written track defined by unit step functions $u(x + \frac{1}{2}W) - u(x - \frac{1}{2}W)$ to obtain the track profile

$$S(x, R, W, \eta) = \frac{\ln\left[\dfrac{\cosh\left[\dfrac{|x| + 0.5(W + R)}{\eta}\right] \cosh\left[\dfrac{|x| - 0.5(W + R)}{\eta}\right]}{\cosh\left[\dfrac{|x| + 0.5(W - R)}{\eta}\right] \cosh\left[\dfrac{|x| - 0.5(W - R)}{\eta}\right]}\right]}{2\ln\left[\cosh\left(\dfrac{R}{\eta}\right)\right]}. \quad (7.3)$$

Hypothetical microtrack profiles showing many details associated with actual microtracks are given in Fig. 7.3 for $W = 0.10\,\mu$m and $R = 1.0\,\mu$m for various side-reading parameter values. The curves are normalized with the relation

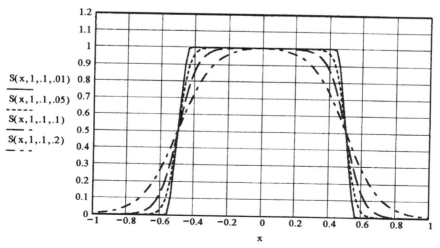

FIGURE 7.3 Hypothetical microtrack profiles $S(x, R, W, \eta)$ for $R = 1.0\,\mu$m and $W = 0.10\,\mu$m for side reading parameters $\eta = 0.01, 0.05, 0.10, 0.20\,\mu$m, respectively.

$2\ln\{\cosh[0.5(W+R)/\eta]/\cosh[0.5(W-R)/\eta]\}$, and one immediately sees the profile skirts increase monotonically with η. For a symmetric read profile, the excess width skirt at either edge is given by the relation

$$\text{Skirt}(W, \eta) = \sqrt{(\tfrac{1}{2}W)^2 + \eta^2} \tag{7.4}$$

and Fig. 7.4 gives plots of the skirt versus η for microtracks of 0.1 and 0.2 μm width.

The three-dimensional analytical equations of shielded MR devices are rather complicated, but Heim (1994), Yuan et al. (1994), and Yuan and Bertram (1994) discuss the use of the three-dimensional Green's function for a shielded MR sensor centered over a magnetic transition in the medium. Yuan and Bertram discuss idealized simplifications to the field equations that nevertheless give useful insight into side-reading behavior and provide a connection between the effective spacing $y_e = d + a + \tfrac{1}{2}\delta$ between the medium and sensor and the side-reading parameter expressed by η. The free-space potential between the head and medium is found by convolving the Green's function with the perfect image of the medium charge distribution in the shields and sensor; the gradient of this potential is the field of interest. The form of the field given by Yuan and Bertram (1994) is

$$H_z(x, y, z) = \frac{1}{2\pi} \int_{-R/2}^{R/2} dx' \int_{-\infty}^{\infty} dz' H_z^S(x', z') \frac{y}{[(x-x')^2 + y^2 + (z-z')^2]^{3/2}}, \tag{7.5}$$

where the field $H_z^S(x', z')$ along the boundary ($y = 0$) equals $+H_0$ and $-H_0$ in the respective gap regions (each of length g) and drops to zero at each read track edge ($x = \pm\tfrac{1}{2}R$). This is roughly equivalent to the "ideal" sensor read profile excited by a

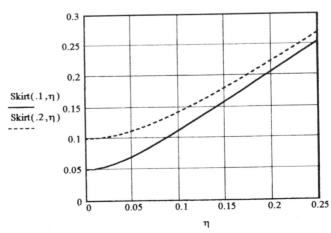

FIGURE 7.4 Plots of the excess skirt width (μm) versus side-reading parameter η (μm) for written widths $W = 0.1, 0.2$ μm.

magnetic point charge at the effective distance y_e from the sensor. Yuan and Bertram take a Fourier transform of (7.5) to obtain the result

$$H_z^{\text{FT}}(k,x,y) = \frac{y}{2\pi i g} \frac{\sqrt{2}}{\Gamma(3.2)} \left\{ \cos\left(\frac{kt}{2}\right) - \cos\left[k\left(g+\frac{t}{2}\right)\right] \right\}$$

$$\times \int_{-R/2}^{R/2} dx' \frac{K_1\left[k\sqrt{(x'-x)^2+y^2}\right]}{\sqrt{(x'-x)^2+y^2}}, \qquad (7.6)$$

where $K_1(z)$ is the first-order modified Bessel function and $k = 2\pi/\lambda$ is the wavenumber for sine wave recording at a wavelength λ. For long wavelengths the argument z for $K_1(z)$ is small and the limiting form for the modified Bessel function becomes $K_1(z) \sim 1/z$ (see Ambramowitz and Stegun, 1964). Using this form, (7.6) reduces to

$$H_z^{\text{FT}}(k,x,y) \simeq \frac{1}{k} \int_{-R/2}^{R/2} \frac{dx'}{(x'-x)^2+y^2} = \frac{1}{ky}\left[\tan^{-1}\left(\frac{x+R/2}{y}\right) - \tan^{-1}\left(\frac{x-R/2}{y}\right)\right], \qquad (7.7)$$

where the factors in g and t in (7.6) are ignored. A normalized form of (7.7) given by the relation $h_z(x,y) = (ky/\pi)H_z^{\text{FT}}(k,x,y)$ is plotted in Fig. 7.5 for several values of effective spacing y. As y approaches zero, the field approaches a step function along x, as expected. In other words, the total spread along the x-direction can be understood as the convolution product of two step functions $u(x'+\tfrac{1}{2}R) - u(x'-\tfrac{1}{2}R)$ with a line charge at the distance y; the line spread function is given by the relation $1/[(x'-x)^2+y^2]$, which is the denominator of the integrand in (7.7). The side-reading results with (7.7) are similar to those obtained with a three-dimensional head field model given by Girschovichus and Girschovichus (1964) for which the horizontal field component at $z=0$ and $y \geq 0$ is written

$$H_z(x,y,z=0) = \frac{H_0}{\pi}\left\{ \tan^{-1}\left[\frac{0.5g(x+0.5R)}{y\sqrt{(x+0.5R)^2+y^2+(0.5g)^2}}\right] \right.$$

$$\left. - \tan^{-1}\left[\frac{0.5g(x-0.5R)}{y\sqrt{(x-0.5R)^2+y^2+(0.5g)^2}}\right] \right\}. \qquad (7.8)$$

This form, which is more appropriate for the side reading of an MR element flush with the edge of a finite width shield, also shows a dependence of side reading on gap length g. The skirt width for a vanishingly narrow microtrack (a line charge along the z-direction) is defined here as the linear extrapolation of the track edge

232 MICROTRACK PROFILES

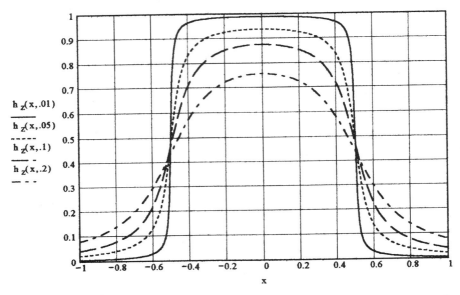

$h_z(x,.01)$ ———
$h_z(x,.05)$ ·······
$h_z(x,.1)$ — — —
$h_z(x,.2)$ — · —

FIGURE 7.5 Plots of normalized ideal sensor read profiles excited by a magnetic point charge at the effective distance $y_e = 0.01, 0.05, 0.1, 0.2\,\mu m$, respectively, from the sensor.

profile inflection point (at $x = \tfrac{1}{2}W$) down to the intercept with the x-axis; this excess width is essentially half of the length of a ramp function whose slope matches the arctangent function at the inflection point. That is,

$$\text{Skirt width} \simeq \tfrac{1}{2}\pi(d + a + \tfrac{1}{2}\delta) \tag{7.9}$$

because the length of the ramp transition is $L = \pi y$ (see Potter, 1970). Other definitions of skirt width are found in the literature. Broadening is a general result of the convolution product, and N repetitive convolutions of a function with itself lead to a total broadening proportional to $\eta\sqrt{N}$, where η is the characteristic broadening for a single convolution.

ANISOTROPIC FLUX PROPAGATION IN A SENSE LAYER

Heim (1989) presented micromagnetic computations showing that injected flux from a point source propagates through an MR layer preferentially in a direction that is perpendicular to the magnetization **M** at each location in the sensor; along the direction of **M** the permeability is 1.0 because $|\mathbf{M}| = M_S$, which is a fixed value. Heim's insight leads directly to fruitful interpretations of left–right asymmetries in microtrack profiles and the dependence of profile asymmetry on bias current magnitude and direction in MR sensors. A more complete analysis of track profiles was published later (Heim, 1994). Wallash et al. (1991) studied experimental microtrack profiles for various stripe heights and interpreted their results with a simple model derived from the concept of anisotropic flux propagation; their work is followed closely here. Figure 7.6 shows microtrack profiles for AMR heads at different stripe heights; a profile is substantially asymmetric for long stripes and becomes symmetric at short stripes. Figure 7.7 shows a sketch of flux propagation through the plane of an MR sense layer for several positions of a microtrack; along the left side the path length $L(x, \theta_b)$ for flux flow increases linearly as $L(x, \theta_b) = x/\sin\theta_b$ to a maximum of $h/\cos\theta_b$ as the microtrack moves from $x = 0$ to $x = h\tan\theta_b$, and the path length remains constant with x until W moves beyond $x = R$ at the right edge where flux propagation ceases. For negative bias angles, the profile is reversed. These idealized plots ignore the skirts at each edge that arise from the sensitivity functions shown in Fig. 7.5; edge rounding that arises from side reading is included later in this discussion. The maximum path length is limited by the device dimensions to $L_{max} = (h^2 + R^2)^{1/2}$. The bias angle θ_b is approximately $\pm 45°$ for a well-biased AMR head and close to $0°$ for a correctly biased GMR spin valve head, but the bias angle depends on the bias current, details of the head design, its construction, and consistency in manufacturing execution. Figure 7.8 shows calculated "ideal" flux path lengths for bias angles θ_b of 0.2, 0.5, 0.75, and 1.0 rad, with h and R equal to 0.5 and 0.9 μm, respectively.

The average ΔR for the AMR effect (which is proportional to $\Delta\rho \sin^2\theta$) is found by assuming the bias angle changes by a small amount when excited with a microtrack of width W; that is, $\theta(x, y) = \theta_b + \Delta\theta(x, y)$ and the resistivity is averaged

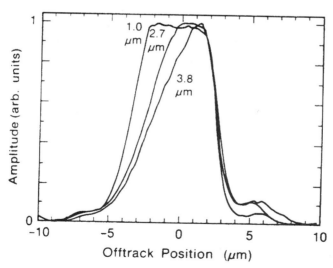

FIGURE 7.6 Experimental AMR microtrack profiles for three heads with stripe heights of 1.0, 2.7, and 3.8 μm. Peak amplitudes have been normalized to 1.0 for ease of comparison. Reprinted with permission from A. Wallash, M. Salo, J. K. Lee, D. E. Heim, and G. Garfunkel, *J. Appl. Phys.*, **69**, 5402. Copyright © 1991 by the American Institute of Physics.

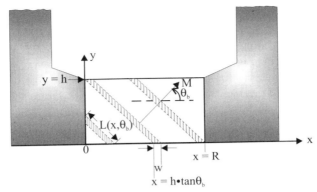

FIGURE 7.7 Geometry for anisotropic flux propagation through a sense layer from a microtrack of width W. Sensor magnetization M is biased at an angle θ_b and signal flux propagates along a direction perpendicular to M.

over the active area. If the stripe height $h \approx \lambda_{sh}$ (the signal flux decay length), then the average change in angle is one-half of the maximum change $\Delta\theta_0$ at the sensor bottom ($y = 0$) [see (5.9) and the discussion in Chapter 5]. That is,

$$\Delta R = \int_0^h \int_0^W \Delta\rho \sin^2 \theta(x, y)\, dx\, dy \simeq \frac{\Delta\theta_0 W \Delta\rho \sin 2\theta_b}{2th^2} L(x, \theta_b), \qquad (7.10)$$

and it is seen that the major features of left/right track edge asymmetry observed in the experimental curves of Fig. 7.6 are described by $L(x, \theta_b)$, which defines a

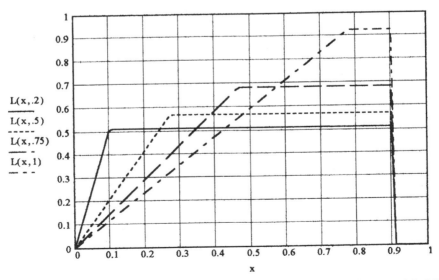

FIGURE 7.8 Calculated flux propagation path lengths for bias angles $\theta_b = 0.2$, 0.5, 0.75, 1.0 rad, with $h = 0.5$ and $R = 0.9\,\mu m$.

microtrack profile for an AMR element with constant sensitivity along the signal input edge. For spin valves with the pinned layer at $\theta_p = \pm\frac{1}{2}\pi$ and the free-layer bias angle given by θ_F, the GMR effect is proportional to $\Delta\rho \sin(\theta_F + \Delta\theta)$, so (7.10) is rewritten with $\Delta\rho \sin(\theta_F + \Delta\theta) \simeq \Delta\rho \cos\theta_F \Delta\theta$ for small θ_F, and the average variation in resistance is

$$\Delta R = \int_0^h \int_0^W \frac{\Delta\rho \cos\theta_F \Delta\theta(x,y)}{th^2} dx\,dy \simeq \frac{\Delta\theta_0 W \Delta\rho \cos\theta_F}{2th^2} L(x, \theta_f). \quad (7.11)$$

Side reading of a shielded sensor is included by forming the convolution product of $L(x, \theta_{b,F})$ and a line spread function, as was done for $h_z(x, y)$ in (7.7). The width of the written microtrack W with sharp edges is included by an additional convolution. Figure 7.9a shows the microtrack profiles of Fig. 7.8 after convolutional broadening with a line spread function at a spacing $y_e = d + a + \frac{1}{2}\delta = 0.025\,\mu\text{m}$ and by a microtrack with $W = 0.20\,\mu\text{m}$. Fig. 7.9b shows profiles for $\theta_b = 0.75\,\text{rad}$ normalized to 1.0 for direct comparison; the two cases are $W = 0.01$ (essentially a line charge) and $W = 0.20\,\mu\text{m}$ (which is somewhat wide, but it demonstrates the extra broadening that interferes with good resolution of the sensor profile).

READ WIDTH AND MAGNETIC CENTER

The microtrack read width W_r (noted on Fig. 7.9b) is taken at the 50% signal level on either side of the "magnetic center" and is less than or equal to the apparent full read width $R + [W^2 + (\pi y_e)^2]^{1/2}$ of the active sensor. In terms of the stripe height and bias angle,

$$\text{Microtrack read width} = W_r = R - \tfrac{1}{2}h \tan\theta_b. \quad (7.12)$$

The physical and magnetic centers are offset by a distance Δ given by the relation

$$\text{Magnetic center offset} = \Delta = \tfrac{1}{2}(R - W_r) = \tfrac{1}{4}h \tan\theta_b, \quad (7.13)$$

which is indicated in Fig. 7.9b. The edges, which in ideal cases would be steps, are broadened by the amount $[W^2 + (\pi y_e)^2]^{1/2}$ and the excess skirt width is half this value; this is seen at the right edges of Figs. 7.9a,b. The distance x_0 along the top of the roughly trapezoidal profile is given by the approximate relation

$$x_0 \simeq W_r - \tfrac{1}{2}\left[\sqrt{W^2 + (\pi y_e)^2} + h\tan\theta_b\right]. \quad (7.14)$$

MICROTRACK PROFILES OF NARROW READ TRACKS

When the width of an ideal read track becomes narrow with respect to the edge profile, the trapezoidal nature of a microtrack profile is lost. Figure 7.10a shows a line spread function (with y_e set at $0.055\,\mu\text{m}$) along with the convolution of this

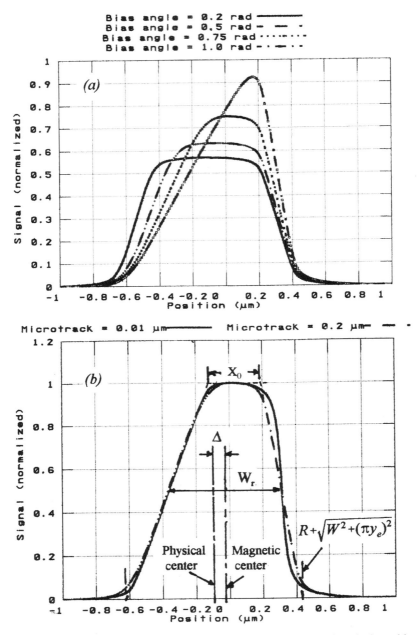

FIGURE 7.9 (a) Microtrack profiles of Fig. 7.8 after convolutional broadening with a line spread function at a spacing $y_e = 0.025\,\mu\text{m}$ and by a microtrack with $W = 0.20\,\mu\text{m}$. (b) Microtrack profiles at $\theta_b = 0.75$ rad after convolutional broadening with $W = 0.01, 0.2\,\mu\text{m}$ at the effective spacing $y_e = 0.025\,\mu\text{m}$.

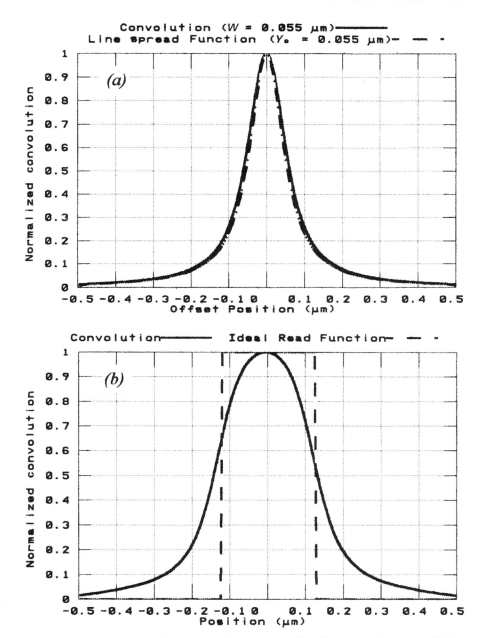

FIGURE 7.10 (a) Spread function of a line charge at the effective spacing $y_e = 0.055\,\mu\text{m}$ and convolution of a written track $W = 0.055\,\mu\text{m}$ with the line spread function. (b) Ideal read profile with $R = 0.25\,\mu\text{m}$ and convolutions of the ideal with a line spread function and a written microtrack with $W = 0.055\,\mu\text{m}$.

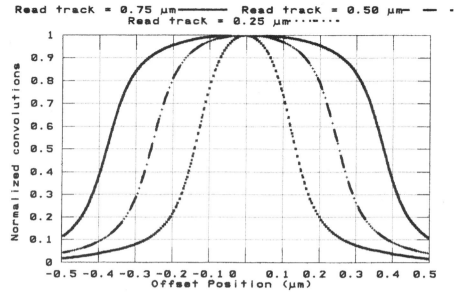

FIGURE 7.11 Convolutions of ideal read profiles for $R = 0.25, 0.50, 0.75\,\mu m$ with spread functions as plotted in Fig. 7.10b.

function with a very narrow but finite-width microtrack ($W = 0.055\,\mu m$); convolutional broadening is apparent. Figure 7.10b is the result of an additional convolution of the broadened line with an ideal step profile for $R = 0.25\,\mu m$. As R goes from 0.25 to 0.50 μm and to 0.75 μm, the trapezoidal nature of the profile emerges; this is shown in Fig. 7.11. The edge width (determined with a ramp function tangent to the track edge slope) is given by the relation

$$\text{Edge width} = \sqrt{W^2 + (\pi y_e)^2} \simeq 0.18\,\mu m, \tag{7.15}$$

and when $R \sim 2$ (edge width) details of the actual profile are not resolved in the experimental microtrack read profile and deconvolution techniques become necessary for extracting precise information about the sensor.

EXPERIMENTAL MICROTRACK PROFILES WITH SPIN VALVES

Microtrack profiles reveal details of sensor behavior that may be pathological and whose explanations require speculation, or the profiles may be well behaved and the explanations follow the theoretical ideas outlined above. The experimental results given below fall into the broad categories of stable, unstable, pathological, and well behaved, and these results are presented to give the reader a feeling for the richness and complexity of microtrack analyses; there can be more than one explanation for a given profile, and those offered here are not necessarily the correct or only interpretation. Figures 7.12a,b, for example, show full and microtrack profiles of a conventional top spin valve for application at an areal density of 3–4 Gbits/in.^{2}; the full track width (1.62 μm or 63.8 μinches) is determined by the width of the write head and the microtrack read width $W_r = 1.38$ μm (54.2 μinches.) The profile is similar to that shown in Fig 7.9b, with the reduction in flux path length appearing on the right edge instead of the left. The interpretation is compatible with a spin valve whose free layer is poorly biased at an angle $\theta_F \sim -0.68$ rad ($-39°$), an estimated stripe height $h \simeq 1.0$ μm, excess skirt width $0.5[W^2 + (\pi y_e)^2]^{1/2} \simeq 0.19$ μm, and $x_0 \simeq 0.78$ μm. In this experiment, $W = 0.25$ μm (10.0 μin.), and the effective spacing is estimated at $y_e = d + a + \frac{1}{2}\delta \simeq 0.09$ μm(3.5 μin.), both of which combine to yield an excess skirt width in good agreement with the experimental left edge. The magnetic center offset Δ is about 0.1 μm experimentally, whereas the calculated value from (7.13) is 0.2 μm; the source of this discrepancy may be in the estimated stripe height.

A somewhat different spin valve head is shown in Figs. 7.13a,b; the full write width is 1.63 μm (64.2 μin.) and the microtrack read width $W_r = 0.89$ μm (34.9 μin.). The profile is stable and asymmetric, but there is no flat region because the read width is small and the sensitivity is not constant. The left edge breadth is

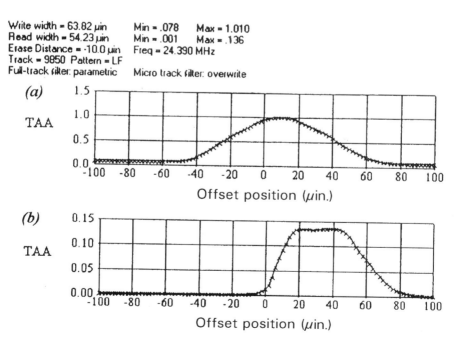

FIGURE 7.12 Experimental profiles for (a) full track and (b) microtrack for a conventional top spin valve showing anisotropic flux propagation. Amplitude is in millivolts (p-p) and offset position is in microinches.

240 MICROTRACK PROFILES

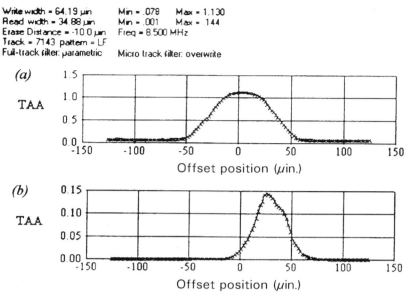

Write width = 64.19 µin   Min = .079   Max = 1.130
Read width = 34.88 µin   Min = .001   Max = 144
Erase Distance = -10.0 µin   Freq = 8.500 MHz
Track = 7143 pattern = LF
Full-track filter: parametric   Micro track filter: overwrite

FIGURE 7.13 Experimental track profiles for (a) full track and (b) microtrack behavior of a conventional top spin valve showing unexplained variations of read sensitivity along the read profile. The full track output is stable. Amplitude is in millivolts (p-p) and offset position is in microinches.

0.50 µm, of which amount only 0.38 µm can be assigned to W and y_e convolutional broadening. This head falls in the category of stable and pathological.

The microtrack profile of a dual synthetic spin valve (DSSV) capable of 36 Gbits/in.^{2} is shown in Fig. 7.14 at a linear density of 48 kfci. The width of the written microtrack is $W = 0.053$ µm and the effective spacing $y_e = d + a + \frac{1}{2}\delta \simeq 0.025 + 0.02 + 0.01 = 0.055$ µm. The left and right edge breadths are nearly equal (0.20 and 0.19 µm), and a theoretical symmetric profile with edges $[W^2 + (\pi y_e)^2]^{1/2} \simeq 0.18$ µm and $R = 0.25$ µm is shown as the dotted curve for comparison with the experimental results. The experimental microtrack read width $W_r = 0.31$ µm at this low density and it narrows to about 0.25 µm at 543 kfci. Part of this reduction can be attributed to the writing width at different densities.

Figure 7.15 presents full track profiles at 48 and 543 kfci for which the apparent written widths are 0.39 and 0.34 µm, respectively. Some of the extra width at 48 kfci could arise from a wavelength-dependent side-reading behavior of the DSSV sensor (see Yuan and Bertram, 1994), but at this time it is assigned to the writing process.

Microtrack profiles may depend on sense current polarity as well as the width of the written track. Figures 7.16a,b show how the read profiles change for W values of 0.125, 0.25, 0.375, and 0.50 µm at sense currents of $+5$ and -5 mA. The head is a conventional top spin valve with a free layer of CoFe/NiFe that has a significant

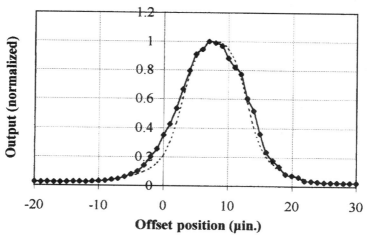

FIGURE 7.14 Experimental microtrack profile of a 36-Gbit/in.^{2} DSSV. The dotted line is computed from an ideal (step) read profile convolved with a written microtrack with $W = 0.053\,\mu\text{m}$ at an effective spacing $y_e = d + a + \frac{1}{2}\delta = 0.025 + 0.02 + 0.01\,\mu\text{m} = 0.055\,\mu\text{m}$.

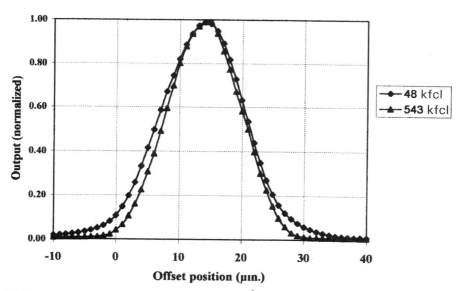

FIGURE 7.15 Full track profiles of a 36-Gbit/in^2 DSSV head at 48 and 543 kfci for which the full widths are 0.38 and 0.34 μm, respectively.

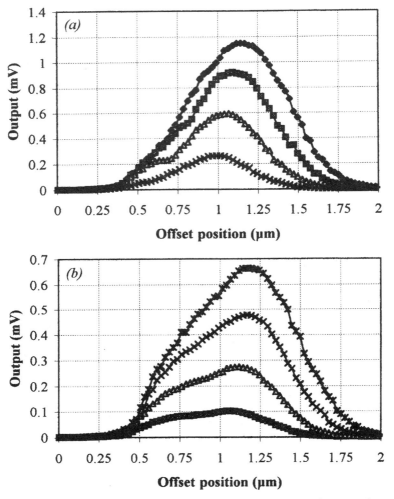

FIGURE 7.16 Microtrack profiles of a top conventional spin valve at written widths $W = 0.125, 0.25, 0.375, 0.50\,\mu m$ for (a) $I = +5\,mA$ and (b) $I = -5\,mA$.

AMR effect, so the output signal is greater for positive than for negative sense currents. (See the discussion of this subject in Chapter 6.) Profile shape is also noticeably distorted with a negative sense current. Figure 7.17a shows a linear increase of output with W as expected from theory, and Fig. 7.17b is a plot of the microtrack read width W_r as a function of W for which W_r increases, but the difference in widths for negative and positive currents is unexplained.

Output level and its stability in MR recording heads are of great interest in the design stage and in failure analysis activities when problems arise. Microtrack profiles are most useful for discovering the loci of unstable regions or for diagnosing issues regarding damage from electrostatic discharge (ESD) or corrosion. Since the subject of reliability is addressed in more depth in Chapter 8, a discussion of microtrack profiles manifesting output instabilities is deferred to that chapter.

The final device for discussion is a conventional spin valve that has been intentionally unpinned and repinned by injecting currents of sufficient magnitude and direction to heat regions of the device beyond the blocking temperature. The instrument used is a human body model (HBM) tester that contains circuitry for controlled simulation of ESD; this subject is addressed in Chapter 8 on reliability

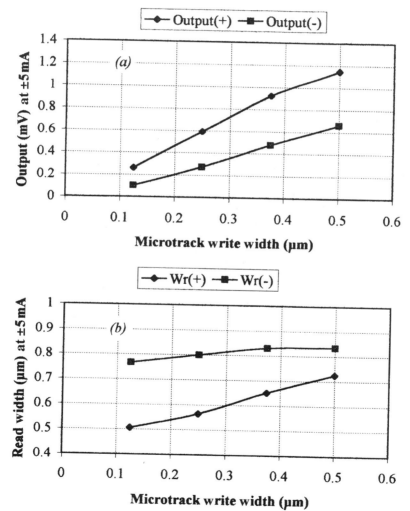

FIGURE 7.17 Data extracted from the microtrack profiles of Figs. 7.16 (*a*),(*b*). (*a*) Output signals for $I = \pm 5$ mA and (*b*) read width versus microtrack written width.

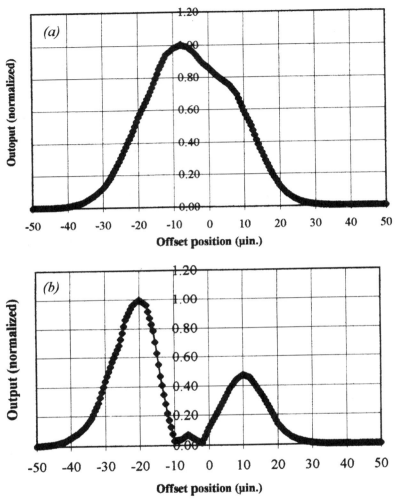

FIGURE 7.18 Microtrack profiles of a conventional spin valve that has been intentionally unpinned by transient heating using an HBM tester: (a) original undamaged head, (b) the head after a voltage pulse (+35 V or +23 mA), and (c) after receiving a "healing" pulse of −35 V or −23 mA). The curves are each normalized to unity response, but repinning an unpinned device seldom (if ever) restores the full amplitude of the head response. Unpublished data courtesy of D. M. Hannon, Western Digital Corp. (1999).

testing, so details regarding ESD damage are deferred to that chapter. The microtrack profile of the original spin value head is given in Fig. 7.18a and the unpinned profile is shown in Fig. 7.18b. The original $W_r = 0.87$ μm and the unpinned device is almost completely insensitive in the center region. The HBM voltage was +35 V, and in the opposition mode the positive current (about 23 mA) was sufficient to heat

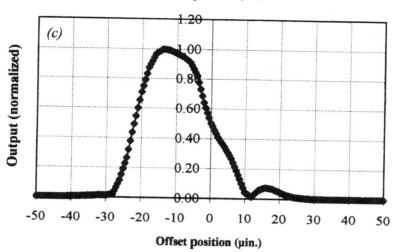

FIGURE 7.18 (*Continued*)

the device to the blocking temperature in the vicinity of the track center. A "healing" pulse of -35 V ($I = -23$ mA) was then discharged through the device, which reheated the stripe to the blocking temperature and reset portions of the pinned layer to a useful direction. The "healed" microtrack profile is shown in Fig. 7.18c for which the active read width $W_r = 0.55$ µm and a small side lobe remains at the right side of the track. These unpublished microtrack profiles are from the courtesy of Hannon (1999). Jang et al. (1999) studied microtrack profiles of AMR and GMR heads with various levels of ESD damage induced by a HBM tester; both types of heads showed loss of sensitivity near the track center. In the AMR heads, loss of sensitivity was attributed to sensor melting, and in GMR heads the loss was assigned to unpinning of the pinned layer.

REFERENCES

Abramowitz, M. and Stegun, I. A., *Handbook of Mathematical Functions,* NBS Applied Math Series 55, U.S. Department of Commerce, Washington, DC, 1964.

Feng, J. S., *Intermag Conf. Proc.,* Brighton, United Kingdom, Digest FP-08, (1990).

Girschovichus, S. Kh., and Girschovichus, I. Kh., *Radiotekhnika*, **19**(4) [*Telecomm. Radio Eng.*, **19**(4), Part 2 (1964)].

Hannon, D. M., unpublished data, Western Digital, San Jose, CA, 1999.

Heim, D. E., *IEEE Trans. Magn.*, **MAG-30,** 1453 (1994).

Heim, D. E., *Intermag Conf. Proc.*, Washington, DC, Digest AA-5, (1989).

Jang, E. K., Kim, W. W., Kao, A. S., and Lee, H. J., *IEEE Trans. Magn.*, **MAG-35**, 2616 (1999).

Kanai, H., Hosono, K., and Takagi, H., *Intermag Conf. Proc.,* Washington, DC, Digest EQ-17 (1989).

Lee, J. K., Wallash, A., and Poon, A. L., *J. Appl. Phys.,* **69**, 5399 (1991).

Potter, R. I., *J. Appl. Phys.,* **41**, 1647 (1970).

Shen, J. X., Chang X., Ding, J., Shultz, A., and Liao, S., *IEEE Trans. Magn.*, **MAG-35**, 2595 (1999).

Wallash, A., Salo, M., Lee, J. K., Heim, D. E., and Garfunkel, G., *J. Appl. Phys.,* **69**, 5401 (1991).

Yuan, S. W. and Bertram, H. N., *IEEE Trans. Magn.*, **MAG-30**, 1267 (1994).

Yuan, S. W., Bertram, H. N., and Bhattacharyya, M. K., *IEEE Trans. Magn.,* **MAG-30**, 381 (1994).

8

CHARACTERIZATION OF MR DEVICE FUNCTION AND RELIABILITY

To retrieve data at acceptable bit error rates, MR read sensors must be accurately positioned over precisely written data tracks. Writing and reading of data require dynamic test equipment composed of a magnetic disk, spindle motor, head micropositioning equipment, data encoding and write circuitry, read electronics, and a decoding channel with error detection logic. Chapter 1 discusses some of the root causes of data errors, and successful analysis of device malfunction requires a cadre of dynamic tests that, when applied in a correct sequence, can isolate the cause (or causes, where multiple causation must be invoked) and assign responsibility to the appropriate component. Some tests produce useful information about an MR device without actually writing and reading on a spinning disk; quasi-static testers (QSTs) are used to characterize finished or unfinished sensors in a wafer (which contains 30,000–40,000 devices). This final chapter discusses dynamic functional tests with MR read/inductive write heads, nonfunctional QST tests, and correlations between QST and dynamic tests. Wang and Taratorin (1999) give unusually broad coverage of noises, nonlinear distortions, and various performance tests in magnetic recording systems, and Ashar (1997) discusses considerations for integrating recording heads with detection electronics and channels.

Devices are easily damaged by electrostatic discharge (ESD) of triboelectrically accumulated charge that is subsequently discharged through the very thin layers of the sensor. Damage may show up as a loss in output signal or total melting of the sensor metallurgy, so it is necessary to relate the knowledge of MR design and analysis gained in Chapters 2–7 to the discipline of ESD and human body model (HBM) testing. Not all damage will be immediately apparent; because MR heads are activated with a sense current, and because volume production creates devices with a distribution in electrical resistances, those devices with high resistance and that are hot relative to the mean temperature of a population could, over extended periods of time (weeks, months, years), gradually accumulate damage from electromigration or layer interdiffusion. For this reason, it is important to leave the reader with an appreciation of the device physics, geometry, and material properties necessary to understand the prediction of sensor lifetime.

DYNAMIC FUNCTIONAL TESTS

Dynamic tests must first verify if a write head can prepare useful written data tracks; the write current is incrementally increased from low to high values and an MR read head is positioned in the center of the written track to obtain signal data for each write current level. The response curves of read signal attributes versus write current are called *Write saturation curves*, and, with a known good MR read head, an appropriate level of write current can be established or writer malfunctions may be diagnosed. Having prepared an acceptable written track with a good writing head, an MR reader can be characterized regarding its output signal, asymmetry, and pulsewidth at any selected bias current. These dynamic tests are normally referred to as read bias curves when the current is varied in small increments between maximum negative and positive limits on the bias current. Additional dynamic tests include output stability and microtrack profiles at various levels of bias current.

Write Saturation Curves

For those readers unacquainted with recording measurements, it is useful to study actual data such as those given in Fig. 8.1 showing output signals (peak to peak) for a conventional bottom spin valve head (~ 6 Gbits/in.^{2}) at low and high frequencies (LF and HF signals, respectively) as a function of the write current (base to peak); the read bias current is 5 mA. At write currents below 10 mA (0-p) the write field is below the medium coercivity and one detects only a constant noise level of about 144 µV peak-to-peak. [*Note:* The test equipment uses a sample-and-hold amplifier for measuring all signals, and thus the peak values of noise are detected. The amplitudes of positive and negative peaks are measured separately (0-p) and later added to present the peak-to-peak results. The root-mean-square noise is (p-p)/(2$\sqrt{2}$) = 50.9 µV.] This excessively high noise may arise from magnetic instabilities such as those studied by Wallash (1998). Above 10 mA write current, the write field switches the recording medium and magnetic transitions are written, thus the LF and HF signals rise above the noise floor. The medium remanence approaches saturation at 16 mA and the LF signal levels off at 20 mA to approximately 2500 µV (p-p) whereas the HF signal reaches a maximum level of 447 µV (p-p) at 16 mA write current, then reduces to 380 µV (p-p) at higher currents. A reduction in HF signal with increasing write current is evidence of write head saturation such that the write field gradient *broadens* the written transition (i.e., the transition sharpness is inversely proportional to the write field gradient) and the pulsewidth increases. If the transitions are crowded at a high linear density, pulse spreading

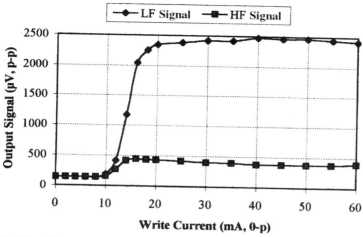

FIGURE 8.1 Write saturation curves of a conventional spin valve designed for ~ 6 Gbits/in.^{2} The low-frequency (LF) and high-frequency (HF) signals (μV p-p) are plotted as a function of write current (mA, 0-p).

increases the intersymbol interference (ISI) and the HF amplitude drops; at lower recording densities the pulses are sufficiently isolated, so any drop in amplitude is small. Figure 8.2 shows the saturation curve for PW50. At 14 mA the medium in not fully switched, so the pulses are poorly formed and the PW50 curve increases to a maximum of 322 nm; at 16 mA, PW50 goes through a *local minimum* of 235 nm, then gradually increases to 259 nm with increasing write current. The minimum PW50 at 16 mA (which coincides with the maximum HF signal) causes the increase in HF signal because narrow pulses reduce the amount of ISI at high recording densities. A pulsewidth of 259 nm is compatible with linear density of about 226 kbpi for a channel density $U = 2.3$ [see (1.12), where it is shown that $U = $ PW50 $\times$ linear density.] The signal resolution (which is the ratio of HF and LF amplitudes) and overwrite are plotted in Fig. 8.3 (note the units for resolution are given as *percent* and OW is in *negative* decibels). Above 20 mA the signal resolution drops to about 16% and remains nearly constant. Overwrite (OW) is a phenomenon with mixed causation and is normally measured in the frequency domain with a spectrum analyzer set to specific frequencies; the data in the plot of Fig. 8.3 cover the range from 0 dB (no erasure of old data at low write current) to -41 dB (a residual LF signal of 0.89%) at 50 mA. Wang and Taratorin (1999) list three factors that contribute to the "residual" overwrite signal:

1. incomplete erasure of the initial LF signal by the overwriting HF signal,
2. side reading at incompletely erased track edges, and
3. hard transition shifting (HTS) of HF transitions arising from the magnetostatic fields of old transitions approaching the writing zone at the LF rate.

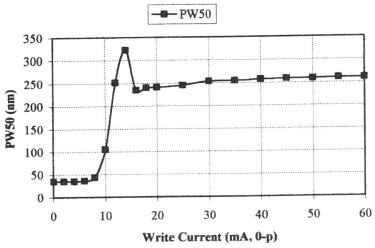

FIGURE 8.2 Write saturation curve of pulsewidth (PW50 in nm) versus write current (mA, 0-p). The device is the same one as presented in Fig. 8.1.

According to Wang and Taratorin, incomplete erasure and side reading seldom contribute significantly to overwrite problems in digital recording systems, and thus HTS is the main source of the residual OW signal. Hard transition shifting is a measure of the accuracy a write head can achieve in placing new transitions at high densities over old transitions. That is, OW is a spectral analysis of pulse position

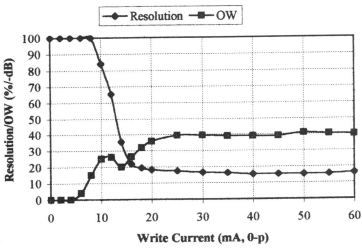

FIGURE 8.3 Write saturation curves of HF/LF signal resolution (in %) and overwrite (OW given as a residual LF signal in −dB) versus write current.

modulation, and the ratio of the final LF amplitude to the original amplitude is a measure of the amount of HTS. That is,

$$\text{OW(dB)} = 20 \log_{10}\left(\frac{\text{residual LF amplitude}}{\text{initial LF amplitude}}\right)$$
$$\simeq 20 \log_{10}\left(\frac{\sqrt{2}\pi}{4T}\Delta\right), \quad (8.1)$$

where Δ is the HTS and T is the time (both in nanoseconds) between HF transitions for the case where HF is twice the LF rate (see Bertram, 1994, p. 257). Most test specifications allow OW $= -30$ dB or less residual, which translates into $\Delta/T \leq 0.0284$; that is, at a rate of 100 Mbits/sec, $T = 10$ nsec and the allowable HTS is 284 psec or less. That HTS is independent of the MR sensor is seen in the relation for Δ given in Bertram (1994, p. 255) and in (1.28) of this book. This relation is repeated here for convenience to the reader:

$$\Delta = \frac{M_r \, \delta(d + \delta/2)}{2x_0 \pi Q H_c} \quad \text{cm}, \quad (8.2)$$

where $d, 2x_0$, and Q refer to the write head spacing, writing zone length, and normalized write field gradient, respectively; the recording medium remanence, thickness, and coercivity are given by M_r, δ, and H_c, respectively. The data in Figs. 8.1—8.3 show acceptable levels of signal, PW50, and OW for the write current set to roughly 30 mA (0-p) or greater. In the neighborhood of 30 mA write current, the readback responses do not change much, so the write current adjustment is not critical in this example.

Read Bias Curves

After successful characterization of the write head behavior, the reading behavior of an MR head is assessed at various levels of positive and negative sense bias current. Read bias curves are prepared from positive, negative, and peak-to-peak signals, and asymmetry is calculated [see (5.12) for the definition]. Each response is plotted as a function of bias current, and these curves are used for defining the best operating current for a device. Figures 8.4 and 8.5 show the read signal and asymmetry behavior, respectively, for the conventional bottom spin valve (~ 6 Gbits/in.^{2}) discussed above. The behavior of this device (of unknown origin) is compatible with a (hypothetical) structure such as Ta 50 Å/AFM/CoFe 35 Å/Cu 25 Å/CoFe 20 Å/NiFe 60 Å/Ta 50 Å similar to the spin valve discussed in Chapter 6 (see Figs. 6.17c,d and the discussion on the AMR effect in spin valves). The signal level saturates for negative bias currents less than about -6 mA, and the asymmetry is large and negative at all bias currents, and thus the device suffers from a significant AMR effect arising from the NiFe portion of the free layer. At positive bias currents greater than about 7 mA, device heating causes

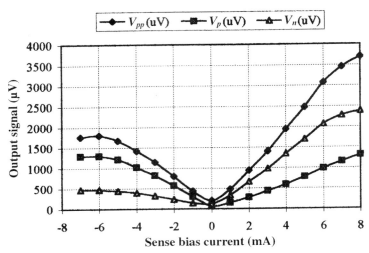

FIGURE 8.4 Read bias curves for a conventional spin valve (~ 6 Gbits/in.^{2}). The high signal at positive bias is compatible with Co "dusting" at both interfaces with the Cu spacer, and the degraded signal at negative bias arises from a significant AMR effect in a CoFe/NiFe free layer.

losses in the GMR and AMR effects, so the signal does not increase linearly with current. The best operating point for this device occurs at a bias current of approximately 5 mA where the peak-to-peak LF signal is 2500 µV, but the very large asymmetry ($\sim -37\%$) could lead to marginal error rate performance in a disk drive.

The read bias curves of dual synthetic spin valves (DSSV) are much improved over the behavior of conventional spin valves. Figures 8.6–8.8 show the signal,

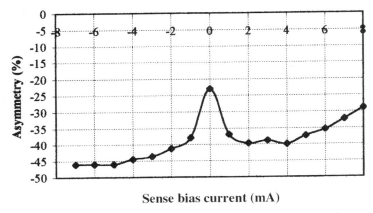

FIGURE 8.5 Asymmetry bias curve of the conventional spin valve shown in Fig. 8.4. Large negative asymmetry is associated with bimetallic free layers such as CoFe/NiFe, where the NiFe layer has a significant AMR effect relative to the GMR effect.

asymmetry, and PW50 responses versus bias current for a DSSV with the structure (excluding Ta capping layers) AFM/CoFe 20 Å/Ru 8 Å/CoFe 20 Å/Cu 25 Å/CoFe 20 Å/Cu 25 Å/CoFe 20 Å/Ru 8 Å/CoFe 20 Å/AFM. The GMR effect is 15% and the AMR effect is nil in the CoFe 20 Å free layer. With a magnetic read width of 0.35 μm and a shield–shield gap $G_{ss} = 0.12$ μm, this device is capable of 36 Gbits/in.² areal density. At 6 mA bias current, the peak-to-peak signal is about 2.0 mV absolute or 5.7 mV/μm when normalized to the read width; the sublinear behavior with sense/bias current is evidence of device heating and loss of GMR effect. Signal asymmetry is small (an average of about 2.5%) over the entire bias curve, and this is one of the benefits of the DSSV structure. The PW50 (taken from the pulses with positive polarity) is about 110 nm (±5 nm for positive bias) and does not change much with bias current; at a channel density $U = 2.3$, this pulsewidth is compatible with a linear density of 531 kfci.

Output Stability Tests

The formation and movement of magnetic domain walls in magnetic recording sensors cause Barkhausen noise spikes that destroy data integrity. The subject of stabilization fields from exchange layers or permanent-magnet layers used for

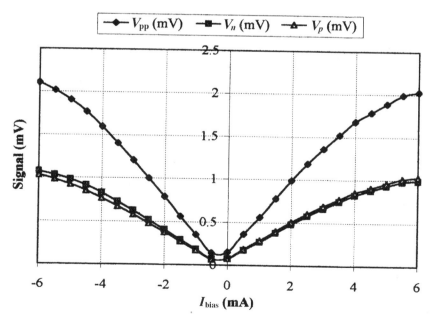

FIGURE 8.6 Read bias curves for a DSSV spin valve (36 Gbits/in.²) showing good symmetric output behavior at positive and negative bias currents. Normalized signal is 5.7 mV/μm.

suppressing Barkhausen noise was discussed in Chapter 4, and the reader is encouraged to review the appropriate sections of that chapter. Various dynamic tests are useful in identifying problems in the stability of MR output signals; among these are tests that examine the coefficient of variation (COV) of track amplitude, instabilities in the baseline around isolated readback pulses [baseline noise (BLN) and baseline popping (BLP)], and microtrack profiles with various levels of disturbance of the bias state.

Coefficient of Variation of Track Average Amplitude In high-volume production of MR recording heads, variations in etching and deposition processes lead to changes in film thickness and composition at the junction between the hard-bias stabilization material and the sensor. These changes may lead to devices whose magnetic domain structures are marginally stabilized, and variations in output signal from these devices could degrade the bit error rate (BER) in a hard disk drive. In the "amplitude COV" test, a full track is written, the amplitude of positive and negative peaks is averaged along the track, the track average amplitude (TAA) is stored and reported, and the write–read cycle is repeated up to a few hundred times. The amplitude COV is then computed from the relation

$$\text{Amplitude COV} = \frac{\text{standard deviation of } N \text{ TAA measurements}}{\text{average TAA of } N \text{ measurements}}. \qquad (8.3)$$

Test specifications depend on customer requirements, but COV $\leq 2\%$ represents an acceptable level of variation. Figures 8.9*a,b* show 300 iterations of TAA and

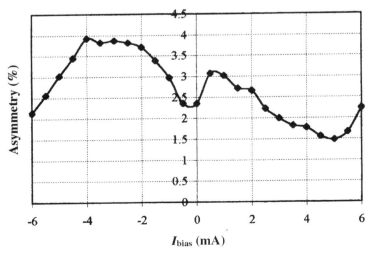

FIGURE 8.7 Asymmetry of a DSSV versus bias current. The asymmetry averages about 2.5% and does not exceed 4% anywhere.

DYNAMIC FUNCTIONAL TESTS   255

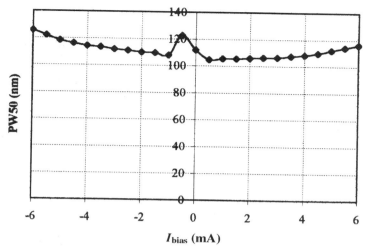

FIGURE 8.8 PW50 (in nm) versus bias current. The average PW50 is about 110 nm, which is compatible with a linear density of 531 kfci at a channel $U = 2.3$.

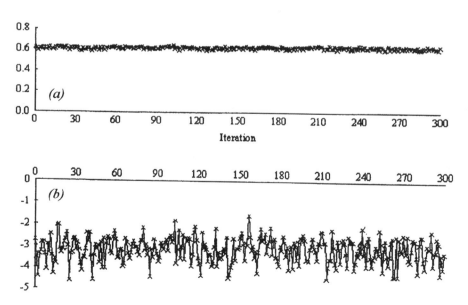

FIGURE 8.9 Output stability tests for an AMR device showing (*a*) track average amplitude (TAA) signal (mV, p-p) and (*b*) TAA asymmetry for 300 iterations of write/read cycles.

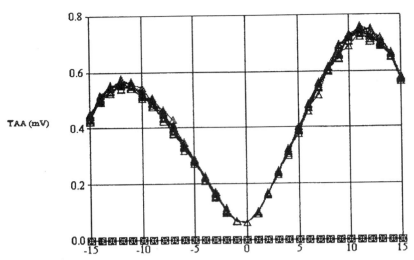

FIGURE 8.10 Read bias curves showing fairly good stability of the TAA signal of the device shown in Fig. 8.9 with writing performed before each read bias cycle. The TAA is in millivolts (p-p) and bias current is in milliamperes.

asymmetry measurements for an AMR head operating at 8 mA bias current; the ranges of the TAA and asymmetry measurements are 0.054 mV (p-p) and 3.045%, respectively. At each writing cycle, the write head field disturbs the AMR sense layer and leaves it in an altered magnetic state. Figure 8.10 shows a number of bias curves (I in milliamperes) for the TAA (millivolts, p-p) of this AMR head; the track was rewritten for each curve. The TAA and asymmetry variations of this head are not excessive and would be regarded as acceptable under most, if not all, circumstances.

Baseline Instabilities: Noise and Popping Severe cases of instability reveal sudden changes in the oscillographic trace of an output signal; the appearance is that of a shift in the baseline, and for this reason, the shifting baseline phenomenon is called "baseline noise" (BLN) or "baseline popping" (BLP). Most of the data analyzing these instabilities show the problems are caused by insufficient hard-bias flux flowing into the free layer. Machining and lapping of MR devices are required to achieve useful stripe heights, and the severe mechanical stresses thus generated can lead to partial demagnetization of the Co-based hard-bias material. (See the section on effective anisotropy field of the free layer in Chapter 6 for a discussion regarding stress anisotropy in layers containing Co or other highly magnetostrictive material.) Remagnetizing the hard-bias layers after final lapping (a process step called "reinitialization") normally eliminates the majority of problems with baseline instabilities. Examples of baseline noise in a conventional spin valve (before and after reinitialization) are shown in Figs. 8.11a and b, respectively. The baseline shift in Fig. 8.11a, which alternates up and down with each cycle of positive and negative pulses, is explained by the presence of hysteresis along the hard axis of the free layer.

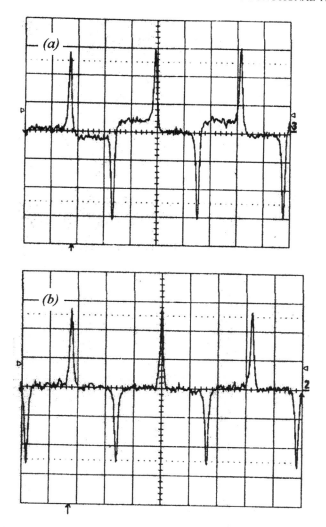

FIGURE 8.11 Baseline instability in a conventional spin valve head (*a*) before reinitialization and (*b*) after reinitialization. Remagnetizing the hard-bias layers normally eliminates baseline instabilities.

Quasi-static testing, which is discussed later in this chaper, is used to verify the presense of hysteresis.

Microtrack Tests for Instability Unstable regions along the read width of a sensor are readily identified and studied by the microtrack technique discussed in the section on microtrack profiles in Chapter 7. Figure 8.12, for example, shows eight repetitions of microtrack profiles (all at $I = +5$ mA) in which a conventional top

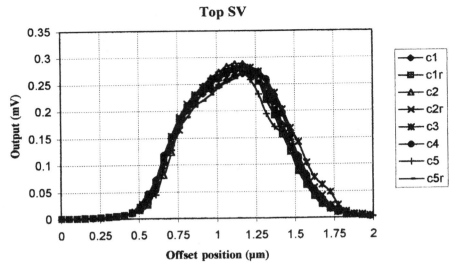

FIGURE 8.12 Microcrack profiles of a conventional spin valve showing evidence of output instability. The microcrack is written once, the head is moved off the microtrack and the bias current is changed to the value listed in the legend. The head is returned to the microtrack and energized at $I = +5$ mA. Unpublished data courtesy of E. Louis, Read-Rite Corp.

spin valve was moved away from the written microtrack ($W = 0.25$ μm), exercised at various sense currents (between 1 and 5 mA), and moved back over the written track. The microtrack read profiles reveal how the bias state of an MR device can be disturbed; in the present case, the read profile maximum amplitudes range between 0.26 and 0.29 mV (p-p) and the read widths W_r vary from 0.74 to 0.80 μm. The stability of this head is marginally acceptable. Five repetitive microtracks (at +5 mA and $W = 0.25$ μm) of a similar device are shown in Fig. 8.13; this top spin valve was exercised at 6–9 mA with the head moved away from the written track; at 9 mA, significant instability was induced along the right edge. Presumably, the instability arises from insufficient hard-bias pinning of the free layer at the junction between the spin valve sandwich and the hard-bias material. Because this head is a top spin valve (see Fig. 6.5a), the free layer (at the bottom of the sandwich) is wider than the top layer and the magnetic remanence of the hard-bias layer may be weak in this region. Severe instability is demonstrated by the profiles shown in Fig. 8.14; all profiles are taken at +5 mA and $W = 0.25$ μm, but the read width W_r is 0.6–0.7 μm for disturb currents of 6, 9, and 10 mA, and W_r varies from 1.0 to 1.2 μm when disturbed at 7 and 8 mA. The normal read width for these devices is about 1.2 μm with an output greater than 1.0 mV, so the unstable output covers a range from roughly 0.5–1.0 mV. The instability appears to arise in the left half of the free layer where the average magnetization has at least two metastable orientations. In one orientation, the free layer is nearly a single domain and the full width is sensitive to the medium flux, but in the other orientation, a domain wall forms down the center of the free layer,

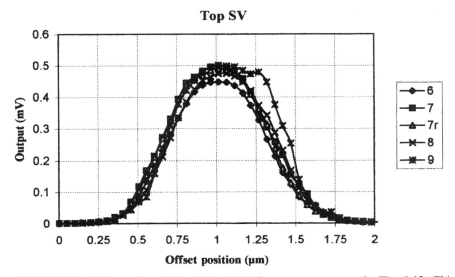

FIGURE 8.13 Microtrack profiles obtained in the same manner as in Fig. 8.12. This conventional spin valve shows significant instability at 9 mA of offtrack disturb current. Unpublished data courtesy of E. Louis, Read-Rite Corp.

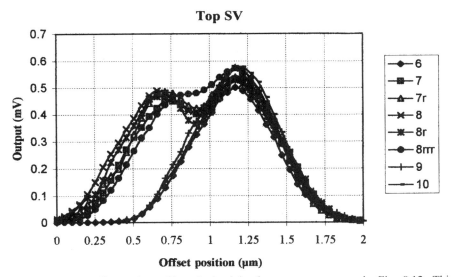

FIGURE 8.14 Microtrack profiles obtained in the same manner as in Fig. 8.12. This conventional spin valve exhibits extreme instability in output signal. The variations are interpreted as arising from a free layer with two metastable domain states. The left domain has an easy direction (presumably from stress anisotropy) perpendicular to the right domain, which has an easy axis along the track width direction. It is likely that hard-bias stabilization is insufficient at the left junction of the free layer. Unpublished data courtesy of E. Louis, Read-Rite Corp.

leaving the magnetization of the left side aligned antiparallel with the pinned-layer magnetization, such that medium flux does not change the angle θ_F of the free-layer magnetization. This could happen in a situation where the free layer has a poorly defined easy direction (arising from high magnetostriction and stress anisotropy in the left half of the film) combined with a weak hard-bias stabilizing field; heads of this nature respond to a process called *reinitialization* where the device is placed in a strong magnetic field (~ 10 kOe) aligned along the track width direction (the intended direction for the easy axis of the free layer) and the hard layer is remagnetized.

Thermal Asperities

Magnetic recording media are not perfectly smooth, and at the very low spacing (~ 25 nm) between the head and medium necessary for high areal density recording, mechanical interference between the flying head and asperities on the the rotating disk is likely. As an asperity slides along the hard air-bearing surface (ABS) of the head, frictional heating rapidly occurs (at 15 m/sec surface velocity, an asperity travels the length of a 2-mm-long ABS in 133 μsec), and if the pathway intersects the MR element, the thermal transient disturbs the resistance and output signal of the MR device. Hempstead (1974, 1975) first introduced and analyzed this phenomenon. Klaassen and van Peppen (1997) analyze signal processing techniques for reducing bits errors associated with thermal transients, and Stupp et al. (1999) discuss the trends and implications of thermal asperities (TA) at increased areal

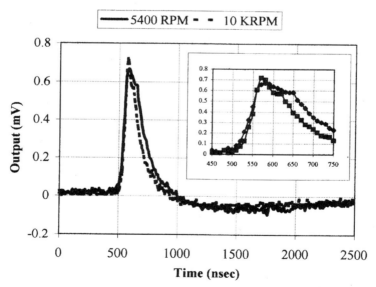

FIGURE 8.15 Thermal asperities from a 40-nm pedestal on a thin-film disk detected with a top spin valve head. Disk rotation was 5400 and 10,000 RPM for the two curves. Data courtesy of J. Dong, Read-Rite Corp. (1999).

densities. Figure 8.15 shows experimental (TA) transients for a GMR head (conventional top spin valve) tested on a disk at 5400 RPM (21.6 m/sec) and 10,000 RPM (40.0 m/sec); the asperity is a 40-nm pedestal. The data are collected directly from the linear readback circuit of the tester, which includes alternating current (AC) coupling through a capacitor, so passage of DC or low-frequency information to the oscilloscope is precluded. The transients decay and go negative at 1000 nsec and greater; this is an artifact of the measuring system and arises from differentiation (high-pass filtering) in the AC-coupled readback circuitry. Nevertheless, two basic aspects of a TA event are clearly seen: first, a rapid (~ 50-nsec) increase in voltage (temperature) occurs as the asperity passes through the sensor region and, second, a slower (~ 300-nsec) exponential decay follows as the heat diffuses from the element to the shields and body of the head slider. The inset in Fig. 8.15 shows the transients on an expanded time scale that has an arbitrary reference for $t = 0$; each pulse peaks at about 570 nsec, and kinks are seen at 620 and 650 nsec for the 10,000 and 5400 RPM cases, respectively. The distances associated with the peak-to-kink intervals are 1.73–2.0 µm and appear to be related to geometrical/mechanical influences, although electrical influences (such as amplifier nonlinearity at high signal levels) may also be involved. The voltage shift, which is roughly the same magnitude as the peak-to-peak signal (~ 1 mV), is equivalent to a resistance change of

$$\Delta R(\text{thermal asperity}) \simeq \frac{\Delta V(\text{thermal shift})}{I_{\text{bias}}}, \quad (8.4)$$

or about 0.2 Ω for a bias current of 5 mA. The temperature rise (*averaged over the entire element*) may be estimated with the relation

$$\Delta T(\text{average}) \simeq \frac{\Delta R(\text{thermal asperity})}{R_0 \alpha}, \quad (8.5)$$

and if the sense element resistance (excluding connections) is $R_0 = 25$ Ω with a thermal coefficient of resistance $\alpha \simeq 2 \times 10^{-3}$ parts/°C, the *average* temperature rise is about 4°C. It is unreasonable to assume the frictional energy of an asperity is uniformly injected over the sensor width or that the heat is instantaneously spread over the sensor height, and thus the actual temperature rise could be much greater than 4°C, and the hot spot could be confined to a region much smaller than the read track width. Unpublished results suggest thermal asperities can unpin a region of the pinned layer, and thus one would infer the temperature approaches the blocking point (~ 250°C) of the AFM material, where a portion of the temperature rise comes from the TA transient and the rest is assigned to the steady-state operating temperature.

QUASI-STATIC DEVICE AND WAFER-LEVEL TESTS

Quasi-static tests do not require flying a head over a magnetic disk. The magnetic field necessary to excite the MR head sense element is provided by an electromagnet capable of producing a nearly uniform magnetic field in the approximate range of −1000 to +1000 Oe, and in some QST equipment, the MR head ambient temperature can be elevated to about 250°C. Magnetic testing can be performed on individual devices or on devices at special test sites within a wafer containing 30,000–40,000 devices. Quasi-static and dynamic test results can be correlated in many cases, but the shields surrounding an MR element are only slightly magnetized by the highly nonuniform magnetic field from a written track on a magnetic disk. Because the uniform field of a QST instrument can magnetize the entire shield structure, domain walls in the shields can sweep over the vicinity of an MR sensor and induce noisy behavior or other instabilities in devices that, if tested dynamically (in a highly nonuniform field), would not necessarily show instabilities. The relation between the response of a shielded MR element to excitation with a uniform field QST and a nonuniform field dynamic test environment is analyzed in Chapter 4 in the section on shield dimensions and sensor field. That analysis is revisited in this present chapter to assist the reader in applying the concepts to specific situations.

QST Transfer Curves

Read signal bias curves from dynamic testing of a spectacularly unstable AMR head are shown in Fig. 8.16; at bias currents I of +17 and −18 mA, the bias curves show discontinuities in output signal that arise from domain wall movement and magnetic hysteresis in the MR layer. To verify the presence of hysteresis or gain further insight into root causes of instabilities, voltage transfer curves may be obtained with a QST. The device is connected to a preamplifier and excited by a uniform field source (an electromagnet). The read bias current is set to the desired level, the magnetic field is swept between positive and negative values, and the sensor voltage is plotted as a function of the uniform field to produce a voltage transfer curve. The QST transfer curves for the device shown in Fig. 8.16 are given in Figs. 8.17a and b for bias currents of +12 and +17 mA, respectively. There is obvious magnetic hysteresis on the +12-mA transfer curve, but the range of hysteresis extends from about −50 to −120 Oe and hysteresis is not elicited with positive (external) excitation. At +17 mA bias current, the additional (internal) bias field from the higher current has shifted the hysteretic behavior to cover the range from −80 to +50 Oe, and thus hysteresis is elicited with positive and negative (external) excitation. This behavior correlates with the dynamic test results in Fig. 8.16, which show erratic behavior in positive, negative, and peak-to-peak signals at bias currents of +17 mA and greater. The QST transfer curves for negative bias currents (not taken at high values) would presumably show approximately the same behavior with a reversal in polarity. The instability in this AMR head would normally stabilize with increased M_r in the hard-bias layers.

Forward and reverse transfer curves for a 20+ Gbits/in.^{2} DSSV are shown in Fig. 8.18; the forward curve is scanned from −750 up +750 Oe, and the reverse curve goes from +750 back to −750 Oe. A number of anomalies appear on these curves, but the most obvious are the Barkhausen jumps in the vicinity of −100 Oe. More details of the hysteretic behavior can be seen on additional transfer curves for which

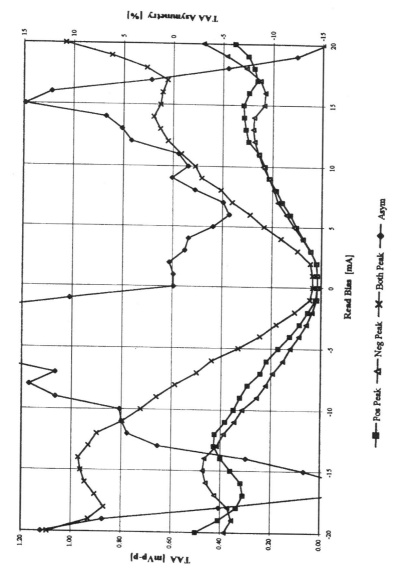

FIGURE 8.16 Read bias curves of a very unstable AMR head. At bias currents of +17 and −18 mA the output curves show discontinuities attributed to magnetic hysteresis along the easy axis of the MR layer.

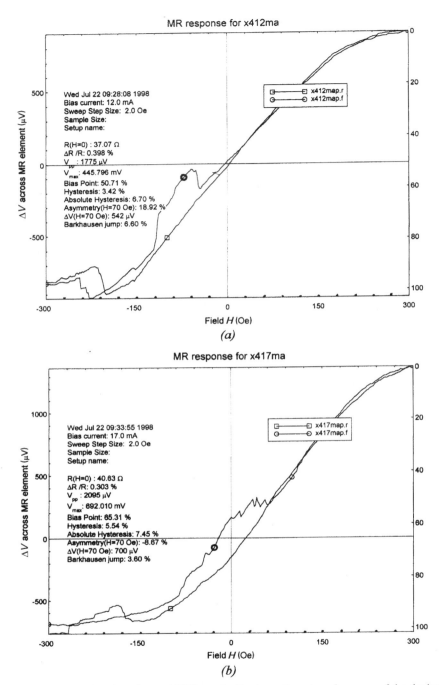

FIGURE 8.17 Quasi-static test (QST) results for the voltage transfer curve of the device in Fig. 8.16: (a) $I = +12$ mA and (b) $I = +17$ mA, showing the magnetic hysteresis shifts with the increased bias field arising from high current (+17 mA). The circled curves are for field sweeping from negative to positive polarities, and the square curves are for the field sweeping from positive to negative polarities.

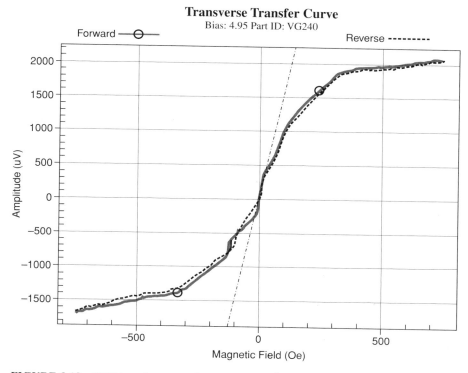

FIGURE 8.18 QST transfer curves for a 20-Gbit/in.² DSSV. Forward field sweep is −750 to +750 Oe, and reverse sweep is +750 back to −750 Oe. Barkhausen discontinuities are seen in the vicinity of −100 Oe. Unpublished data courtesy of K. Stoev, Read-Rite Corp.

the scale is expanded; this is shown in Fig. 8.19 for scanning between ±500 Oe. The hysteresis loop spans a field range over the interval −180 to −74 Oe. The Barkhausen noise persists in a third scan between limits of ±250 Oe, although the hysteretic behavior has narrowed to a field range between −115 and −75 Oe; this is shown in Fig. 8.20.

Correlation between Quasi-Static and Dynamic Tests

Figure 8.21 shows the dynamic (spin stand) reading behavior of the DSSV head discussed in Figs. 8.18–8.20. The sense bias current is 5.0 mA. Baseline popping (shifting) occurs for each positve-to-negative transition; the positive pulses are distorted by positive baseline shift. About 40–50 nsec after a positive peak, the baseline shifts abruptly in the negative direction and leaves well-formed undistorted negative pulses. Hannon (in press) presents correlations between QST and dynamic tests for a top spin valve with hard-bias stabilization; these correlations are shown in Figs. 8.22*a,b*, 8.23, and 8.24. Hysteretic transfer curves at bias currents of 4.0 and 3.2 mA are shown in Figs. 8.22*a,b*; at 4.0 mA bias, the hysteresis loop occurs for negative excitation only (−80 to −50 Oe), and at 3.2 mA bias, the hysteresis loop

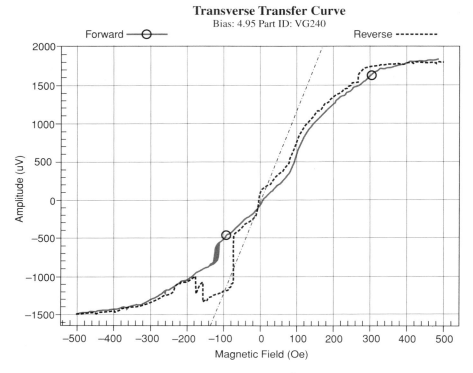

FIGURE 8.19 QST transfer curves for the 20-Gbit/in.^{2} DSSV shown in Fig. 8.18. Field scanning is between ±500 Oe and hysteresis occurs between −180 and −74 Oe. Unpublished data courtesy of K. Stoev, Read-Rite Corp.

covers positive and negative excitations (−50 to +20 Oe). Dynamic test results for bias currents of 3.2 and 5.0 mA are shown in Fig. 8.23. The isolated pulses show large baseline shift (popping) at a bias current of 3.2 mA, which correlates with the transfer curve hysteresis loop covering positive and negative excitations. At 5 mA bias, the baseline returns to zero after positive or negative pulses are detected because the hysteresis loop is beyond the range of positive excitation. The dynamic microtrack profiles in Fig. 8.24 show normal behavior at 5 mA and domain structure instabilities at 3.2 mA. In other words, the instabilities found in dynamic testing can be predicted from the hysteresis seen on the QST transfer curves.

Signal and Asymmetry Correlations

Quasi-static testing of MR heads can replace dynamic testing in many cases. Quasi-static testers are cheaper and root causes of device malfunction can be found quickly. Because a QST voltage transfer curve directly reveals the operating characteristsic curve of the device, the main task in correlating signal and asymmetry measure-

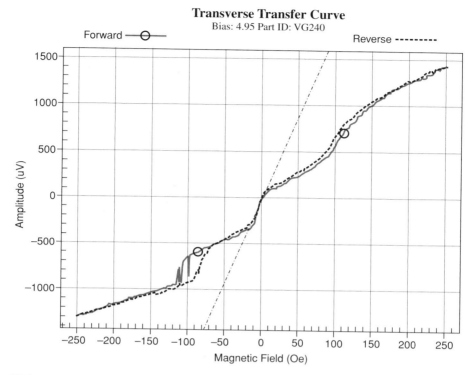

FIGURE 8.20 QST transfer curves for a 20-Gbit/in.² DSSV shown in Figs. 8.18 and 8.19. Field scanning is between ±250 Oe and hysteresis covers a range of −115 to −75 Oe. Unpublished data courtesy of K. Stoev, Read-Rite Corp.

ments with dynamic results is to determine the relationship between excitation fields of each system. Correlations will normally be empirical in nature such that a matrix of heads and media are tested dynamically at known flying heights and recording densities; the heads are retested quasi-statically and the signal (and asymmetry) responses are noted for each level of uniform field excitation. As a rule of thumb, the QST excitation falls in the range of ±70 to ±100 Oe to achieve output signals equivalent to the dynamic results. Yuan et al. (1995) studied this problem theoretically and experimentally with shielded AMR heads, and the results of their analysis are repeated here for convenience to the reader. Figure 8.25 shows the simplified geometry for a uniform field of magnitude H_0 (Oe) exciting a pair of magnetic shields separated by a distance G_{ss}. The MR sensor, located at the centerline between shields, experiences a field at its bottom edge ($y = 0$) given by the relation

$$H_y(y=0) = \frac{H_0}{2\sqrt{2}t_1} F(t_1, h_1) + \frac{H_0}{2\sqrt{2}t_2} F(t_2, h_2), \quad (8.6)$$

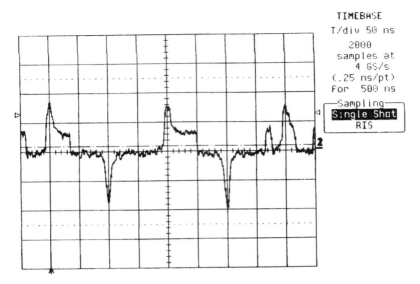

FIGURE 8.21 Read signal from dynamic testing of the DSSV head shown in Figs. 8.18–8.20. Positive pulses are distorted by baseline shifting (popping). Negative pulses are undisturbed. Unpublished data courtesy of K. Stoev, Read-Rite Corp.

where t_1, h_1 and t_2, h_2 are the thickness and height of each shield, respectively. The function describing the normalized field of an individual shield is

$$F(t, h) = \left(t + \frac{G_{ss}}{2}\right)\sqrt{1 + \left(\frac{h}{t + G_{ss}/2}\right)^2} + 1 - \frac{G_{ss}}{2}\sqrt{1 + \left(\frac{h}{G_{ss}/2}\right)^2} + 1. \tag{8.7}$$

The shields magnetize and increase the field at the sensor location, and thus one can think in terms of a field enhancement ratio $R_H(t, h, G_{ss}) = H_y(y = 0)/H_0$ that depends on the shield design and spacing between them. These simple equations ignore the presence of a write head structure built on top of the second shield. A write head would behave like a thick, short shield placed on top of the actual shield layer (2), so its effect would be to reduce the field enhancement calculated with (8.6) and (8.7). The field enhancement ratio $R_H(t, h)$ for identical shields* with $G_{ss} = 0.15$ μm is calculated and plotted in Fig. 8.26; one can estimate that for any reasonable shield thickness between 2 and 4 μm and shield height between 50 and 100 μm, the uniform excitation field is increased by a factor of approximately 3, with a range between 2 and 4. This estimate agrees well with the corresponding values of excitation from recording media in dynamic testing. The vertical field from

QUASI-STATIC DEVICE AND WAFER-LEVEL TESTS   269

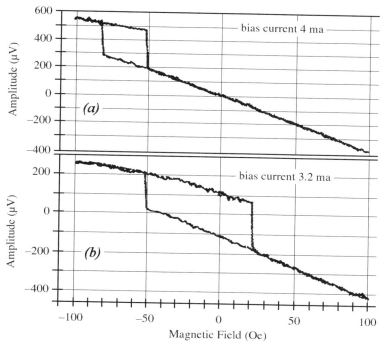

FIGURE 8.22 QST transfer curves for a conventional top spin valve at (*a*) bias current = 4.0 mA with hysteresis between −80 and −50 Oe and (*b*) bias current = 3.2 mA with hysteresis between −50 and +20 Oe. Data courtesy of D. M. Hannon, Western Digital Corp.

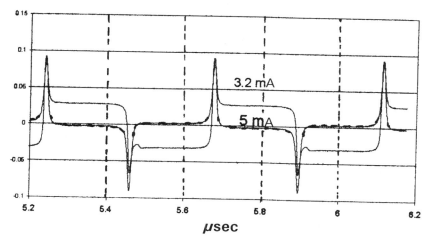

FIGURE 8.23 Read signal from dynamic testing of the top spin valve shown in Figs. 8.22*a,b*. Baseline popping occurs at 3.2 mA bias current and disappears at 5 mA. Data courtesy of D. M. Hannon, Western Digital Corp.

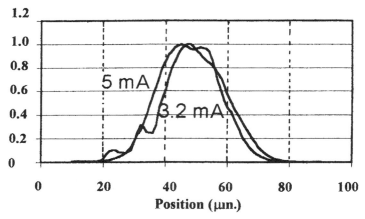

FIGURE 8.24 Microcrack profiles from dynamic tests of the top spin valve shown in Figs. 8.22a,b and 8.23. The profile shows instabilities at 3.2 mA and is normal at 5.0 mA. Data courtesy of D. M. Hannon, Western Digital Corp.

a magnetic transition is estimated with (1.21), which is repeated here for convenience:

$$H_y(x=0, y=d+\delta/2) = 4M_r \log_e\left(\frac{d+a+\delta}{d+a}\right) \simeq \frac{4M_r \delta}{d+a} \quad \text{Oe}. \quad (8.8)$$

The field from a recorded arctangent transition with $d = 25$ nm, $a = 20$ nm, and $M_r \delta = 0.35$ memu/cm^2 is approximately 311 Oe, which, if divided by $R_H = 3$, correlates with a QST field of about 100 Oe.

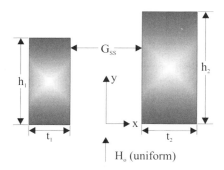

FIGURE 8.25 Simplified geometry for analyzing the field at a sensor when the entire shield structure is excited by a uniform external field from a QST instrument.

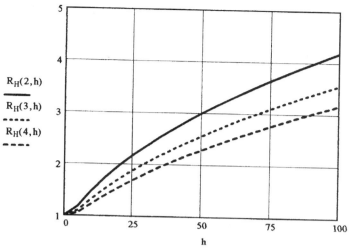

FIGURE 8.26 Plot showing the ratio of the field at the bottom edge of a sensor to the uniform field magnitude of a QST instrument. The ratio $R_H(t, h, G_{ss}) = H_y(y=0)/H_0$ is a function of shield thickness t, shield height h, and distance between shields G_{ss}. In these plots, the shields are identical and $R_H(t, h, G_{ss})$ is calculated as a function of h for $G_{ss} = 0.15$ μm and for $t = 2, 3, 4$ μm.

ELECTROSTATIC DISCHARGE AND SENSOR DAMAGE

The passage of a brief current pulse through an MR sense element is a good example of an adiabatic process: the electrical energy rapidly heats the metallic layers, and the heat slowly diffuses through insulation layers to the rest of the head structure. Controlled testing of MR heads for damage arising from ESD current transients is routinely performed with a HBM system. A HBM tester is essentially a capacitor ($C = 100$ pF) charged to a known HBM voltage and the charge is discharged through a 1500-Ω resistor in series with the device under test. The current transient is closely described by the simple exponential equation for a series RC circuit,

$$I_{ESD}(t) \simeq I_{peak} \exp\left(-\frac{t}{\tau_{HBM}}\right), \qquad (8.9)$$

where $I_{peak} = V_{HBM}/(1500 + R_{MR})$, R_{MR} is the resistance of the device, and $\tau_{HBM} = (1500 + R_{MR})C \simeq 150^+$ nsec is the electrical time constant of the HBM setup. An HBM current pulse is shown in Fig. 8.27; the peak current is 23 mA flowing through a 49-Ω spin valve head. Wallash (1996), Lam et al. (1997), Wallash and Kim (1997, 1998), and Takahashi et al. (1998) can be consulted for discussions of ESD test methods and damage to MR heads. The damage threshold for pinned-layer reversal by ESD heating is lower than that required for electrical breakdown and melting, which cause permanent changes in the electrical resistance. The

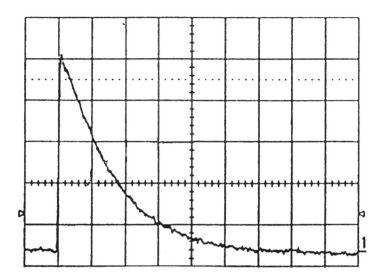

FIGURE 8.27 Experimental current pulse of an HBM tester exciting a 489-Ω conventional spin valve head with $V_{HBM} = 35.0$ V. The vertical scale is 5 mA/division and the horizontal scale is 100 nsec/division.

blocking temperature T_B and melting point T_m of GMR magnetic materials are in the vicinity of 150–300°C and 1450°C, respectively, so it is possible to correlate HBM test results with known properties of the MR head design and materials. The thermal analysis given in Chapter 2 is directly applicable to the ESD current equation (8.9) above; the melting current I_m [given in (2.50)] was derived in Chapter 2 from the power $P = I^2 R$ (W), thermal resistance $R_\tau \simeq g_1 g_2/[\kappa_0(g_1 + g_2)Wh]$ °C/W, and the melting temperature. The relation for damage current (melting or unpinning) can thus be expressed as

$$I_{melt} \simeq \left[\frac{T_m \kappa_0 (g_1 + g_2) t_f h^2}{g_1 g_2 \rho_0 (1 + \alpha T_m)}\right]^{1/2}, \quad (8.10a)$$

$$I_{unpin} \simeq \left[\frac{T_B \kappa_0 (g_1 + g_2) t_f h^2}{g_1 g_2 \rho_0 (1 + \alpha T_B)}\right]^{1/2}, \quad (8.10b)$$

where κ_0 is the thermal conductivity of the gap insulation material, g_1 and g_2 are the insulation gap thicknesses, t_f is the total thickness of the sensor layers, h is the stripe height, ρ_0 is the average room temperature resistivity of the conducting layers, and α is the average thermal coefficient of resistivity of those layers. Note the damage current is independent of the track width W because the area for power dissipation ($A = Wh$) and the sensor resistance ($R = R_S W/h$, where R_S is the sheet resistance) each scale with track width. With $T_m = 1450$°C, $\kappa_0 = 1.0$ W/m-K for alumina, $g_1 = g_2 = 650$ Å, $t_f = 455$ Å (a conventional top spin valve with a PtMn AFM layer

350 Å thick), $\rho_0 = 70$ $\mu\Omega$-cm, and $\alpha = 2 \times 10^{-3}$ parts/°C, the melting current (in milliamperes) becomes $I_{melt} \simeq 27.3h$ (with h in micrometers). For a blocking temperature $T_B = 300°C$ and a HBM voltage of the correct polarity for current flow in the opposing direction, unpinning damage would occur at $I_{unpin} \simeq 19.4h$ (with h in micrometers) according to (8.10b). In other words, a typical top spin valve with $h = 0.7$ μm and $R_{MR} = 35$ Ω would suffer permanent electrical damage at a current of about 19.1 mA and unpinning damage at 13.6 mA. The equivalent HBM voltages can be estimated by equating the damage currents (I_d) with the peak current (adiabatic heating) in (8.9) and solving for the equivalent HBM voltages:

$$V_{HBM} = I_d(1500 + R_{MR}) \simeq 1500 I_d \quad V, \tag{8.11}$$

and in the example given above, $V_{HBM} = (1500\ \Omega)(0.0191\ A) \simeq 28.7$ V for melting damage and 20.4 V for the threshold of unpinning damage. It is useful to look at (8.10a,b) from the viewpoint of damage current threshold $I_d(T_d)/h$ (in milliamperes per micrometer) as a function of temperature (T_d) at which damage may occur. This relation is plotted in Fig. 8.28 using the device parameters given above; note, for the temperature ratio $T_m/T_B \sim 5$, the ratio of damage currents $I_{melt}/I_{unpin} \sim 1.5$, but with low blocking temperatures (150°C for FeMn) $T_m/T_B \sim 10$ and $I_{melt}/I_{unpin} \sim 2.0$. This observation is compatible with the experiments of Wallash and Kim (1998) with GMR heads whose structure is given by Ta 50 Å/NiFe 75 Å/Co 20 Å/Cu 30 Å/Co 20 Å/NiFe 5 Å/FeMn 100 Å/Ta 50 Å. The sheet resistance of these devices would be fairly low ($R_S \simeq 13$ Ω/square), thus with $g_1 = g_2 = 700$ Å, the estimated damage currents are $I_{unpin} = 16.6h$ and $I_{melt} = 32.6h$. Wallash and Kim place the nominal stripe height at 1.3 μm for their devices, and their data in Figs. 8.29a,b,c show unpinning damage occurs at a threshold of 35 V_{HBM}, which is

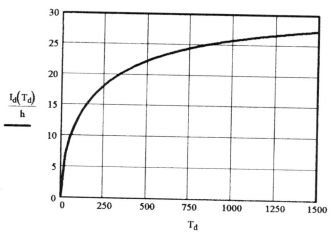

FIGURE 8.28 Plot of the damage threshold current per unit stripe height $I_d(T_d)/h$ computed as a function of the damage temperature T_d using (8.10) with device parameters given in the text.

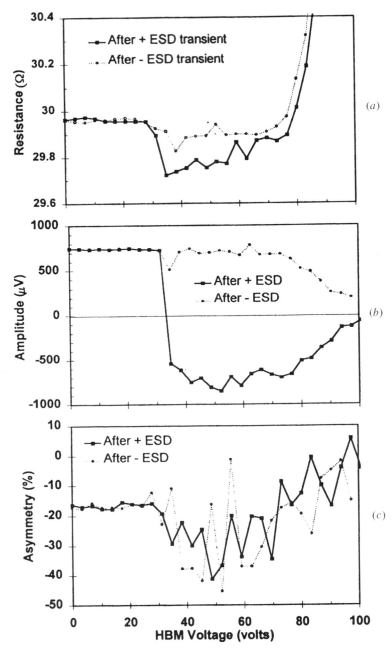

FIGURE 8.29 (*a*) Resistance, (*b*) amplitude, and (*c*) asymmetry for a GMR sensor as a function of HBM voltage. At each voltage, measurements were taken after a +ESD transient (dark line) and a −ESD transient (light line). Reprinted with permission from A. Wallash and Y. K. Kim, *IEEE Trans. Magn.*, **MAG-34**, 1519. Copyright 1998 by IEEE.

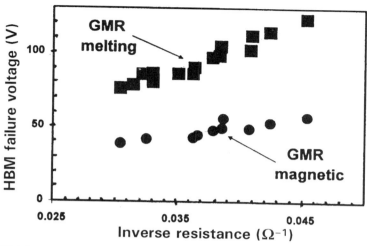

FIGURE 8.30 HBM failure voltage versus inverse resistance for over 100 different GMR sensors. Both melting and magnetic failure levels are shown. Reprinted with permission from A. Wallash and Y. K. Kim, *IEEE Trans. Magn.*, **MAG-34**, 1519. Copyright 1998 by IEEE.

equivalent to $I_{unpin} = 35\,V/1500\,\Omega = 23.2$ mA; this compares favorably with the calculated value $I_{unpin} = (16.6\,mA/\mu m)(1.3\,\mu m) = 21.6$ mA. The failure voltages for melting and unpining threshold for many devices are shown in Fig. 8.30 as a function of $1/R_{GMR}$; the ratio of voltages for melting and unpinning is about 2.0, which is expected for FeMn AFM material with $T_B \simeq 150°C$. Since $1/R_{GMR} = h/R_S W$, the linear behavior is expected from the foregoing theory using (8.10a,b) and (8.11).

Figures 8.31a,b show typical HBM test results with the polarity set to assist the pinning direction and produce permanent damage from melting. The device is a top spin valve with an IrMn AFM layer and at 45 V (30 mA) the QST signal drops and the device resistance increases. Figure 8.32 shows collected HBM test results from Chang et al. (1999) for various GMR designs, all plotted as a function of stripe height; the criterion for damage was an irreversible 10% drop in QST output voltage with the HBM voltage at a positive polarity (which assists pinning). The 61 devices used in this experiment came from five GMR designs: two different conventional top spin valves (CSV-1 and CSV-2), dual synthetic (DSSV), bottom synthetic (BSSV), and a top synthetic (TSSV) spin valve. Indeed, as expected from the foregoing analysis, the data are essentially linear with stripe height and the devices with long stripes have proportionally greater damage threshold voltages (melting currents.) Linear trendlines are plotted through each of the data sets; the trendline equations are tabulated in Table 8.1 along with theoretical calculations based on estimates of the device properties for the CSV-1, DSSV, and TSSV designs. The agreement between theory and experiment is quite good considering the level of complexity involved in estimating the melting current and in measuring damage thresholds. Damage threshold is $V_{HBM} = Kh$ volts with stripe height h in micrometers. The AFM

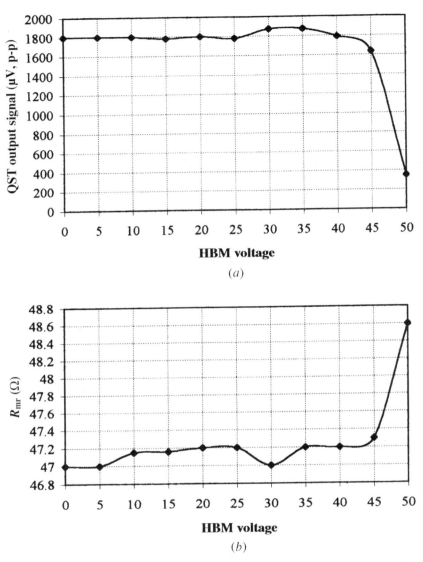

FIGURE 8.31 QST results for (a) signal and (b) resistance versus HBM voltage for a top spin valve with an IarMn AFM layer. Permanent damage from melting occurs at a threshold $V_{HBM} \simeq 45$ V (30 mA). Unpublished data courtesy of C. Lam, Read-Rite Corp.

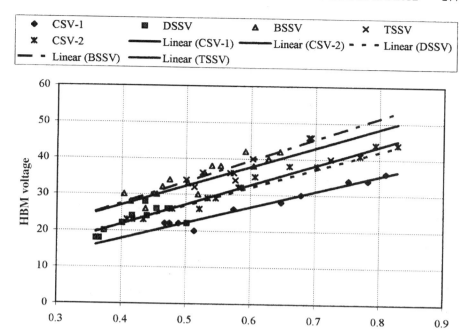

FIGURE 8.32 QST tests results for a permanent (melting) 10% signal loss: V_{HBM} damage threshold voltage plotted as a function of stripe height. The experiment used 61 devices from a variety of spin valve designs. Linear trendlines are fit to the data from each of the device types. Unpublished data courtesy of C. Chang, S. Chim, and C. Lam, Read-Rite Corp. (1999).

layers are quite different in these designs; the CSV-1 and TSSV designs use thick PtMn layers (350–400 Å) while the DSSV design uses thin IrMn layers (70 Å each side) for pinning. In addition, the TSSV design centers the free layer between the shields, thus g_1 is thin to compensate for the thick PtMn AFM layer. These design choices directly lead to the HBM damage voltage threshold. At high areal densities, stripe height h must reduce in proportion to the track width W, and thus the damage

TABLE 8.1 Comparison of Experimental and Calculated HBM Damage Voltages

Design	Experimental Damage, V_{HBM}	Calculated Damage, $V_{HBM} = I_d \cdot 1500\,\Omega$	t_f (Å)	g_1 (Å)	g_2 (Å)	ρ_0 (μΩ-cm)	R_S (Ω/square)
CSV-1	43.47h + 0.15	41.0h	455	650	650	70.0	15.36
DSSV	51.18h + 1.2	53.9h	310	400	500	40.3	12.98
CSV-2	54.43h + 0.01	52.5h	475	700	300	68.8	14.5
TSSV	53.40h + 5.6	55.2h	535	650	200	96.2	17.98
BSSV	60.34h + 3.25	60.3h	600	200	585	92.4	15.4

threshold reduces accordingly. At 35 Gbits/in.^{2}, a DSSV must be lapped to a stripe of about 0.35 μm, which, according to the analysis given above, leads to $V_{HBM} \simeq 53.9h$ or approximately 19 V. Improvements in thermal conductivity would raise the damage threshold accordingly; aluminum nitride, for example, has a thermal conductivity $\kappa_0 \simeq 3$ W/m-K, and thus the damage threshold for a DSSV design could improve to about 33 V.

OPERATING TEMPERATURE AND LONG-TERM RELIABILITY

Hard disk drives are regarded as archival storage devices supporting a lifetime greater than 5 years. Ongoing reliability tests (ORTs) require retrieval of data (at an acceptable error rate) from drives that have been turned on and left running with the MR heads energized for long periods of time. The implications for head design, materials, processing, storage, shipping, and handling are quite challenging. This book ends with a discussion of thermally activated metallurgical and structural changes in MR heads that progess over long time periods. These long-term changes fall into the categories of interdiffusion between the various layers of materials in MR heads and electromigration of metals in conducting thin films. Difffusion between very thin layers of magnetic and nonmagnetic materials may lead to changes in magnetic properties such that the device function degrades. The output signal may decrease with time or the electrical resistance may increase by an unacceptable amount. Electromigration can lead to structural changes at the junction between two conductors of differing properties with changes in resistance that become unacceptable or even catastrophic. The treatment of these issues will be brief and are offered to give the reader an appreciation of some of the physical mechanisms and material properties that must be considered in the design and analysis of MR recording heads.

Interdiffusion in Thin-Metal Films

Baglin and Poate (1978) declare that interdiffusion in thin metal films is distinguished by large-scale mass transport at low temperatures. They observe that diffusion times can be short because the films are thin and contain many defects such as grain boundaries and dislocations and that defects control the interdiffusion rate in thin films. The mathematical theory of diffusion is thoroughly covered in Crank (1975), and some of the fundamental relations are introduced here for those readers unacquainted with diffusion theory. Fick's first law for one-dimensional diffusion is given by the relation

$$F = -D \frac{\partial C}{\partial x} \qquad \text{atoms/cm}^2\text{-sec,} \qquad (8.12)$$

where F is rate of transfer per unit area normal to the x-direction, D is the diffusion constant or diffusivity (in square centimeters per second), and C is the concentration of the diffusing material (number of atoms or vacancies/per cubic centimeter). In analogy with the theory of heat flow, Fick's second law (for one dimension) is

$$\frac{\partial C}{\partial t} = D\frac{\partial^2 C}{\partial x^2} \tag{8.13}$$

when the diffusion coefficient D is constant. The solution to (8.13) for constant D, given in Crank (1975, p. 11), is the well-known relation

$$C = \frac{A}{t^{1/2}} \exp\left(-\frac{x^2}{4Dt}\right), \tag{8.14}$$

where A and D are experimentally determined constants. Most experimental work shows the diffusion constant depends on temperature, according to an Arrhenius equation of the form

$$D = D_0 \exp\left(-\frac{E}{k_B T}\right), \tag{8.15}$$

where E is the activation energy (in joules), $k_B = 1.381 \times 10^{-23}$ J/K is Boltzmann's constant, T is the absolute temperature (in kelvins), $D_0 = va^2$ (in centimeters squared per second) is the frequency factor, v is the atomic vibrational frequency (in reciprocal seconds), and a is the lattice constant (in centimeters) of the material [see, e.g., Glasstone et al. (1941) or Fowler (1966.)]

It is known that the resistance of many alloys follows Matthiessen's rule [see Mott and Jones, (1958, p. 286) or Wilson (1965, p. 310ff)] where the resistance of a metal A increases due to a small concentration of metal B in solid solution with A. For this reason, studies of metal–metal interdifussion often use resistivity as a measure of the diffusion process. For example, Madakson (1991) correlated changes in resistivity with interdiffusion between Cr and Au thin films and was able to calculate diffusion coefficients and activation energies from resistivity measurements. Christou and Day (1973) studied Au–Ta interdiffusion using resistance, x-ray diffraction, and electron microscope techniques. They concluded that Au diffuses into Ta along grain boundaries and the process occurs below 350°C with an activation energy $E_b \simeq 0.41$ eV. Macchioni et al. (in press) demonstrated that during lifetime testing of 3.2-Gbits/in.^{2} AMR heads, the resistance of leads and sense layers increased at different rates while step annealing the structures at 275°C up to 32 h. The AMR sensor layers and hard-bias/lead structures were NiFeRh/Ta/NiFe/Ta and Cr/CoCrPt/Ta/Au/Ta, respectively. Their data for sheet films on Si substrates are given in Fig. 8.33; the percent change in resistance is plotted as a function of (time)$^{1/2}$, and the portion that is linear is compatible with a diffusion model. Diffusion of Ta into the sensor layers appears to level off in about 2 h, whereas in the much thicker lead structure, diffusion of Au into Ta required about 9 h. The lead resistance continued to increase (at a reduced rate) with annealing times up to 32 h, at which point the experiment was terminated.

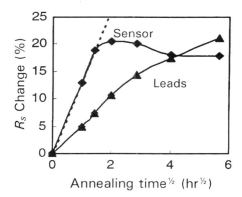

FIGURE 8.33 Resistance of AMR head leads and sense layers as a function of annealing (time)$^{1/2}$ at a temperature of 275°C for 32 h. Diffusion of Ta into the AMR sense layer occurred in about 2 h, whereas Au diffusing into Ta in the leads required about 9 h. Reprinted with permission from C. Macchioni, V. Doan, and Q. He, to be published in *Conf. Proc. of 3M Conference*, San Jose (1999).

Two groups of individual finished AMR heads were also tested; group A2 was tested for 200 h at 150°C with 10 mA bias current (a current density of 7.6 × 10^7 A/cm^2), whereas group B2 was preannealed at 275°C for 168 h and then tested in the same manner as group A2. The preannealed group showed less than 0.4% change in resistance as compared to an average resistance increase of 18.8% for the unannealed group. The data for both groups are shown in Fig. 8.34. It is clear that preannealing AMR heads stabilizes resistance changes arising from grain boundary diffusion (a fast process) such that other, slower changes can be discovered and studied.

Giant MR heads require annealing in a magnetic field to establish a unidirectional exchange anisotropy that aligns the magnetization of the pinned (or reference) layer. The annealing temperature is near the blocking temperature of the AFM layer (250–275°C), and depending on the design and materials used, the time is on the order of several hours or more. In addition to pinning the reference layer, this annealing cycle stabilizes resistance changes that occur rapidly due to low-activation-energy grain boundary diffusion, and thus high-activation-energy, slow interdiffusion processes can be detected. Doan (1999) has shown a gradual reduction with time of the signal amplitude for synthetic spin valve structures where the devices were tested on a QST hot plate at modest temperatures (100–160°C) and at bias currents between 5 and 8 mA. The signal was measured every 2 h with the uniform QST field amplitude at ±100 Oe. Selected data for signal versus (time)$^{1/2}$ and resistance versus time for a type B spin valve at 6 mA and an ambient temperature T_{amb} = 120°C are shown in Figs. 8.35a and b, respectively; at 6 mA, the sensor temperature rises about 40°C above T_{amb} and thus the hot resistance is about 43.8 Ω. The signal reduction follows a $t^{1/2}$ dependence, which is indicative of

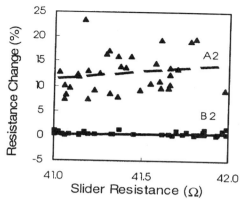

FIGURE 8.34 Resistance of finished AMR heads from two test groups. Group A2 tested for 200 hours at 150°C at 10 mA bias (7.6 × 10⁷ A/cm² through the NiFe layer of the devices) without preannealing. Group B2 preannealed at 275°C for 168 h and then tested in the manner of group A2. Reprinted with permission from C. Macchioni, V. Doan, and Q. He, to be published in *Conf. Proc. of 3M Conference*, San Jose (1999).

a diffusion process, and the resistance is essentially constant up to the end of the experiment at 240 h. Signal and resistance data from a selected type C spin valve at 6 mA and 130°C are plotted in Figs. 8.36a and b; the behavior is qualitatively similar to the type B design.

Saito et al. (1998) have studied the activation energies of interdiffusion at the film interfaces for top spin valve systems whose structures are given by CoZrNb/NiFe/CoFe/Cu/CoFe/FeMn/Ta and CoZrNb/NiFe/CoFe/Cu/CoFe/IrMn/Ta. Their work is reviewed here because it offers plausible explanations for signal loss in spin valves, but interpretation of these results requires speculations about the interdiffusion mechanisms that may be operating in these devices. For brevity, their structures will be called CoFe–FeMn and CoFe–IrMn SV (spin valves.) All of the spin valves were annealed in a magnetic field at 240°C for 1 h to establish exchange field pinning of the reference layers. The initial values of the GMR ratios $(\Delta R/R)_0$ were measured with a four-point probe and the structures were annealed at various temperatures (210–285°C) in a vacuum furnace with no external field for 20–200 h. After each annealing process, the structures were magnetic field annealed (240°C for 1 h) to reestablish exchange pinning such that any changes in GMR ratio could be assigned to atomic interdiffusion at film interfaces and not have the $\Delta R/R$ measurements confounded with changes in the pinned-layer magnetic alignment. Structural changes were analyzed by grazing-incidence x-rays at room temperature after each annealing cycle.

In all of the samples studied, the GMR ratio decreased with annealing time and temperature. At 270°C the CoFe–FeMn SV dropped to 75% of the initial GMR ratio in 100 h, the pinning field decreased from about 300 Oe to about 200 Oe, the interlayer coupling field increased, and the GMR ratio dropped another 3.5% after

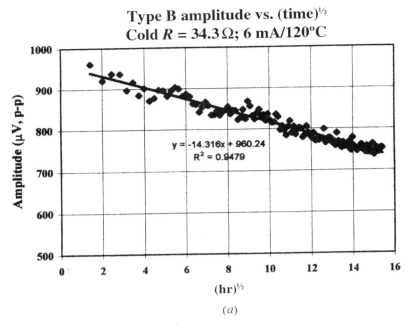

(a)

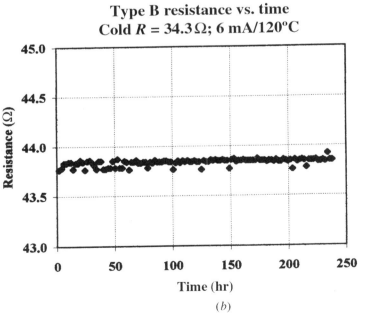

(b)

FIGURE 8.35 QST test results as a function of (time)$^{1/2}$ for a bottom synthetic spin valve (BSSV) at ambient temperature of 120°C and 6 mA bias current: (*a*) signal and (*b*) resistance. Unpublished data courtesy of V. Doan, Read-Rite Corp. (1999).

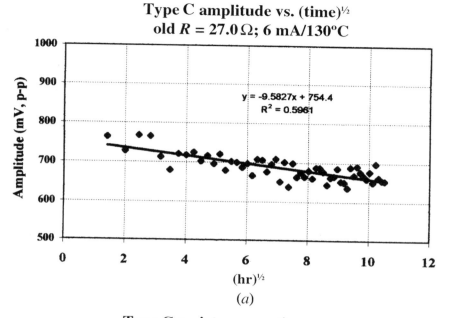

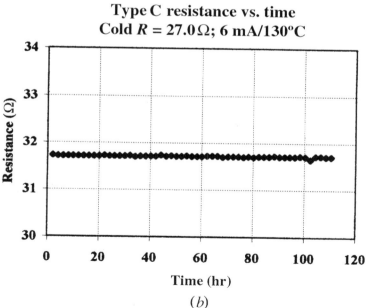

FIGURE 8.36 QST test results as a function of $(time)^{1/2}$ for a dual synthetic spin valve (DSSV) at ambient temperature of 130°C and 6 mA bias current: (*a*) signal and (*b*) resistance. Unpublished data courtesy of V. Doan, Read-Rite Corp. (1999).

annealing a total of 200 h. At 285°C the CoFe–IrMn SV GMR ratio dropped to 91.5% of the initial value in 100 h, and very little change was detected after annealing for a total of 200 h. According to (6.8), these losses in GMR ratio directly imply signal losses of 29 and 8.5% for the FeMn SV and IrMn SV systems, respectively. The data are plotted in Figs. 8.37a,b. Saito and co-workers fitted exponential curves to their data and found the CoFe–FeMn SV approximately followed a relation given by

$$\left(\frac{\Delta R(t)}{R}\right) = \left(\frac{\Delta R}{R}\right)_0 \left[1 - \Delta_1 \exp\left(-\frac{t}{\tau_1}\right) - \Delta_2 \exp\left(-\frac{t}{\tau_2}\right)\right], \quad (8.16)$$

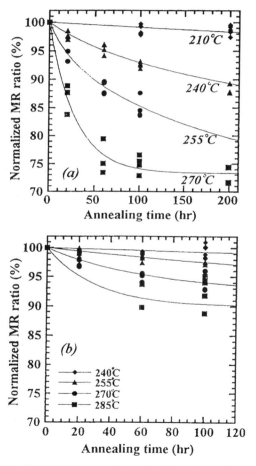

FIGURE 8.37 GMR effect $\Delta R/R$ versus time at different annealing temperatures for two top spin valve systems studied at the film level: (a) CoFe–FeMn SV system and (b) CoFe–IrMn SV system. Reprinted with permission from A. T. Saito, H. Iwasaki, Y. Kamigchi, H. N. Fuke, and M. Sahashi, *IEEE Trans. Magn.*, **MAG-34**, 1420. Copyright 1998 by IEEE.

and the CoFe–IrMn SV approximately followed

$$\frac{\Delta R(t)}{R} = \left(\frac{\Delta R}{R}\right)_0 \left[1 - \Delta_3 \exp\left(-\frac{\tau}{\tau_3}\right)\right]. \quad (8.17)$$

The relaxation time constants were thermally activated and followed Arrhenius relations of the form

$$\frac{1}{\tau_i(T)} \propto D_i \exp\left(\frac{-E_{a_i}}{k_B T}\right). \quad (8.18)$$

Arrhenius plots for the temperature dependence of the time constants are given in Figs. 8.38a,b; the estimated activation energies are shown on each plot. The time constants follow an exponential relation proportional to the inverse of (8.18),

$$\tau(T) = \tau_0 \exp\left(\frac{E_a}{k_B T}\right), \quad (8.19)$$

and from the Arrenhius plots, one can extract the prefactor τ_0 for each of the relaxation processes.

The experimental results and related information for the SV systems of Saito and co-workers are tabulated in Table 8.2. Saito and co-workers (1998) assigned the deterioration of GMR ratios in the CoFe–FeMn SV system to two different mechanisms: The Δ_1 loss at $E_{a1} = 2.16$ eV was likely caused by interface mixing between CoFe and Cu layers, and the Δ_2 loss at $E_{a2} = 1.57$ eV probably arose from atomic diffusion of Mn or Ni atoms to the Cu layer through CoFe grain boundaries at the CoFe–FeMn or NiFe–CoFe interfaces. In the CoFe–IrMn SV system, they detected one diffusion mechanism (Δ_3 at $E_{a3} = 2.49$ eV) and thus the GMR loss was assigned to interfacial intermixing between CoFe and Cu. Table 8.3 shows the as-deposited and annealed (270°C, 100 h) film and interface (δ) thicknesses (in angstroms) for the magnetic and Cu layers of the CoFe–IrMn SV system. While all of the interfaces broadened with annealing, the nonmagnetic spacing between pinned and free CoFe layers increased from 36 to 38.3 Å, approximately. The work of Dieny et al. (1992), which was reviewed in Chapter 6 and summarized with (6.7), showed the GMR ratio varied with nonmagnetic spacer thickness (t_S) as $A_0 \exp(-t_S/\lambda_S)$, where the diffusion length for electrons through Cu is $\lambda_S \simeq 47$ Å. The ratio of spin valve signals before and after diffusion broadening of t_S is thus estimated as $\exp(-38.3/47)/\exp(-36/47) \simeq 0.95$, or about a 5% loss. Considering the accuracy of estimates for film and interface thicknesses before and after annealing, a calculated loss of 5% agrees reasonably well with the 8.5% experimental result. With small amounts of spacer broadening Δt_S, the signal would follow a trend according to

$$\text{Signal} \simeq S_0 \left(1 - \frac{\Delta t_S}{\lambda_S}\right) \quad (8.20)$$

where S_0 is the initial signal level and a loss of 8.5% is nearly equivalent to $\Delta t_S = 4.0$ Å.

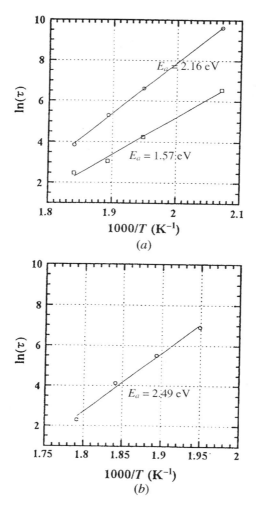

FIGURE 8.38 Arrhenius plots for the GMR ratio decay time plotted versus inverse temperature (K) for two top spin valve systems at the film level: (a) CoFe–Mn SV system and (b) CoFe–IrMn SV system. Reprinted with permission from A. T. Saito, H. Iwasaki, Y. Kamiguchi, H. N. Fuke, and M. Sahashi, *IEEE Trans. Magn.*, **MAG-34**, 1420. Copyright 1998 by IEEE.

TABLE 8.2 Experimental Results for Interdiffusion in CoFe SV Systems

System	Initial $(\Delta R/R)_0$	Final, $(\Delta R(t)/R$, $t = 200$ h	Loss Factor, Δ_i (%)	Activation Energy, E_{ai} (eV)	Time Constant Prefactor, τ_0 (sec)
CoFe–FeMn	0.06–0.07	0.043–0.05	$\Delta_1 = 23.5$	$E_{a1} = 2.16$	$\tau_{01} = 1.66 \times 10^{-15}$
			$\Delta_2 = 5.5$	$E_{a2} = 1.57$	$\tau_{02} = 1.08 \times 10^{-10}$
CoFe–IrMn	0.075–0.08	0.069–0.073	$\Delta_3 = 8.5$	$E_{a3} = 2.49$	$\tau_{03} = 1.47 \times 10^{-18}$

TABLE 8.3 Intediffusion in CoFe–IrMn SV: Layer Thicknesses (Å)

	NiFe	δ	CoFe	δ	Cu	δ	CoFe	δ	IrMn
Initial	13.3	5.4	25.5	5.6	23.5	6.9	12.3	8	45.3
Final	8.8	9.0	22.5	7.1	21.2	10	9.9	9.5	41.8

This work is directly applicable to testing of energized devices at different ambient temperatures. The absolute temperature of an energized sensor is $T(K) = T_{amb} + T_{rise}$, where the temperature rise is estimated with the relations (2.28), (2.29), and (2.30). A nominal MR device operates about 50 K above ambient and a disk drive internal ambient might be 330–350 K, so the approximate lower temperature of an operating head is about 380–400 K. The resistance and temperature of a device increase rapidly as the stripe height is reduced, and thus it useful to estimate the relaxation time constant given by (8.19) over a practical range of temperatures. For example, the resistance and temperature of a BSSV designed for use at 7 Gbits/in.^{2} can be estimated using (2.28), (2.29), and (2.30); for stripe heights h of 0.6, 0.5, and 0.4 μm, the cold resistance and temperature rise at 6 mA bias would be about 27.4, 32.9, and 41.1 Ω with the corresponding temperature rises of 34, 51, and 84 K, respectively. At $h = 0.4$ μm and $I = 7$ mA, the temperature rise jumps to 122 K, and thus the absolute temperature of the sensor in a disk drive at an ambient of 343 K (70°C) would be 465 K. It is instructive at this point in the discussion to estimate $\tau_i(T)$ given by (8.19) for $i = 1, 3$ (in Table 8.3) over a range of operating temperatures, say from 400 to 600 K. The results are plotted in Fig. 8.39. It is very sobering to observe relaxation times plummet from 10^{10} h (1.1 million years) at 400 K down to 10^4 h (1.1 year) at 500 K, and the implications of this understanding regarding the lifetime of MR recording heads is explored in the section on prediction of sensor lifetime, which follows shortly.

Electromigration

Electromigration is a grain boundary diffusion process in which mass is transported primarily by electrical forces. Gradients in alloy concentrations and in temperature along a conductor can also lead to mass transport, but d'Heurle and Ho (1978) assert the driving force in electromigration is normally attributed to two main effects. There is the direct electric field force on ionized atoms stripped of their valence electrons, and there is a frictional force in the opposite direction (the "electron wind") arising from momentum exchange between moving electrons and the ionic atoms. At high current densities in metals, the electron wind force dominates the electric field force and noticeable mass transport occurs. In the studies with long conducting stripes, damage in the form of voids is found around the cathode region. Voids lead to increasing current density, high resistance, hot spots, thermal runaway, and ultimately catastrophic failure from melting. In films without dielectric overlayers,

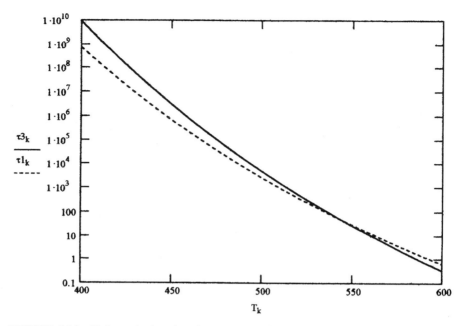

FIGURE 8.39 Estimated relaxation time constants (h) versus absolute temperature (K) for two different decay processes: $E_{a1} = 2.16$ eV with $\tau_{01} = 1.66 \times 10^{-15}$ sec and $E_{a3} = 2.49$ eV with $\tau_{03} = 1.47 \times 10^{-18}$ sec.

hillocks can also grow from buildup of material, and these defects can lead to shorting between closely spaced electrodes.

Blech (1976) suggested that pressure and concentration gradients could lead to a reverse flow of atoms that just balances the forward electromigration (electron wind) flow. His idea allows one to estimate the threshold in current density at which electromigration damage occurs. Blech and Herring (1976) measured stress in Al stripes 4300 Å thick, 35 µm wide, and 150–600 µm long deposited on Si substrates. At room temperature, the Al stripe was in tension and at 330°C the stress became compressive as a result of the greater thermal expansion coefficient of Al relative to Si. With current flow at 2.7×10^4 A/cm^2 (below the threshold for electromigration) for 26 h, the compressive stress was eliminated at the negative end of the stripe, and the stress gradient along the stripe increased with time. After about 100 h of current flow, stress was tensile near the cathode and compressive near the anode. After turning off current flow, the stress gradient relaxed and the film stress became uniform along the stripe length. With sufficiently high current density and temperature, mass transport from electromigration begins and the atomic flux can be written

$$J = \frac{ND}{k_B T} Z^* e E \quad \text{atoms/cm}^2\text{-sec,} \tag{8.21}$$

where N is the atomic (or vacancy) density, $D = D_0 \exp(-E_a/k_B T)$ is the mass transport diffusion coefficient, Z^*e is the effective charge of atomic ions, and E is the electric field. For concentration or stress gradients, Blech and Herring write Fick's law as

$$F = \frac{ND}{k_B T} \frac{\partial(\mu_a - \mu_v)}{\partial x}, \qquad (8.22)$$

where μ_a and μ_v are the chemical potentials of atoms and vacancies, respectively. Blech and Herring equate the difference in chemical potentials to the stress normal to a grain boundary under the assumption that local equilibrium is reached between adding or subtracting lattice defects at grain boundaries. That is,

$$\mu_a - \mu_v = \mu_0 + \Omega \sigma_n \qquad \text{ergs} \qquad (8.23)$$

where μ_0 is a constant, Ω is the atomic volume ($\sim 10^{-23}$ cm^3), and σ_n is the stress (in dynes per square centimeter) normal to a grain boundary. At equilibrium between the electromigration flow and the backwards "healing" flow from stress, Blech and Herring equate (8.21) and (8.22) to obtain a relation for the balance in forces,

$$Z^*eE = \Omega \frac{\partial \sigma_n}{\partial x} \qquad \text{dyn.} \qquad (8.24)$$

The effective ion charge Z^*e is difficult to estimate accurately, but Herzig and Cardis (1975) found that $Z^*e \simeq 1.12 \times 10^{-18}$ C ($Z^* = -7$) for Au at 1200 K ($T_m \simeq 1336$ K), and Christou (1994, p. 305) shows $Z^*e = 1.44 \times 10^{-18}$ C ($Z^* = -9$) for molten Al droplets ($T_m \simeq 933$ K). D'Heurle and Ho (1978) tabulated various estimates of Z^* for several diffusing species, and the average value is about -10.2 (Cu in Al, Al in Al, Ag in Au, and Au in Au). The threshold current density (J_{th}) for electromigration can be estimated directly from (8.24) by noting that J (A/cm^2) $= E$ (V/cm)/ρ (Ω-cm) and that 1.0 dyn-cm $= 1.0$ erg $= 10^{-7}$ J. Therefore

$$J_{th} = \frac{E}{\rho} = \frac{10^{-7} \Omega}{\rho Z^*e} \frac{\partial \sigma_n}{\partial x} \simeq \frac{10^{-7} \Omega}{\rho Z^*e} \frac{\sigma_n}{W} \qquad \text{A/cm}^2. \qquad (8.25)$$

The stress gradient is approximated by σ_n/W, where W is the track width (i.e., stripe length) of an MR sensor [see Tang, 1994, p. 38]. A numerical estimate of J_{th} for a high-stress, short-stripe-length film will be useful at this point. Assuming $\sigma_n = 10^{10}$ dyn/cm^2 (about 0.5% of Young's modulus for NiFe), $\Omega = 10^{-23}$ cm^3, $\rho = 30$ μΩ-cm, $Z^*e = 1.6 \times 10^{-18}$ C ($Z^* = -10$), and $W = 1.0$ μm, the electron wind force would overcome a stress-driven back flow at a current density greater than $J_{th} \sim 2 \times 10^6$ A/cm^2. Since all of the AMR and GMR devices analyzed in this book operate at current densities *more than an order of magnitude greater than the threshold level*, electromigration must be studied and understood well enough to predict time to failure (TTF) under a variety of operating conditions.

The resistance method is routinely used in studying electromigration, and in some MR heads electromigration might occur in the lead structure (Au diffusion into Ta). Changes in lead resistance do not impact reliability in the same manner as changes in the sensor layers, and thus it is necessary to correct test data for any confounding trends not related to sensor lifetime. Lam et al. (1998) studied electromigration in AMR heads designed for 2-Gbit/in.^{2} applications; the approximate geometry is $W \simeq 1.8$ μm, $h \simeq 0.7$ to $h \simeq 1.2$ μm, $G_{ss} \simeq 0.20$ μm, NiFeRh/Ta/NiFe/Ta thicknesses $\simeq 80$ Å/75 Å/120 Å/50 Å, and dielectric gaps $g_1 = g_2 = 850$ Å. The devices, all of which were finished head gimbal assemblies (HGAs), were placed in ovens and energized. Two groups of devices were selected based on the cold slider resistance (slider resistance excludes the wire resistance of 4.6 Ω). Group A slider resistances ranged from 27.3 to 35.4 Ω with a mean resistance of 31.6 Ω, and group B ranged from 34.7 to 44.1 Ω with a mean resistance of 40.6 Ω. The resistance of the devices was monitored and recorded for the duration of the test. Tests were performed on group A at 120°C and 11 mA bias (estimated $J \simeq 6.4 \times 10^7$ A/cm^2 average in the NiFe layer) and on group B at 150 °C and 10 mA (estimated $J \simeq 6.5 \times 10^7$ A/cm^2 average in the NiFe layer). Results for one head each from group A (slider resistance = 33.5 Ω at 23°C) and group B (slider resistance = 39.7 Ω at 23°C) are shown in Figs. 8.40 and Fig. 8.41, respectively. The increase of resistance in each group is proportional to (time)$^{1/2}$, but in 1000 h the group A head increased less than 2%, while the group B head increased 5% in just 64 h. The

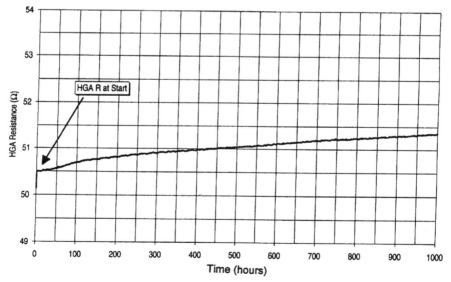

FIGURE 8.40 Experimental electromigration results for 2-Gbit/in.^{2} AMR head with cold resistance $R_0 = 23.5$ Ω tested at 120°C with 11 mA bias current ($J = 6.4 \times 10^7$ A/cm^2 and 100% duty cycle). Unpublished data courtesy of C. Lam, I. C. Barlow, and S. Chim, Read-Rite Corp. (1999).

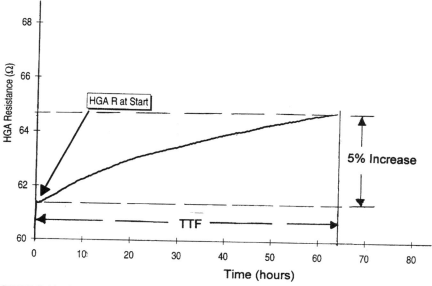

FIGURE 8.41 Experimental electromigration results for a 2-Gbit/in.^{2} AMR head with cold resistance $R_0 = 39.7\,\Omega$ tested at 150°C with 10 mA bias current ($J = 6.5 \times 10^7$ A/cm^2 and 100% duty cycle). Time-to-failure (TTF) is defined at the point where the device resistance increases from 5% from the initial value. Unpublished data courtesy of C. Lam, I. C. Barlow, and S. Chim, Read-Rite Corp. (1999).

implications of these results for device lifetime are explored in the next (and final) section.

Prediction of Sensor Lifetime

With long periods of use, MR sensors can fail from a number of causes such as corrosion, frictional wear, interdiffusion between thin films (which leads to signal loss), and electromigration (which leads to device burn-out). This book ends with a brief discussion of lifetime prediction based on signal loss or increased resistance using the methodology of accelerated testing at elevated temperatures. The information derived from Arrhenius plots is used to define acceleration factors that facilitate estimation of sensor lifetimes under various operating conditions. While the definition of TTF varies within the magnetic recording industry, in the electromigration studies of Lam et al. (1998) a criterion of 5% increase in the initial resistance was used to define TTF, and this is shown in Fig. 8.41. If the device resistance has not increased 5% in 500 h, the test is normally discontinued and the $t^{1/2}$ curve is extrapolated to estimate the 5% TTF point. At the beginning of the test, the group A head temperature was about 485 K and the group B head was at 544 K, approximately, and these differences in temperature were fundamental to the rate at which the electromigration/diffusion process proceeded. Using the 5% increase in resistance as the TTF threshold, Lam et al. plotted the results from 58 heads of the combined groups A and B, and an Arrhenius-type plot of the data is given in Fig. 8.42; the ordinate is ln(TTF) and the abscissa is $1/T$, where T is the absolute temperature of each head at the beginning of the test. There is significant scatter, but the low- and high-resistance groups give sufficient separation in results that a trendline can be drawn. Since an Arrhenius equation follows the form $t = t_0 \exp(E_a/k_B T)$, then $\ln(t) = \ln(t_0) + E_a k_B T$ and the trendline $y = 17768x - 29.351$ identifies the important constants of the experiment. That is, the intercept $\ln(t_0) = -29.351$ or $t_0 = 1.32 \times 10^{-13}$ h, and $E_a/k_B = 17{,}768$ or $E_a = 1.53$ eV. [Boltzmann's constant k_B (in electron-volts per kelvin) = $(1.38 \times 10^{-23}$ J/K$)/(1.6 \times 10^{-19}$ C$) = 8.625 \times 10^{-5}$ eV/K.]

The TTF is linked directly to the cold resistance of a device, and for a given design, the resistance is well correlated with the stripe height. The TTF data of the 58 group A and B devices are plotted versus cold resistance of a slider in Fig. 8.43. Because of the connecting wires, the HGA resistance is about 4.6 Ω greater than the slider value. The equivalent stripe height h (in micrometers) for a given resistance is found with the relation $R = R_S W/h$ with $R_S = 17.6$ Ω/square and $W = 1.8$ μm. A plot of ln(TTF) versus the approximate equivalent h is given in Fig. 8.44.

Electromigration studies for each type of MR device show the general features of diffusion processes; trendlines of TTF versus $1/T$ conform to Arrhenius-type processes from which one derives the activation energy E_a [the slope of the ln(t) plot] and the time factor t_0 [the intercept $\ln(t_0)$ of the Arrhenius plot]. Zolla (1997) discussed the reliability of dual-stripe MR (DSMR) heads based on accelerated life

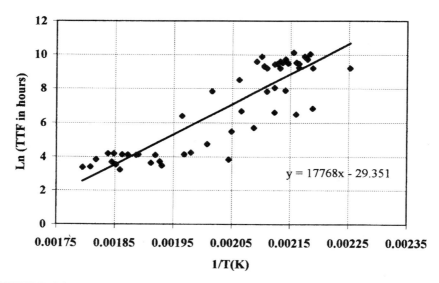

FIGURE 8.42 Arrhenius plot of ln(TTF) versus $1/T$ (K) for 58 AMR devices. The trendline $y = Ax + B$ identifies the important parameters $A = E_a/k_B$ and $B = \ln(\tau_0)$. Unpublished data courtesy of C. Lam, I. C. Barlow, and S. Chim, Read-Rite Corp. (1999).

OPERATING TEMPERATURE AND LONG-TERM RELIABILITY   293

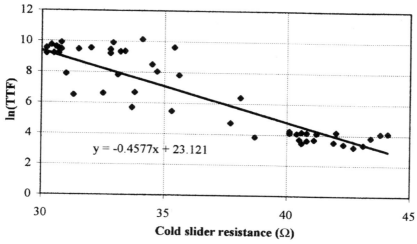

FIGURE 8.43 Plot of ln(TTF) versus R_0 (the cold resistance of each test device) for 58 AMR devices. Unpublished data courtesy of C. Lam, I. C. Barlow, and S. Chim, Read-Rite Corp. (1999).

tests at elevated temperatures and currents. At his criterion for end of life (4–6% change in resistance), the data from 39 devices were scattered but conformed to an Arrhenius plot from which he deduced an activation energy $E_a = 1.99 \pm 0.11$ eV and a time factor $t_0 = 8.96 \times 10^{-17} \pm 8.04 \times 10^{-16}$ h. The TTF measurements had about an order-of-magnitude scatter that Zolla attributed to uncertainties in current

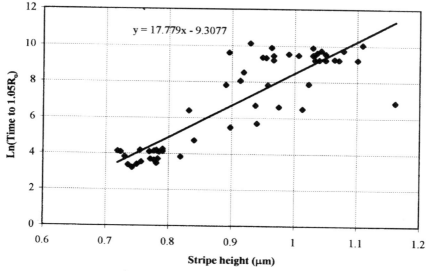

FIGURE 8.44 Plot of ln(TTF) versus stripe height for 58 AMR test devices. Unpublished data courtesy of C. Lam, I. C. Barlow, and S. Chim, Read-Rite Corp. (1999).

density and in temperature of individual devices. Gangulee and d'Heurle (1974) studied electromigration and diffusion in 200-Å-thick NiFe films 125 μm long and 5 μm wide deposited on Si substrates held at 200°C. Resistometric tests were performed at 200–253°C and 0.7–1.4 × 10^7 A/cm^2, and the time to reach electromigration failure was estimated with an equation attributed to Black (1969),

$$t_f \simeq AJ^{-n} \exp\left(\frac{E_a}{k_B T}\right), \tag{8.26}$$

where A is a constant, J is the current density, and the exponent $n \sim 2-3$. [*Note:* The value of n appears to depend on details of the test structures and on the experimental method. Other workers find that $n < 2$ gives a better fit to their results. D'Heurle and Ho (1978) point to the difficulties in estimating A and n in experiments having complex thermal factors.] Gangulee and d'Heurle (1974) found an activation energy $E_a = 0.7 \pm 0.1$ eV, which is compatible with a grain boundary diffusion mechanism but is significantly lower than the value of 1.57 eV found by Lam et al. (1998) or the value of 2.3 eV for self-diffusion of Ni in NiFe given by Walsoe de Reca and Pampillo (1967). Doan et al. (1998) studied electromigration in 3.0-Gbits/in.^{2} AMR heads and fit their data with $E_a = 0.99$ eV and $n \simeq 1.5$ for Black's (1969) equation. Since device temperature is a function of current density, many workers do not seperate the J and T dependencies, and they express their results as a simple Arrhenius equation like (8.19) instead of using Black's equation.

Because of the wide range of activation energies and time factors from different experiments with various designs and materials, general statements about lifetime prediction must be qualitative. Quantitative predictions can be made for given designs and under specific operating conditions. For example, the test results in Fig. 8.42 give a TTF trendline that follows

$$\text{TTF} = (1.32 \times 10^{-13}) \exp\left(\frac{1.53 \text{ eV}}{k_B T}\right) = (1.32 \times 10^{-13}) \exp\left(\frac{17768}{T}\right) \quad \text{h} \tag{8.27}$$

where T is the absolute temperature for a device. This equation does not say anything about the device resistance, operating current, or current density, but it does say the TTF for a device at $T = 470$ K is about 3460 h, but at $T = 490$ K, the device TTF drops to about 740 h. In other words, the *Acceleration Factor* AF for testing at a higher temperature is

$$\text{AF} = \frac{\text{TTF}_1}{\text{TTF}_2} = \exp\left[\frac{E_a}{k_B}\left(\frac{1}{T_1} - \frac{1}{T_2}\right)\right], \tag{8.28}$$

or the lifetime improves by this same factor by reducing the operating temperature accordingly. The accumulation of electromigration damage occurs only with current flow, and thus the TTF (arising from increased resistance) can be increased by reducing the duty cycle. This knowledge is reflected in many test specifications for MR heads where the lifetime (TTF) must be 10,950 h at a bias current duty cycle of 100% and a test ambient temperature of 60°C. Under these conditions at 25% duty cycle for bias current, sensor lifetime would be 43,800 h or 5 years, which is regarded as adequate for archival data storage in today's technology environment. The temperature rise of a sensor, which was analyzed in Chapter 2, may be estimated by the expresssion

$$\Delta T \simeq \frac{R_\tau I^2 R_0}{1 - \alpha R_\tau I^2 R_0}, \quad (8.29)$$

where R_τ is the thermal resistance, $R_0 = \rho_0 W/(th)$ is the cold resistance of the sensor, I is the bias current, and α is the temperature coefficient of resistance of the sensor element. If one ignores heat flow from the sensor through the connnecting lead structures on the wafer, the thermal resistance can be approximated by the simple relation

$$R_\tau = \frac{g_1 g_2}{\kappa(g_1 + g_2)Wh} \quad \text{K/W}, \quad (8.30)$$

and using (8.27) one can conservatively estimate TTF based on specific details of a device to find $T = T_{amb} + \Delta T$.

Sensor life may be limited by amplitude loss instead of increased resistance. This situation is exemplified by the type B and type C spin valve behaviors shown in Figs. 8.35a,b and 8.36a,b. Doan (1999) defined TTF at the point where the signal dropped 10% from the initial value. For the type B devices studied at 6 mA and 120°C ambient, the Arrhenius plot gave $E_a = 1.28$ eV and $t_0 = 1.76 \times 10^{-13}$ h, so the TTF is estimated from the relation

$$\text{TTF} = (1.76 \times 10^{-13}) \exp\left(\frac{14{,}805}{T}\right). \quad (8.31)$$

The device test temperatures ranged from 408 to 471 K, and the portion arising from I^2R dissipation ranged from $\Delta T = 15$ K to $\Delta T = 78$ K for the 68 devices of this study. Figure 8.45 shows various lifetime results scaled to 60°C ambient and 25% duty cycle from the accelerated tests of Doan (1999). Time-to-failure results are plotted as a function of the cold slider resistance for 5-Gbit/in.^{2} type A spin valves, 7-Gbit/in.^{2} type B spin valves at 6 mA, and 10-Gbit/in.^{2} type C spin valves at 5 and 6 mA. For a given design, the resistance of finished devices will depend primarily on the final stripe height, so the devices with short stripes and high resistance will, on average, have short lifetimes. This is shown in Fig. 8.46 for the 10-Gbit/in.^{2} type C sensors. Table 8.4 gives the estimated activation energy for each type of device. At a cold slider resistance of 25 Ω, the type C sensors have a projected lifetime of 1000 years at 5 mA sense current, but this value drops to 100 years at 6 mA. As a consequence, any short-term improvements in bit error rate achieved by increasing

296 CHARACTERIZATION OF MR DEVICE FUNCTION AND RELIABILITY

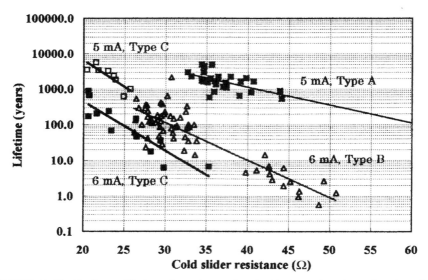

FIGURE 8.45 Projected lifetimes for various spin valve designs plotted as a function of the cold slider resistance. Results are scaled from the accelerated test conditions (120°C, 100% duty cycle) to the operating conditions of 60°C and 25% duty cycle. Unpublished data courtesy of V. Doan, Read-Rite Corp. (1999).

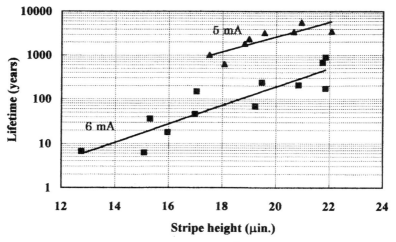

FIGURE 8.46 Projected lifetime for a 10-Gbit/in.^{2} DSSV design plotted as a function of stripe height. Results are scaled from the accelerated test conditions (120°C, 100% duty cycle) to the operating conditions of 60°C and 25% duty cycle. Unpublished data courtesy of V. Doan, Read-Rite Corp. (1999).

TABLE 8.4 Activation Energies for Lifetime Tests

Device Identification	Activation Energy, E_a (eV)
Type A	1.47
Type B	1.28
Type C	1.46

the signal with higher sense currents must be carefully weighed against the risks of reduced failure times that find their cause in interlayer diffusion processes.

REFERENCES

Ashar, K. G., *Magnetic Disk Drive Technology*, IEEE Press, New York, 1997.

Baglin, J. E. E. and Poate, J. M., "Metal-Metal Interdiffusion," Chap. 9 in *Thin Films—Interdiffusion and Reactions*, Poate, J. M., Tu, K. N., and Mayer, J. W. (Eds.), Wiley-Interscience, New York, 1978.

Bertram, H. N., *Theory of Magnetic Recording*, Cambridge University Press, Cambridge, 1994.

Black, J. R., *Proc. IEEE*, **57**, 1587 (1969).

Blech, I. A., *J. Appl. Phys.*, **47**, 1203 (1976).

Blech, I. A. and Herring, C., *Appl. Phys. Lett.*, **29**, 131 (1976).

Chang, C., Chim, S., and Lam, C., unpublished internal results, Read-Rite, 1999.

Christou, A. (Ed.), *Electromigration and Electronic Device Degradation,"* Wiley-Interscience, New York, 1994.

Christou, A. and Day, H., *J. Appl. Phys.*, **44**, 3386 (1973).

Crank, J., *The Mathematics of Diffusion*, Oxford University Press, Oxford, 1975.

D'Heurle, F. M. and Ho, P. S., "Electromigration in Thin Films," Chap. 8 in *Thin Films—Interdiffusion and Rections*, Poate, J. M., Tu, K. N., and Mayer, J. W. (Eds.), Wiley-Interscience, New York, 1978.

Dieny, B., Humbert, P., Speriosu, V. S., Metin, S., Gurney, B. A., Baumgart, P., and Lefakis, H., *Phys. Rev. B*, **45**, 806 (1992).

Doan, V., unpublished internal data, Read-Rite, Milpitas, CA, 1999.

Doan, V., Cheng, K., and Barlow, I. C., unpublished internal data, Read-Rite, Milpitas, CA, 1998.

Dong, J., unpublished internal data, Read-Rite, Milpitas, CA, 2000.

Fowler, R. H., *Statistical Mechanics*, Cambridge University Press, Cambridge, 1966.

Gangulee, A. and d'Heurle, F. M., *Proc. 6th Int'l. Vacuum Congr. Japan J. Appl. Phys. Suppl.*, **2**, Pt. 1, 625 (1974).

Glasstone, S., Laidler, K. J., and Eyring, H., *The Theory of Rate Processes*, Mc-Graw-Hill, New York, 1941.

Hannon, D. M., *INTERMAG* (in press).

Hempstead, R., *IBM J. Res. Devel.*, **18**, 547 (1974).

Hempstead, R., *IEEE Trans. Magn.*, **MAG-11**, 1224 (1975).
Herzig, Ch. And Cardis, D., *Appl. Phys.*, **5**, 317 (1975).
Klaassen, K. B. and van Peppen, J. C. L., *IEEE Trans. Magn.*, **MAG-33**, 2611 (1997).
Lam, C., Sahli E., and Chim, S., *Proc. 19th EOS/ESD Symp.*, 386 (1997).
Lam, C., Barlow, I. C., and Chim, S., unpublished internal data, Read-Rite, Milpitas, CA, 1998.
Macchioni, C., Doan, V., and He, Q., *3M Conf. Proc.*, San Jose (in press).
Madakson, P., *J. Appl.Phys.*, **70**, 1380 (1991).
Mott, N. F. and Jones, H., *The Theory of the Properties of Metals and Alloys*, Dover, New York, 1958.
Saito, A. T., Iwasaki, H., Kamiguchi, Y., Fuke, H. N., and Sahashi, M., *IEEE Trans. Magn.*, **MAG-34**, 1420 (1998).
Stupp, S. E., Baldwinson, M. A., McEwen, P., Crawford, T. M., and Rogers, C. T., *IEEE Trans. Magn.*, **MAG-35**, 752 (1999).
Takahashi, M., Maeda, T., Inage, K., Sakai, M., Morita, H., and Matsuzaki, M., *IEEE Trans. Magn.*, **MAG-34**, 1522 (1998).
Tang, P. F., Chap. 2 in *Electromigration and Electronic Device Degradation*, Christou, A. (Ed.), Wiley-Interscience, New York, 1994.
Wallash, A., *IEEE Trans. Magn.*, **MAG-32**, 49 (1996).
Wallash, A., *IEEE Trans, Magn.*, **MAG-34**, 1450 (1998).
Wallash, A. and Kim, Y. K., *J. Appl. Phys.*, **81**, 4921 (1997).
Wallash, A. and Kim, Y. K., *IEEE Trans. Magn.*, **MAG-34**, 1519 (1998).
Walsoe de Reca, E. and Pampillo, C., *Acta Metall.*, **15**, 1263 (1967).
Wang, S. X. and Taratorin, A. M., *Magnetic Information Storage Technology*, Academic, San Diego, 1999.
Wilson, A. H., *The Theory of Metals*, Cambridge University Press, Cambridge, 1965.
Yuan, S. W., Kobayashi, T., and Liao, S. H., *IEEE Trans. Magn.*, **MAG-31**, 2627 (1995).
Zolla, H. G., *IEEE Trans. Magn.*, **MAG-33**, 2914 (1997).

INDEX

Accelerated life tests, 292
Acceleration factor (AF) (in lifetime tests), 291, 294
Activation energy
  defined, 279
  for grain boundary diffusion of Au-Ta, 279, 294
  and interdiffusion, 281
  for self diffusion of Ni in NiFe, 294
Active region
  of a conventional pinned layer, 181
  of a saturated SAL, 136
  of a synthetic pinned layer, 209
Adiabatic heating by an ESD pulse, 271
Adjacent track reading, 226
Aluminum nitride (AlN), 47
Aluminum oxide (Al_2O_3), 47
Ambient temperature (inside a disk drive), 49, 287
Ampere's circuital law, 66, 86
  and field inside a conductor, 86
Amplitude asymmetry, see Asymmetry
Amplitude coefficient of variation (COV), 254
Anisotropic magnetoresistance (AMR) effect, 20
  parabolic nature of, 31
  presence in GMR free layer, 40f, 160
  and ratio (defined), 30, 116
  and skew angle, 30
  and temperature dependence, 46–48
Anisotropic propagation of signal flux, 226, 233
Anisotropy
  barrier height, 24
  energy, 64

exchange, see Exchange coupling
field, 24f
  and effective value of a free layer, 166, 171
  induced, 23–25
  and influence of transfer curves, 31, 80
  magnetocrystalline, 26
  pair-ordering, 23–25, 171. See also Directional ordering
  shape, 26, 80, 135
  stress, 24, 26, 260
    and effective field, 172
    and patterning effect, 25
Annealing
  in magnetic field, 36
  temperature and limits on upper values, 36f, 98, 190, 192, 205
Antiferromagnetic (AFM) films, 160
  criteria for selection, 37
  CrPtMn, 37
  FeMn, 38, 168, 183, 188
  FeMnRh, 37f
  IrMn, 38, 170, 174, 184, 190, 203
  NiMn, 38, 190
  NiO, 209, 211, 216
  PdPtMn, 38, 190, 203, 216
  PtMn, 37
  RhMn, 37
Antiferromagnetic (AFM) systems
  disordered, 38, 203
  ordered, 38, 203

299

300 INDEX

Areal density (AD), 226. *See also* Scaling
  defined, 1, 4
  and dependence on magnetic spacing, 9
  and read signal sensitivity, 18, 205, 207
Arrhenius
  equation, 279
  plots, 285, 291*f*
  trendline (intercept and slope), 292
Aspect ratio, *see* Bit cell ratio
Asperities, *see* Thermal asperities
Asymmetric side reading, *see* Microtrack profile
Asymmetry
  computational artifacts, 185, 191
  defined, 114, 185
  influence of AMR effect in GMR devices, 184
  relation to bias angle, 114
  relation to bias current polarity, 160
  of signal pulse amplitudes for DSMR heads, 128
  of signal pulse amplitudes for single layer, 114, 117
  of a spin valve head, 185
Atomic
  flux (in electromigration), 288
  vibrational frequency, 279
  volume, 289

Bandwidth
  noise, 13
  signal, 226
Barber-pole MR heads, 108
Barkhausen noise, 82, 262, 265
  and dynamic recording tests, 254*ff*
  origin of, 82, 96
  and QST transfer curve hysteresis, 262
  suppression with longitudinal field, *see* Stabilization
Baseline
  noise (BLN), 254, 256, 265*f*
  and reinitialization of hard bias layers, 256
  popping (BLP), 254, 256, 265*f*
  shift of PM stabilized layers, 256, 265*f*
Bessel function (first order, modified), 231
Bias angle, 98, 233
  and influence on sensor H_D, 98
  of a magnetic layer, 82
  for SAL/MR head, 112, 133, 136, 142, 151
  at zero asymmetry, 151
Bias compensation layer (BCL), *see* Spin valves
Bias current
  estimate for operating point, 152
  polarity, 160
Bias curves, *see* Read bias curves
Bias field
  on free layer of a DSSV, 219

  for DSMR head, 125
  for self-biased head, 109
  for shunt-biased head, 120*f*
  for SAL/MR head (conducting spacer), 141
  for SAL/MR head (insulated spacer), 133
Bias point
  of DSSV free layer and interlayer coupling fields, 219, 221
  of DSSV free layer and pinned layer thickness, 221
  and optimization of free layer thickness, 176, 184
  shifting with gap asymmetry, 137*f*
  shifting with net moment of SSV, 209, 214
  shifting with sense current, 168, 170, 182
  on a transfer curve, 109, 149
Biasing techniques, for shielded sensors, 108*f*, 139
  self biasing, 108*f*
  shunt biasing, 108
  SAL biasing, 108
Bit cell (aspect) ratio (R_{bc}), 3, 226
Bit error rate (BER), 2, 247
  of archival data storage systems, 2, 295
  and complementary error function, 2
  and signal variations, 2, 254, 260
  and SNR, 3
Black's equation, 294
Blocking temperature, *see also* Temperature
  definition, 37
  distribution and Fermi–Dirac statistics, 38
  tabulation of values, 37, 191
Boltzmann
  constant, 47, 69, 279
  distribution, 39
  transport equation, 42, 168, 216
    and thin film resistivity, 168
    and Fuchs–Sondheimer theory, 42
Bottom synthetic spin valve (BSSV), *see* Spin valves
Boundary conditions
  for coupled magnetic films, 65
  and dependence on signal flux, 76, 134
  for a free layer, 175
  for thermal model, 54
Broadening of microtrack profile, 232
Buffer layer, 217
  and promotion of [111] texture by Ta, 35, 160
Bulk nature
  of AMR effect, 42. *See also* Scattering
  of GMR effect, 32. *See also* Scattering

Capping layer, 161, 217
Channel density (U), 8, 249, 253

INDEX **301**

Characteristic decay length
 for domain wall, 71
 for heat flow, 54f, 188
 for flux propagation 65, 121
  exchange energy ignored, 74
  exchange energy included, 74
  for nonidentical films, 68, 127
  for shielded films, 90, 121, 127, 134, 141, 174f, 210
Charge (effective) of atomic ions (electromigration), 289
Chemical potential of atoms and vacancies, 289
Co dusting for enhanced GMR ratio, 35, 164, 211
Coefficient of variation (COV) of signal, 254
Coercivity, 10, 22
 remanent value, 10
Columnar crystalline [111] structure, 21, 35, 47, 122, 161, 217. See also Buffer layer
Comparison between experimental signals and model
 self-biased MR head, 120
 DSMR head, 132
 SAL/MR (conducting spacer), 147–149
 top spin valve, 187f
Complementary error function, see Bit error rate
Composite free layer (CoFe/NiFe), see Free layer
Concentration (of diffusing material), 279
Conducting spacer, see SAL/MR
Conductivity, 43, 45, 50. See also Resistivity.
Contiguous junction (CJ) hard bias design, 83, 100, 102
Conventional spin valve (CSV) head, 161, 166, 178
Convolution product
 and side reading of a shielded sensor, 229, 231, 235
 of write and read sensitivity functions, 227
Correlation
 between HBM results and MR head properties, 277
 between QST and dynamic tests, 269f
 coefficient (product moment), 155, 198, 214
 length (cross-track) for medium jitter, 14
Cost/megabyte for hard disk drives, 1
Coupled film analysis, compared with uniform rotation model, 91, 167
Coupled FM films and irrelevance of demagnetizing factors, 91
Cr underlayer for hard bias materials, 23
Curie temperature, 23, 69
Current
 aiding and opposing modes in spin valves, 170, 177, 217

density
 and temperature rise, 53, 56
 and threshold for electromigration, 288f
 and melting of MR sensor, 58, 272
 sense/bias, 4
 and unpinning of a pinned layer, 199ff
Damage in sensors
 criterion for HBM tests, 275
 current, melting, or unpinning, 272f
 electrostatic discharge (ESD), 275
 interdiffusion between layers, 194
 threshold for electrical breakdown and melting, 271f
 threshold for pinned layer reversal, 272f
 by void formation (electromigration), 287
Decay length, see Characteristic length
Deconvolution techniques (and microtrack profiles), 238
Defects (grain boundaries and dislocations), 278
Demagnetizing (stray flux) energy, 171
 in region between two films, 93
Demagnetizing factors, 25
Demagnetizing (stray) field
 effective value
 of an isolated film, 27, 98
 of a pinned layer, 166, 214f
 of a shielded film, 91
 for flat ellipsoidal films, 96
Density
 areal, 1, 4
 bit, 1
 current, see Current density
 linear, 1, 8
 track, 4
Design parameters
 for 25 Gbits/in.^{2} DSSV head, 220
 for SAL biased MR heads, 146
Diamond-like carbon (DLC) overcoat, 9
Diffuse scattering of spin polarized electrons, see Scattering
Diffusion, 49, 160, 205, 278. See also Reliability.
 of Au into Ta, 279
 broadening of spacer layer in spin valves, 285
 constant, 278
 length, electrons in Cu, 162, 285
 of Mn or Ni through grain boundaries, 285
 process, and time dependence, 279f
 rate, 278
 theory, 278
Diffusivity, 278
Directional ordering (of Fe-Fe pairs), 23
Disordered AFM materials, see AFM systems

Domain wall
  Bloch and Néel types, 70
  in sensors at regions of PM field reversal, 258
  thickness, 70f
Dual spin valve (DSV), 217
Dual stripe MR (DSMR) head, 108, 123
Dual synthetic spin valve (DSSV) head, 217
Dusting of GMR interfaces with Co, 35, 206
Duty cycle (of current flow) and lifetime tests, 295
Dynamic range (of spin valve output signal), 184
Dynamic tests for MR heads, 247

Easy axis (EA), 24
Edge broadening of a track profile, 227ff
Edge curling wall, 62
Edge effect (field reversals near PM edges), 101ff
Edge width (of a microtrack), 227
Effective spacing for MR readback signal, see Spacing, magnetic
Efficiency of readback flux in shielded MR sensor, 94
Electromigration, 49, 248, 278. See also Reliability
  effective charge of atomic ions, 289
  electric field force, 287
  electron wind, 287f
Electron reflection at interfaces, 160
Electrostatic discharge (ESD)
  current transient, 271
  damage thresholds, 243
  HBM testing, and damage, 243, 247
End effects (of finite height films), 89
Energy
  anisotropy 64
  budget in coupled films, 72f
    of a spin valve, 174
  demagnetizing (stray field), 64, 171
  exchange, 64. See also Exchange coupling
  magnetostatic potential 64
  stress (pattern effect) anisotropy, 25
Equalization for partial response detectors, 15
Euler's equation, 65
Exchange coupling
  anisotropy of CoFe/IrMn systems, 191
  anisotropy of NiFe/FeMn systems, 32
  anisotropy of NiFe/NiMn systems, 38, 191
  biasing, with patterns on sensors, 96, 100
  constant for Co, Fe, Ni, NiCu, NiFe, 6, 22, 174
  coupling (RKKY) and AFM layers, 35, 179, 207
  coupling suppression with thin Cu layer, 216
  energy
    of Co/Ru multilayers, 207
    and dependence on temperature, 38, 205
    and in-plane rotation of magnetization, 72
    and out-of-plane tilting of magnetization, 71
  field, 168
    and distribution parameters with temperature, 38, 190
    at operating temperature, 190
  integral and values for materials 69
  penetration distance of a field, 71
  stiffness, see Exchange constant
Excitation field (flux), 267. See also Signal flux
  at a free layer, 181
  at a pinned layer, 181
  of an MR sense element, 5, 12, 95, 111
Experimental results
  AMR effect and skew angle, 28f
  AMR effect versus temperature, 47f
  AMR read bias curves, 149
  AMR self-biased head, 120
  AMR signal and asymmetry, 148
  AMR transfer curves, 98
  Arrhenius plots
    GMR ratio decay time, 286
    lifetime of AMR heads, 292
  bias field for sensor stabilization, 97
  current pulse of HBM tester, 272
  distribution of blocking temperatures, 37, 39, 191, 205
  exchange field versus temperature, 191
  exchange and SSV saturation fields versus Ru thickness, 208
  GMR amplitude and resistance versus life test time, 282f
  GMR effect versus layer thickness, 163, 165
  GMR effect versus temperature, 48f, 163
  GMR parameters for Co, NiFe, and Ni free layers, 163, 165
  GMR ratio versus annealing time, 284
  GMR ratio versus free layer thickness, 163
  GMR resistance, signal, and asymmetry versus HBM voltage, 274, 276
  HBM failure voltage (GMR sensors) versus resistance, 275
  HBM failure voltage (GMR sensors) versus stripe height, 277
  interlayer coupling energy versus Cu thickness, 179f
  lifetime
    AMR heads versus head resistance, 293
    AMR heads versus operating temperature, 292
    AMR heads versus stripe height, 293
    GMR heads versus head resistance, 296
    GMR heads versus stripe height, 296
  magnetic and physical track width, 105
  magnetization and GMR effect versus field, 33

INDEX **303**

microtrack profiles
  for AMR heads, 233, 239f
  for CSV head, 242, 258–260, 270
    unpinned and repinned, 244f
  for DSSV head, 241
  for various stripe heights, 233
read bias curves
  for AMR head with instabilities, 263
  for BSV head, 187, 252, 256
  for DSSV head, 222, 224, 253–255
resistance change versus annealing time, 280, 290f
resistance change versus initial resistance, 281
resistivity and AMR effect versus film thickness, 46
signal and asymmetry
  of AMR (self-biased) head, 120
  of BSSV head, 212
  of BSV head, 187
  of DSMR head, 132
  of DSSV head, 222, 224
  of SAL biased AMR heads, 148f, 255
  of SSV head, 212
  of TSV, 188
signal pulses
  CSV at different bias points, 269
  CSV head before and after reinitialization, 257
  DSSV head with BLP, 268
  thermal asperities, 260
transfer curves
  AMR head with instabilities, 264
  CSV head with Barkhausen noise, 269
  DSSV head with Barkhausen noise, 265–267
write saturation curves for BSV head, 249f

Face-centered cubic (fcc) texture, 21f
Failure causes in MR sensors, 199, 291
Failure voltage (melting or unpinning) of HBM, 275
Fermi–Dirac statistics, 38
  and blocking temperature, 38
  and basic hypotheses, 38f
Fermi surface, 30
Ferromagnetic (FM) materials, electrical and magnetic properties
  Co, 22, 160
  CoFe, 22, 160
  CoFeB, 22, 216
  CoNbTi, 22
  CoZrCr, 22, 154
  CoZrMo, 23
  NiCo, 22, 47
  NiCu, 47f

NiFe, 22f, 47, 144, 160f, 168
NiFeRh, 22, 144
NiFeZr, 22
Fick's laws (first and second) of diffusion, 278f, 289
Field
  of an arctangent transition, 12
  for annealing FM/AFM systems, 36f
  ferromagnetic coupling (topological), see Interlayer coupling
  from current flow in a film, 85f
Fluctuations, see Superparamagnetism
Flux (magnetic)
  closures, 62, 93
  conduction through pinned layers of SSV, 209
  decay, see Characteristic length
  decay length between free layers and shields, 175
  propagation, 63
Flying height and magnetic spacing, see Spacing, magnetic
Four-point probe, 30, 281
Fraction of current flowing in each layer of a sandwich, 169
Free (F) layer, 160
  Co and CoFe, 32, 35, 206
  composite (CoFe/NiFe) and effective magnetization, 35, 170, 176, 185, 252
  NiFe, 32
  thickness, and SV performance, 162, 176, 205
Free space potential (between head and medium), 230
Frequencies (LF and HF) for magnetic recording, 248
Frequency factor (in atomic vibration), 279
Frictional heating (at head-disk interface), see Thermal asperity
Fuchs–Sondheimer theory (film resistivity), 42, 168, 216
Full width at half-maximum (FWHM) of track profile, 227

Galvanomagnetic effect (ordinary), 30
Gap
  asymmetry, 147, 197
  shield-to-shield, 93, 95, 109, 147, 173
  symmetry, 124
Gaussian statistics
  electrical noise, 2
  and Monte Carlo analysis, 154
Gauss's law 67
Giant magnetoresistance (GMR), 20
  description of effect, 32, 161
  ratio (response or figure of merit), 21, 161f

304 INDEX

Giant magnetoresistance (GMR) (*Continued*)
    and dusting with Co alloys, 35, 217
    and empirical nature, 160*ff*
    and enhancement with specular reflection, 216
    and free layer materials, 32, 162
    and resistance states (high and low), 161*f*
    and sheet conductivity, 162
    and temperature dependence, 163*f*
    and thickness of FM layers, 162*f*
    and thickness of spacer layer, 162
Gradient
    in alloy concentration (electromigration), 287*f*
    in magnetic field of write head, 10. *See also* Transition
    in stress (or pressure), 288*f*
    in temperature (electromigration), 287
Grain boundary (diffusion), 279*f*
Grain size (and medium noise), 13
Green's function, 230

Hard axis, 24
    permeability, 24
Hard bias. *See also* Permanent magnet (PM) materials
    collimated deposition (no overspray), 101
    material properties
        CoCr, 22
        CoCrPt, 22*f*
        CoCrTa, 22
        CoCrTaPt, 22
        CoNiCr, 22
    stabilization, 99, 254, 265
        contiguous junction (CJ) design, 166
        field at a free layer, 166
        field attenuation by shields, 104
        field and magnetic width of sensor, 105
        overlay lead design, 166
Hard transition shifting (HTS) and OW, 16, 249
Head field model (three-dimensional), 231
Healing
    flow from stress (electromigration), 289
    pulse of a HBM tester, 245
Heat damage, 51. *See also* Damage
Heat sink (for MR sensors), 49
High frequency (HF) signal and dependence on write current, 248
High pass filtering (in AC coupled circuit), 261
Hot spot (of a thermal asperity), 261, 287
Human body model (HBM) test, 247
    and controlled ESD, 247
    and current transient, 271
    and damage voltage (or current), 273
    and polarity of voltage, 273–275

Hysteresis
    along hard axis of free layer, 256
    and transfer curves, 262, 265

Ideal bias point (of free layer), 41, 176
Images (magnetic) of a layer in a shield, 88
Incomplete erasure of LF signal and OW, 249
Induced uniaxial anisotropy, *see* Anisotropy
Instability
    and insufficient hard bias pinning, 258
    magnetic, and overspray of HB material, 101
    in microtrack profiles, 258*f*
    and root causes, 102, 262
    and uncontrolled stress in free layer, 259
Insulated spacer, *see* SAL/MR
Insulation (electrical) of sensor layers, 47
Interaction energy of correlated waviness (orange-peel effect), *see* Topological coupling
Interaction field and relation to surface energy, 34
Interdiffusion, 248, 278
    Au-Cr, 279
    Au-Ta, 279
    and signal loss in spin valves, 285
    of thin films, 281
Interface
    broadening with annealing, 192, 285
    dusting with Co, 35
    mixing between CoFe and Cu layers, 285
Intergranular coupling (and medium noise), 14
Interlayer coupling
    field, 32, 34, 175, 178, 198
    between free and pinned FM layers, 166, 177, 213
Intersymbol interference (ISI), 2, 17, 249
    and HF amplitude reduction, 249
Ion charge (effective) in electromigration, 289

Jitter, position and width variances, *see* Noise and transition jitter
Junction angle of CJ stabilized sensors, 51, 101, 206

Karlquist head, 10
    and analysis of shielded MR sensor, 5
    magnetic field of, 10

Laminated magnetic films 62, 69
Lattice
    body-centered cubic (bcc), 69
    constant, 69, 279
    defects at grain boundaries, 289
    simple cubic (sc), 69
Length (of magnetic transition), *see* Transition, length

Lifetime, 58, 295
   and amplitude loss of a sensor, 295
   prediction for sensors, 248, 287, 291, 295
   of spin polarized electrons, 216
Line broadening (from convolution), 227
Line charge (magnetic), 231
Line spread function, 231, 235
Longitudinal
   fields at a free layer, 167
   recording of magnetic transition, 12
Lorentz number, 5
   and thermal conductivity, 47
Lorentzian pulse, 5
   and comparison with accurate calculation, 6
   and sum of infinite sequence, 7
Low frequency (LF) signal, 248
   and approach to saturation level, 248

Magnetic
   center of a microtrack, 235
      offset of a microtrack, 239
   domain walls, and Barkhausen noise, 253
   field
      of a current sheet, 78
      at a free layer, 166f, 175
      at a pinned layer, 167, 173
   force microscopy (of medium jitter), 15
   instabilities from CJ overspray, 83
   oxide layers (MOL) for specular reflection, 217
   relaxation effects, *see* Superparamagnetism
   spacing, 6, 9. *See also* Spacing, magnetic
      and areal density, 9, 109
      and basis for scaling parameters, 6
      and relation to flying height, 10
   susceptibility
      of isolated film, 89, 91
      of shielded film, 91
   transitions, *see* Transitions, magnetic
   viscosity, *see* Superparamagnetism
   width, and physical width of sensor, 105
Magnetization
   angle of free layer, 184
   profiles
      for coupled films 68
      for DSMR head, 126
      for free layers, 174, 174
         of a BCL head, 215
         of a DSSV free layer, 219
         of a SSV free layer, 209
      for pinned layer, 173f
      for SAL/MR head, 134, 141
      for self-biased AMR head, 109f
      for shunt-biased AMR head, 121
Magnetostriction, 256
   coefficient, various materials, 21, 24f
   and influence on anisotropy field, 260
Mass transport (in electromigration), 287
Matthiesen's rule, 279
Mean free path
   of electrons (bulk scattering), 32, 42f, 45, 160
   of experimental parameters (electrons), 168
   of spin polarized electrons in NiFe, 32, 161
Mean square error (MSE), 2
Medium (magnetic)
   coercive squareness, 10
   coercivity, 10
   moment ($M_r\delta$) and head design, 111, 116, 183
   noise, *see* Noise
   remanence, 10, 248
   remanent coercivity, 10
   thickness, 10, 109
Melting current, 194, 272f
   dependence on geometry and properties, 58, 272
   for SAL/MR biasing (conducting spacer), 152
Melting point (of NiFe), *see* Temperature
Metastable orientations of sensor magnetization, 258f
Microstructure and magnetic properties, 21, 23, 35, 160
Microtrack profile, 226
   asymmetry, 226, 232, 234
      and bias current, 232
      and stripe height, 233
   of CSV head, 239
   of DSSV head, 240
   with ESD damage, 243
   experimental categories, 238
   output instability, 242
   nonuniformity from unpinning, 193
   read width defined, 227, 235
   of unpinned and repinned sensor, 243–245
Model calculations and specification of variables, 146
Model parameters
      for low density self-biased AMR head, 120
      for 1.0 Gbit/in.^{2} DSMR head, 132
      for 3.0 Gbit/in.^{2} NiFeRh SAL/MR head, 132
      for 3.0 Gbit/in.^{2} CoZrCr SAL/MR head, 147
      for 6.0 Gbit/in.^{2} BSV head, 159
      for 25 Gbit/in.^{2} DSSV head, 220
      for Monte Carlo analysis
         of BSSV with NiO AFM layer, 213
         of CSV with IrMn AFM layer, 195
         of CSV with PdPtMn AFM layer, 196
         of SAL biased AMR heads, 154
Modulation, of AMR resistance by signal flux, 117
Moment ratio formula for SAL/MR biasing, 98, 147, 150f

Monte Carlo analysis, 109, 154, 160

Néel model of correlated roughness, *see* Topological coupling
Noble metals (Ag, Au, Cu), 32, 161
Nonmagnetic (NM) spacer layers, Ag, Au, Cu, 32, 161
Nonmagnetic oxide layers for specular reflection, 216
Noise, 13, 247
  bandwidth, 13
  budget (in a recording system), 17
  common mode (thermal asperity), 123
  head, 13
  medium, 13f. *See also* Transition jitter
  power in magnetic medium (jitter), 14
  power spectrum for transition noise, 14
  preamplifier, 13
  root-mean-square (RMS), 15, 248
  voltage, 13
Noise-equivalent resistance, 13
Noise spectral density (NSD), 13
  combined value for head and preamplifier, 13
  head, 13
  preamplifier, 13
Nonlinear distortion (NLD), 3, 16, 247
Nonlinear transition shift (NLTS), 3, 16
Nyquist's theorem, 13

Ohm's law, and size effect in thin metal films, 50
Ongoing reliability tests (ORT), 278
Operating point (on SV transfer curve), 116
  and shift by AMR effect, 42
  and shift by sense current, 168ff
Orange-peel effect, *see* Topological coupling
Ordered AFM materials, *see* AFM systems
Oscillatory exchange interaction (RKKY coupling), 35, 177, 179, 207f
Overspray of CJ materials, *see* Instability
Overwrite (OW), 16
  and BER, 16
  in frequency and time domains, 249
  and mixed causation, 249
Oxide layers (for specular reflection), 216

Parallel resistor analysis of multilayered sensors, 168
Partial erasure, 3, 16
Partial response maximum likelihood (PRML), 2
  and PR4 channels, 8, 15
Patterning effect, *see* Anisotropy, stress
Penetration distance for magnetic fields, 71f
Percolation, transition, *see* Partial erasure

Performance tests in magnetic recording, 248ff
Permanent magnet (PM) material, *see* Hard bias
Permeability, magnetic, 24, 65
Pin holes (and layer thickness), 34, 164, 177, 205
Pinned (P) layer, and thickness tradeoffs, 164
Pinning field, *see* Exchange field
Planck's constant, 231
Point charge (magnetic), 231
Polarity reversal of peak-to-peak signal, 120
Polarity of sense current (and device function tradeoffs), 160, 177
Pole tip recession (PTR), 9
Preannealing and resistance changes in sensors, 280
Probability of device unpinning (and sense current), 199
Process control tolerances and influence on MR signal, 109, 198, 214
Production quantities of MR heads (in, 1998–1999), 1, 109
Product moment correlation, *see* Correlation coefficient
Pseudorandom sequence (PRS), 2, 15
Pulses and baseline shift, 266, 268f
Pulsewidth (PW50), 109, 248f, 253
  shielded DSMR head 6, 131
  shielded single-element head, 6
  unshielded sensor, 5

Q-parameter for write heads, *see* Write field gradient
Quasistatic tests (QST), 247, 257
  and correlation with dynamic tests, 247, 262
  and excitation level, 95, 262

Read bias curves, 248, 251
  for 6 Gbits/in.^{2} BSV device, 186f, 252
  for 12 Gbits/in.^{2} BSSV device, 212
  for 36 Gbits/in.^{2} DSSV device, 222, 252f
  for DSMR head, 124, 130f
  for SAL/MR (conducting) head model, 145f, 150
  model and experiment, 149
  for SAL/MR (insulated) head, 139
  for self-biased AMR head
    heating, 119
    no heating, 16
  for shunt biased AMR head, 122
Read sensitivity function, 229
Read width, physical and magnetic (with hard bias), 206. *See also* Track pitch
Reflection layers in spin valves, *see* Specular reflection

INDEX 307

Reinitialization of hard bias layers, 260
Relaxation
  effects, *see* Superparamagnetism
  processes, 285
  time (thermally activated), 172, 285, 287
Reliability, 49
  of DSMR heads, 292
  and long term changes, 290
Remagnetizing hard bias layers, *see* Reinitialization
Remanence of hard bias layer, 22
Remanence of medium and approach to saturation, 248
Repinning an unpinned layer, 191f
Resistance
  active and inactive regions in GMR structures, 161
  SAL/MR head with conducting spacer, 140
  noise-equivalent value of MR head, 13
Resistivity, 44
  antiferromagnetic (AFM) films, 37, 168, 217
  bulk values, and use in thin film analysis, 42, 168
  change with magnetic field (AMR effect), 30
  dependence on film thickness, 44
  effective value for multilayered films, 188
  ferromagnetic (FM) thin films, 22
  limiting forms (thick and thin films), 44
  mean free path parameters for spin valves, 161f
  Ru thin films, 45
  Ta thin films, 46
Resolution (of readback signal), 250
Root-mean-square (RMS) noise, and relation to peak-peak, 248
Ruderman–Kittel–Kasuya–Yosida (RKKY) coupling, *see* Oscillatory interaction

Sample-and-hold amplifier (MR test equipment), 248
Saturation
  current
    for a BSSV head, 212
    for a DSMR head, 128
    and dependence on stripe height, 78ff
    for SAL/MR head, 134, 141
    for self-biased AMR head, 113
    for shunt biased AMR head, 123
  curves, *see* Write saturation
  field of FM/Ru/FM trilayers, 207
  of pole tips in a write head, 12, 248
  region (location) in coupled films, 75, 77, 173
  threshold field for coupled films, 76, 78, 112
Scaling
  parameters, 277
  rules for areal density, 3, 99, 159, 226, 277
Scattering
  diffuse, 42, 215
  of electrons (bulk, surface, grain boundary), 42, 161
  of electrons at interfaces, 160, 216
  probability, and mean free path of electrons, 162
Self-biasing, influence of sense current, 88, 162
Sense current, 4
  aiding or opposing the pinning field, 174
  optimal value, 117, 176
Sensitivity, 17, 206f
  function, shielded MR head, 226
  of a pinned layer to signal flux, 181, 209
  readback of AMR heads, 17f
  readback of DSSV heads, 18, 221, 253
  of a transfer curve, 167
Shape anisotropy, of sensor height/width, 25
Sheet conductance, for thin films, 45, 162, 168, 186
Sheet resistance, 45, 168
  bottom spin valve, 184, 196
  CSV sandwich with IrMn AFM layer, 196
  CSV sandwich with PdPTMn AFM layer, 196
  of DSSV sandwich with IrMn AFM layer, 217, 220
  metallic thin films, 45
  SSV sandwich with NiO AFM layer, 211, 214
Shields, 5
  affect on pulsewidth, 5, 7, 82
  attenuation of hard bias field, 103
  field enhancement ratio for sensor, 268
  images of free layer (self-biasing) in spin valves, 173, 175
  magnetization of, in uniform and nonuniform fields, 91, 94, 267f
Shield-to-shield gap in spin valves, *see* Gap
Side reading
  behavior of sensors, 226
  influence of overwrite (OW), 249
  parameter in track profiles, 227, 229f
Signal
  amplitude of MR devices, 114, 122, 138
  flux division between magnetic layers, 127
  flux for shielded MR element, 5, 93, 111. *See also* Excitation flux
  loss in spin valves (interdiffusion process), 285
  resolution (HF/LF amplitudes), 249
  voltage of a GMR head, 185
Signal-to-noise ratio (SNR), 15, 17
  defined, 3
  medium-limited, 14

Signal-to-noise ratio (SNR) (*Continued*)
  of *N*-bit PRS sequence, 15
Single domain state 62
Size effect, 50. *See also* Fuchs–Sondheimer theory
Skew angle (between current and easy axis), 30*f*
Skirt (and width of a microtrack profile), 227, 230*f*
Soft adjacent layer (SAL)
  biasing of MR heads, 108*f*
    with conducting spacer, 138*f*
    with insulating spacer, 133, 138
  ideal properties, 21
  materials
    CoNbTi, 22
    CoZrCr, 22, 154
    CoZrMo, 22
    NiFeCr, 22, 144
    NiFeRh, 22
    NiFeZr, 22
Solid solution of metals, 279
Spacer (S) layer in spin valves, 160, 215
  and GMR ratio versus thickness, 162
Spacing
  efffective value for readback, 5, 181, 230
  magnetic, 9
Spectral analysis of pulse position modulation, 249
Spectrum analyzer and overwrite (OW) measurement, 249
Specular reflection (and enhanacement of GMR ratio), 216
Spin dependent scattering, 32
Spin polarized electrons, lifetime extension, 216
Spin quantum number, 69
Spin valves
  bias compensation layer (BCL), 159, 207, 215
  bottom conventional (BSV), 160, 166, 170
  bottom synthetic (BSSV), 208
  conventional (CSV), 161, 166
  dual synthetic (DSSV), 159
  effect, 21
  free (F) layer, 160, 168, 170, 174, 176
  hysteresis loop, 33
  magnetoresistance versus field, 33
  pinned (P) layer, 160, 168, 170
  spacer (S) layer, 168, 170
  specular reflection layers, 159
  synthetic (SSV), 207
  top conventional (TSV), 39, 160, 166
  top synthetic (TSSV), 207
Spread function, 231
Stability
  coefficient (SC) of hard-biased MR heads, 101
  of output signal, 22, 242
Stabilization
  comparison of CJ and overlaid designs, 101
  with exchange pinning, 253
  with hard bias (PM) field, 101
  trade-offs between exchange and hard bias, 99
Step-annealing and resistance changes, 281
Stoner–Wohlfarth model (uniform rotation), 26, 109, 167
  and application to GMR spin valves, 40, 170
Stray field, *see* Demagnetization field
Stress, 21, 256, 288
  anisotropy, and patterning effect, 25, 256
  residual, 24*f*, 288
Stripe height, 64*f*, 277
  control, 195, 214
  damage current threshold, 199
  for minimum saturation current, 80, 135*f*
Superlattice structures (Co/Ru, Co/Cr, Fe/Cr), 35
Superparamagnetism (and relaxation effects in media), 14
Surface roughness and interlayer coupling mechanisms, 177
Susceptibility (magnetic), 30
Switching of magnetic medium with write field, 249
Synthetic spin valves (SSV), 12 Gbit/in.^{2} design parameters, 159, 211, 355
Synthetic trilayers (FM/Ru/FM), 207
  and exchange energy, 207
  net moment and free layer biasing, 207, 219
  temperature dependence of exchange energy, 209

Ta layers, 21, 35, 47, 122, 161, 217. *See also* Buffer and capping layers
Temperature
  annealing, *see* Annealing temperature
  average over a sensor width, 56
  blocking, 37, 38, 272
  distribution of exchange energy, 38
  melting point of AMR and GMR layers, 58, 272
  Néel (point), 37
  profile across a sensor, 55, 189, 193
  rise (in a sensor), 51*f*, 117
  unpinning, 38, 191*f*
Temperature coefficient
  for AMR effect, 47*f*
  for GMR effect, 47, 49
  of resistance, 47, 261, 295
    average value for multiple layers, 188, 272
  of thermal conductivity (Cu and Fe), 47
Thermal
  analysis of MR heads
    and boundary conditions, 54

INDEX **309**

one dimension, 50*ff*
two dimensions, 52*ff*
asperities (TA), 123, 260
   and resistance change of sensor, 261
   and temperature rise of sensor, 261
   and unpinning of a pinned layer, 261
conductivity, 188, 272
   of AFM layer, 46, 188
   and Lorentz number, 188
   of various materials, 118, 123
fluctuations, *see* Superparamagnetism
resistance, 295
   defined, 50, 58
   values for specific head designs, 118, 123, 130
   transient (and frictional heating), 260
Thermally activated changes in MR heads, 278
Time
   constants for thermally activated processes, 285
   dependence of diffusion processes, 279*ff*
   factor, in Arrhenius plots, 285*f*
Time to failure (TTF) in reliability testing, 289
Topological coupling between magnetic layers, 34, 177
Top synthetic spin valve (TSSV), *see* Spin valves
Track average amplitude (TAA), 254
Track pitch, 4
   and read width, 4, 226, 277
   and scaling rules for areal density, 4
   and write width, 4, 226
Track profiles (read signal), 226
Transfer curve
   AMR effect, 31, 111, 113, 143*f*
   and Barkhausen noise, 264–267, 269
   GMR, 179, 183*f*, 211, 220
   GMR + AMR, 179, 182, 251
   and hysteresis, 264–267, 269
   operating point, 109. *See also* Bias point
   SAL/MR head, 144
Transition (magnetic), 10, 248
   a-parameter, 10
   intermediate and final parameters, 10*f*
   jitter in position and width, 13–15. *See also* Noise
   and dependence on linear density, 14
   length, and relation to *a*-parameter, 6, 109
   magnetic field of, 13, 270
   sharpness, and write field gradient, *see* Write field gradient
Transmission line model (temperature rise), 53
Transverse fields at a free layer, 166
Triboelectric accumulation of charge, 247

U-parameter, *see* Channel density
Uniform rotation analysis, *see* Stoner–Wohlfarth model
Unit step function, 229, 231
Unpinning, 38, 187, 190, 197
   and evolution with increasing temperature, 190, 203
   and influence on signal and asymmetry, 187, 191, 193, 199–203
   and location in the pinned layer, 244
   and reversal of signal polarity, 203, 206
   and reversal of transfer curve slope, 203, 206
   and sensitivity to stripe height, 199
   and threshold in sense current, 197, 199

Vacancies (lattice), 289
Variances in magnetic transition, *see* Noise
Viterbi detector, 2
Voids, 287
Voltage
   HBM and relation to current, 273
   transfer curve
      and definition of output signals, 209
      and QST, 262

Wavelength (of sinewave recording), 231
Wave number (of sinewave recording), 14, 231
Waviness (correlated) between magnetic films, *see* Topological coupling
Wiedemann–Franz law (and Lorentz number), 47
Williams–Comstock model (of magnetic transition), 10
Write current and determination of optimal level, 248*f*
Write head saturation, 12, 248
Write field
   of a Karlquist head, 10
   disturbance of a sensor, 256
   gradient
      normalized value (*Q*-parameter), 10, 251
      and transition broadening, 11, 248
Write precompensation for NLTS, 16
Write saturation curves, 248*f*
Write width, *see* Track pitch
Writing zone length (size), 251

X-ray diffraction (grazing incidence) test, 279, 281

Young's modulus (NiFe), 289

Zr layers (and crystalline texture), 21